Additional Volumes in Preparation

ELECTRICAL ENGINEERING-ELECTRONICS SOFTWARE

1. Transformer and Inductor Design Software for the IBM PC, *Colonel Wm. T. McLyman*
2. Transformer and Inductor Design Software for the Macintosh, *Colonel Wm. T. McLyman*
3. Digital Filter Design Software for the IBM PC, *Fred J. Taylor and Thanos Stouraitis*

Two-Dimensional Digital Filters

Wu-Sheng Lu
Andreas Antoniou

University of Victoria
Victoria, British Columbia, Canada

Marcel Dekker, Inc.   **New York • Basel • Hong Kong**

Library of Congress Cataloging-in-Publication Data

Lu, Wu-Sheng
    Two-dimensional digital filters / Wu-sheng Lu, Andreas Antoniou.
        p.   cm. -- (Electrical engineering and electronics ; 80)
    Includes bibliographical references and index.
    ISBN 0-8247-8434-0 (acid-free paper)
    1. Electric filters, Digital.   2. Signal processing--Digital
techniques.   I. Antoniou, Andreas.   II. Title.   III. Series.
TK7872.F5A65   1992
621.3815'324--dc20   92-18745
                                                                  CIP

This book is printed on acid-free paper.

MARCEL DEKKER, INC.
270 Madison Avenue, New York, New York 10016

Current printing (last digit):
10 9 8 7 6 5 4 3 2 1

PRINTED IN THE UNITED STATES OF AMERICA

To Catherine and Rosemary

Preface

The theory of two-dimensional (2-D) digital filters has been a subject for study during the past two decades and has matured so that it now forms an important part of multidimensional digital signal processing. The objective of this book is to present basic theories, techniques, and procedures that can be used to analyze, design, and implement 2-D digital filters. Applications of 2-D digital filters in image and seismic data processing are also presented that demonstrate the importance of using 2-D filters in real-world signal processing.

The prerequisite knowledge is an undergraduate mathematics background of calculus, linear algebra and complex variables, and 1-D digital filters. Sufficient background information on 1-D digital filters can be found in the book *Digital Filters: Analysis, Design, and Applications*, Second Edition, by A. Antoniou.

Chapter 1 introduces the 2-D digital filter as a linear, discrete system. Characterization, flow-graph and network representations, and an elementary space-domain analysis are then introduced. The chapter also includes a brief discussion on stability and on the realization of 2-D digital filters.

The Givone–Roesser model is introduced in Chapter 2 and is then used as a tool in a more advanced space-domain analysis. The concepts of controllability and observability are then defined. The chapter concludes with a section on state-space realization.

The 2-D z transform is studied in Chapter 3. The complex convolution is introduced as a tool for obtaining the z transforms of products of space-domain functions and for the derivation of Parseval's formula. The 2-D sampling theorem is then introduced as the fundamental link between 2-D continuous and discrete signals.

Chapter 4 introduces the important concept of the transfer function as an alternative, z-domain representation of 2-D digital filters. Some of the properties of the discrete transfer function and its application to space-domain and frequency-domain analysis are then described.

Chapter 5 is devoted to a detailed stability analysis in both frequency and state-space domains. Also included in this chapter is a 2-D Lyapunov stability theory developed recently. It is shown that the problem of stability in 2-D digital filters, unlike its 1-D counterpart, can get quite involved.

Various design techniques are described in Chapters 6 to 9. Chapter 6 deals with the approximation problem for nonrecursive filters. Methods described in the chapter include the window method and a method based on the McClellan transformation. Chapter 7 presents several transformation methods for the design of recursive filters. In Chapter 8 a least pth and a minimax optimization method are used to design recursive filters. A design approach based on two-variable network theory that leads to stable filters is then presented. The chapter concludes with a design method based on the singular-value decomposition (SVD). The design of nonrecursive filters by optimization is considered in Chapter 9.

Several direct and indirect realization methods are considered in detail in Chapter 10. These include nonrecursive realizations based on the McClellan transformation, nonrecursive and recursive realizations based on matrix decompositions, and recursive realizations based on the concept of the generalized-immittance converter.

The finite wordlength effects of 2-D digital filters are considered in Chapter 11. An analysis on quantization errors and their computation is first discussed. State-space structures with minimized roundoff noise are then examined. Two types of parasitic oscillations, namely, quantization and overflow limit cycles, are examined and conditions for their elimination are presented.

Chapter 12 introduces the 2-D discrete Fourier transform and methods for its efficient computation in terms of the 1-D and 2-D fast Fourier transforms as tools in the implementation of 2-D digital filters. Several systolic structures for the implementation of 2-D nonrecursive and recursive filters are then examined.

Chapter 13, which concludes the book, introduces applications of 2-D filters to image enhancement, and restoration and seismic data processing. Most of the concepts and techniques are illustrated by examples and a selected set of problems is included at the end of each chapter. The book can serve as a textbook for a one-semester course on 2-D digital filters for first-year graduate students. The book should also be of interest to filter designers and engineers who intend to utilize 2-D filters in dealing with a variety of digital signal processing problems.

We wish to thank Len T. Bruton of the University of Calgary for providing materials that have been used to write Secs. 8.4 and 13.4; Chris Charalambous for allowing us to use the design data presented in Examples 8.2 and 9.1; Majid Ahmadi for reading parts of the manuscript and for supplying useful comments; Catherine Chang for her assistance in the preparation of the artwork and for typing Chapters 8 to 13 of the manuscript; Eileen Gardiner and Ruth Dawe for their persistent encouragement and involvement as the acquisition editors and Joseph Stubenrauch for his assistance as the production editor of Marcel Dekker, Inc.; the Natural Sciences and Engineering Research Council of Canada and Micronet, Networks of Centres of Excellence Program, for supporting the research that led to some of the new results presented in Chapters 8 to 10; and the University of Victoria for general support. In addition, we wish to thank our wives Catherine Chang and Rosemary C. Antoniou for their sacrifices and constant support.

Wu-Sheng Lu
Andreas Antoniou

Contents

Introduction

The field of two-dimensional (2-D) digital signal processing has been growing rapidly in recent years. Images such as satellite photographs, radar and sonar maps, medical X-ray pictures, radiographs, electron micrographs, and data from seismic, gravitational, and magnetic records are typical examples of 2-D signals that might need to be processed. The types of processing that can be applied may range from improving the quality of signals to extracting certain useful features from them. For example, a picture degraded by wideband noise might be improved by removing the noise without blurring the edges, or a seismic record might be made more readable by removing a certain large-amplitude, low-frequency signal known as ground-roll interference.

A *continuous 2-D signal* is a physical or contrived quantity that is a continuous function of two real independent variables. An example of such a signal might be the light intensity in the case of a photograph or image as a function of distances in the x and y directions. A *discrete 2-D signal* is a sampled version of a continuous 2-D signal and is normally in the form of a 2-D array of numbers. Like a one-dimensional (1-D) discrete signal, a 2-D discrete signal can be represented by a frequency spectrum that can be modified, reshaped, or manipulated through filtering. This type of processing can be carried out by using 2-D digital filters.

Two-dimensional digital filters, like their 1-D counterparts, are *discrete systems* that can be linear or nonlinear, shift-invariant or shift-dependent, causal or noncausal, and stable or unstable. They can be characterized in terms of difference equations or state-space equations in two independent

1

variables and in terms of transfer functions or matrices of transfer functions, which are rational functions of polynomials in two variables. Time-domain analysis in 1-D digital filters is replaced by *space-domain analysis* in 2-D digital filters, since neither of the two independent variables needs to be time but frequency-domain analysis continues to be referred to by the same name, although frequency is sometimes referred to as *spatial frequency* to emphasize that frequency may not bear an inverse relation with time. As in 1-D digital filters, the transfer function yields the amplitude and phase responses, which are surfaces over a 2-D frequency plane rather than curves plotted over a frequency axis.

Two-dimensional digital filters can be classified as *recursive* or *nonrecursive*, depending on whether the output of the filter depends on previous values of the output; alternatively, they can be classified as *infinite-impulse response* (IIR) or *finite-impulse response* (FIR) filters, depending on whether their impulse response is of infinite or finite duration. These types of 2-D digital filters are consistent with their 1-D counterparts and have analogous properties. For example, FIR filters can be designed to have linear phase, whereas IIR filters can be designed to be more economical in terms of the amount of computation required for a given degree of selectivity.

The design of 2-D digital filters, like that of 1-D digital filters, involves several steps, as follows:

- Approximation
- Realization
- Implementation
- Study of quantization effects

Approximation is the process of generating a rational transfer function that satisfies required specifications imposed on the amplitude, phase, or space-domain response. It can be accomplished by applying transformations to 1-D analog or digital filters, by using optimization methods, or by applying transformations in conjunction with optimization methods. A prerequisite property here is that the transfer function generated should represent a stable digital filter.

Realization is the process of converting the transfer function obtained through the approximation step into a signal-flow graph, digital-filter network, or state-space representation.

Implementation is the process of convering the signal-flow graph, digital-filter network, or state-space representation into a computer program or a dedicated piece of equipment. In this way software and hardware digital-filter implementations can be obtained.

When a 2-D digital filter is implemented in terms of either software on a general-purpose computer or dedicated hardware, numbers representing transfer-function coefficients and signals must be stored and manipulated in registers of finite length. When the approximation step is carried out, transfer-function coefficients are calculated to a high degree of precision; consequently, they must be quantized before the implementation of the digital filter. The net effect of *coefficient quantization* is to introduce inaccuracies in the amplitude response of the filter that tend to increase as the word length of the hardware is reduced. On the other hand, internal signals generated as products when signals are multiplied by coefficients are almost always too long to fit in the available registers and must again be quantized. The effect of *signal quantization* is to introduce noise at the output of the filter, which degrades the signal-to-noise ratio. Signal quantization can lead to other problems as well, such as the generation of spurious parasitic oscillations, known as limit cycles. While the effects of coefficient and signal quantization are insignificant if a general-purpose computer is to be used, owing to the high precision of the hardware, particular attention must be paid to these effects when fixed-point arithmetic or specialized hardware with reduced word length is to be employed. In such applications, the design process is not considered complete until the effects of quantization are studied in detail.

The characterization, properties, and design of 2-D digital filters are usually simple extensions or generalizations of the characterization, properties, and design of 1-D digital filters. Nevertheless, notable exceptions arise where the extension or generalization is not simple, and it may on occasion be quite complicated. An example in this regard is *stability analysis*. In the 1-D case, the stability of the digital filter is linked to the poles of the transfer function, which are isolated points in the 1-D complex plane. As a result, powerful mathematical tests that have been used in the past to determine whether the zeros of a polynomial in z are located in a specified region of the z plane have been used to develop stability criteria, such as the Jury–Marden stability criterion, which simplify the stability analysis of 1-D digital filters. In the 2-D case, on the other hand, the stability of the digital filter is closely linked to contours in the 2-D complex plane for which the denominator polynomial of the transfer function is zero and may also on occasion be linked to contours in the 2-D complex plane for which the numerator polynomial is zero. These difficulties arise because polynomials in two variables are not generally factorable into first-order polynomials, owing to the fact that the Fundamental theorem of algebra does not extend to polynomials in two variables.

A major consideration in the case of 2-D digital filters relates to the issue of *computational complexity*. With one more dimension added to the field of data, the curse of dimensionality is brought to bear and the amount of computation tends to increase as the square of the order of the filter. Consequently, in 2-D filters it is particularly important to design the lowest-

order filter that will satisfy the specifications. Another important consideration in 2-D digital filters relates to the ultimate receiver of the processed
2-D signal. In the case of hearing, phase distortion is quite tolerable and
1-D filters are often designed without regard to *phase-response linearity*.
On the other hand, phase distortion tends to be as objectionable as amplitude distortion when the final receiver is the human eye. Particular
attention must, therefore, be paid to the phase response when 2-D digital
filters are to be designed for image processing applications.

1
Fundamentals

1.1 INTRODUCTION

A 2-D digital filter is a discrete system that can be used to process 2-D discrete signals. Like any other system, it can be linear or nonlinear, shift-invariant or shift-dependent, causal or noncausal, and stable or unstable. It can be characterized by a difference equation or by a transfer function and can be analyzed in the space domain or frequency domain.

In this chapter, 2-D continuous and discrete signals are introduced as extensions of their 1-D counterparts. The characterization and fundamental properties of 2-D digital filters are then introduced. Space-domain analysis, the process of finding the response of the filter to a given input signal, is examined and is then used to develop the concepts of causality and stability in 2-D digital filters. The chapter concludes with a brief introduction to the realization of 2-D digital filters, which is the process of converting a mathematical description of a filter into a network.

1.2 2-D DISCRETE SIGNALS

A 2-D continuous signal is a physical or contrived quantity that depends on two independent continuous variables t_1 and t_2. It can be represented by a function $x(t_1, t_2)$ and each of the two variables t_1 and t_2 may represent time, distance, or any other physical or contrived variable. Two examples of 2-D continuous signals are the light intensity of an image as a function of distance in the x and y directions and the depth of an ocean as a function of distance in the east and north directions.

A 2-D discrete signal, on the other hand, is a physical or contrived quantity that depends on two real independent integer variables n_1 and n_2. It can be represented by a function $x(n_1 T_1, n_2 T_2)$, where T_1 and T_2 are constants and $n_1 T_1$ or $n_2 T_2$ may represent time, distance, etc. A shorthand notation for a 2-D discrete signal, which will be used frequently in this book, is $x(n_1, n_2)$. Although $x(n_1, n_2)$ can in principle be complex, it is normally real and can be represented in terms of a 3-D plot, as depicted in Fig. 1.1. Two-dimensional discrete signals are usually obtained by sampling corresponding continuous signals. Two such examples are the arrays of numbers representing a digitized image and the depth of an ocean at discrete points in the north and east directions.

In many applications, 2-D signals arise that are continuous with respect to one variable and discrete with respect to the other variable. Such signals are sometimes said to be mixed. An example of a 2-D mixed signal is the set of acoustic waveforms that might be produced by an explosion, as measured by transducers placed at discrete intervals near the surface in

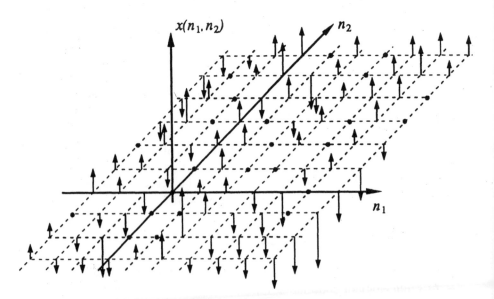

Figure 1.1 A 2-D discrete signal.

the ground along a given direction. In such a signal the continuous variable is time and the discrete variable is distance.

1.2.1 Region of Support

A useful concept in the description of 2-D discrete signals is the region of support. Consider a signal $x(n_1, n_2)$ and let A be a connected region in the (n_1, n_2) plane such that $x(n_1, n_2) = 0$ for all points in A. The complement of A with respect to the (n_1, n_2) plane, represented by S, is said to be a region of support of the signal. If A is the largest connected region in the (n_1, n_2) plane such that $x(n_1, n_2) = 0$ for all points in A, the complement of A is said to be the minimum region of support. If a 2-D signal is zero everywhere except the first quadrant, it is said to have first-quadrant support, and if a 2-D signal is zero everywhere in the left half plane, it is said to have right half-plane support. A 2-D signal is said to be of finite extent if its minimum region of support is finite.

1.2.2 Quantization

As in the 1-D case, the amplitude of a 2-D discrete signal can be quantized by representing it in terms of a finite number of distinct levels so as to facilitate the storage of the signal in a digital memory or mass-storage device. Signal quantization depends on the type of arithmetic used for the representation of the signal (e.g., fixed-point or floating-point) and can be carried out by rounding or truncating the numbers that represent the levels of the signal.

1.2.3 Periodicity

A 2-D discrete signal can be periodic with respect to either or both of the space-domain variables n_1 and n_2. A 2-D signal that is periodic with respect to both space-domain variables satisfies the relation

$$x(n_1 + k_1 N_1, n_2 + k_2 N_2) = x(n_1, n_2)$$

for any pair of integers k_1 and k_2. The constants N_1 and N_2 are the periods of the signal in the two directions. A periodic signal is illustrated in Fig. 1.2a.

   If a 2-D signal is periodic with respect to both space variables with periods N_1 and N_2, then any connected domain of $x(n_1, n_2)$ comprising exactly $N_1 \times N_2$ points is said to be a 2-D period of the signal. Although the 2-D period of a signal is usually rectangular, as in Fig. 1.2a, many geometrical shapes are possible, as illustrated in Fig. 1.2b.

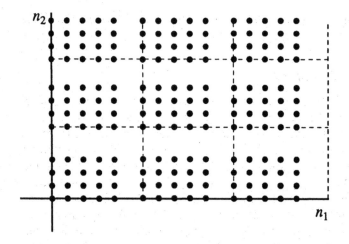

(a)

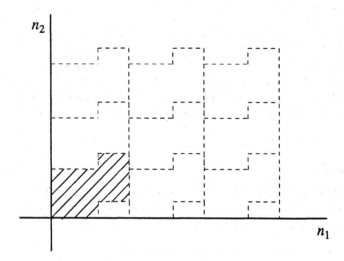

(b)

Figure 1.2 Two-dimensional periodic discrete signals: (a) with rectangular period; (b) with nonrectangular period.

1.2.4 Separability

A 2-D discrete signal is said to be separable if it can be expressed as a product of two 1-D discrete signals; that is,

$$x(n_1, n_2) = x_1(n_1)x_2(n_2)$$

The importance of separability arises from the fact that if the output of a 2-D digital filter can be expressed as a product of two 1-D discrete signals, then the design of the filter can be broken down into the design of two 1-D filters.

Although 2-D signals are usually nonseparable, in certain circumstances it is possible to express a 2-D signal in terms of a linear combination of products of 1-D signals; that is,

$$x(n_1, n_2) = \sum_{i=N_L}^{N_H} \alpha_i x_{1i}(n_1)x_{2i}(n_2) \tag{1.1}$$

where α_i for $i = N_L, \ldots, N_H$ are constants. In such a case it would be possible to design the 2-D filter by simply designing a set of 1-D filters and then interconnecting them to form the 2-D digital filter. In the case where the 2-D signal has a finite support, the decomposition in Eq. (1.1) is always possible, as will now be demonstrated. Let

$$x(n_1, n_2) = 0$$

for any pair (n_1, n_2) not in the rectangle defined by

$$A = \{(n_1, n_2): N_{L1} \leq n_1 \leq N_{H1}, N_{L2} \leq n_2 \leq N_{H2}\}$$

Such a signal can be expressed in terms of Eq. (1.1) by letting

$$x_{1i}(n_1) = x(n_1, i)$$

and

$$x_{2i}(n_2) = \begin{cases} 1 & \text{for } n_2 = i \\ 0 & \text{otherwise} \end{cases}$$

as can be easily verified.

The preceding decomposition of a 2-D signal, though simple to understand, is not the most effective for the applications to be considered later. An alternative and more efficient decomposition can be obtained by expressing the $(N_{H1} - N_{L1} + 1) \times (N_{H2} - N_{L2} + 1)$ matrix defined by

$$\mathbf{X} = \{x(n_1, n_2)\}$$

for $(n_1, n_2) \in A$ as a singular-value decomposition [1] of the form

$$\mathbf{X} = \sum_{i=1}^{r} \sigma_i \mathbf{u}_i \mathbf{v}_i^T \qquad (1.2)$$

where σ_i for $i = 1, 2, \ldots, r$ are said to be the singular values of $\mathbf{X}$, r is the rank of A, $\mathbf{u}_i$ and $\mathbf{v}_i$ are column vectors, and $\mathbf{v}_i^T$ is the transpose of $\mathbf{v}_i$. Evidently, $\mathbf{u}_i$ and $\mathbf{v}_i$ can be regarded as 1-D signals and Eq. (1.2) can be considered to be equivalent to Eq. (1.1).

The issue of separability will be reexamined later, in Chapters 8 and 9, when the design of 2-D digital filters is undertaken.

1.3 THE 2-D DIGITAL FILTER AS A SYSTEM

A 2-D digital filter is a discrete system that will receive an input signal $x(n_1, n_2)$ and produce an output signal $y(n_1, n_2)$, as depicted in Fig. 1.3. The output is related to the input by some rule of correspondence. This fact can be represented mathematically by the relation

$$y(n_1, n_2) = \mathcal{R}x(n_1, n_2) \qquad (1.3)$$

where $\mathcal{R}$ is an operator. This relation can be used to define the fundamental properties of a 2-D digital filter as a system, such as linearity, shift invariance, and causality.

1.3.1 Linearity

A 2-D digital filter is linear if its response satisfies the principles of homogeneity and additivity given by

$$\mathcal{R}\alpha x(n_1, n_2) = \alpha \mathcal{R}x(n_1, n_2)$$

and

$$\mathcal{R}[x_1(n_1, n_2) + x_2(n_1, n_2)] = \mathcal{R}x_1(n_1, n_2) + \mathcal{R}x_2(n_1, n_2)$$

for all possible values of α and all possible input signals $x_1(n_1, n_2)$ and $x_2(n_1, n_2)$. The above two conditions can readily be combined into the

Figure 1.3 The 2-D digital filter as a system.

general superposition relation given by

$$\Re[\alpha_1 x_1(n_1, n_2) + \alpha_2 x_2(n_1, n_2)] = \alpha_1 \Re x_1(n_1, n_2) + \alpha_2 \Re x_2(n_1, n_2) \quad (1.4)$$

If Eq. (1.4) is violated for some signals or constants of proportionality, the 2-D digital filter is nonlinear.

1.3.2 Shift Invariance

An initially relaxed 2-D digital filter in which $x(n_1, n_2) = 0$ and $y(n_1, n_2) = 0$ for all $n_1 < 0$ or $n_2 < 0$ is said to be shift-invariant if

$$y(n_1 - k_1, n_2 - k_2) = \Re x(n_1 - k_1, n_2 - k_2) \quad (1.5)$$

for all input signals $x(n_1, n_2)$ and all possible integers k_1 and k_2. If the condition for shift invariance is not satisfied, then the filter is said to be shift-dependent.

1.3.3 Causality

A 2-D digital filter is said to be causal if its output for $n_1 \leq k_1$ and $n_2 \leq k_2$ is independent of the input for values of $n_1 > k_1$ or $n_2 > k_2$ for all possible values of k_1 and k_2. In mathematical terms, a 2-D digital filter is causal if and only if

$$\Re x_1(n_1, n_2) = \Re x_2(n_1, n_2) \quad (1.6)$$

for $n_1 \leq k_1$ and $n_2 \leq k_2$ for all possible signals $x_1(n_1, n_2)$ and $x_2(n_1, n_2)$ such that

$$x_1(n_1, n_2) = x_2(n_1, n_2)$$

for $n_1 \leq k_1$ and $n_2 \leq k_2$. Conversely, if Eq. (1.6) is not satisfied for some pair of possible signals $x_1(n_1, n_2)$ and $x_2(n_1, n_2)$ such that

$$x_1(n_1, n_2) = x_2(n_1, n_2)$$

for $n_1 \leq k_1$ and $n_2 \leq k_2$ and

$$x_1(n_1, n_2) \neq x_2(n_1, n_2)$$

for some pair of values of n_1 and n_2 such that $n_1 > k_1$ or $n_2 > k_2$, then the filter is noncausal. If Eq. (1.6) holds only with respect to variable n_1 (n_2), the 2-D digital filter is said to be causal only with respect to variable n_1 (n_2).

Causality has physical significance only in the case where $n_1 T_1$ or $n_2 T_2$ represents real time. Nevertheless, the concept of causality is useful even in the case where neither of the two variables is time.

1.4 CHARACTERIZATION

Two types of 2-D digital filters can be identified, namely, recursive and nonrecursive, and like their 1-D counterparts, they can be characterized in terms of difference equations.

1.4.1 Nonrecursive Filters

In a nonrecursive 2-D filter the output $y(n_1, n_2)$ is a function of $x(n_1 - i, n_2 - j)$ for $-\infty < i, j < +\infty$. Assuming linearity and shift invariance, the output of a nonrecursive 2-D digital filter can be expressed as a weighted sum of all the possible values of the input; that is

$$y(n_1, n_2) = \sum_{i=-\infty}^{\infty} \sum_{j=-\infty}^{\infty} a_{ij} x(n_1 - i, n_2 - j)$$

where a_{ij} are constants. Assuming causality, the output for $n_1 \leq k_1$ and $n_2 \leq k_2$ is independent of the input for values of $n_1 > k_1$ or $n_2 > k_2$ for all possible values of k_1 and k_2, and therefore we have

$$a_{ij} = 0 \qquad \text{for } -\infty \leq i \text{ or } j < 0$$

and thus

$$y(n_1, n_2) = \sum_{i=0}^{\infty} \sum_{j=0}^{\infty} a_{ij} x(n_1 - i, n_2 - j)$$

Now if the input is assumed to have a quarter-plane support, that is, $x(n_1, n_2) = 0$ for $-\infty < n_1$ or $n_2 < 0$, and $a_{ij} = 0$ for $N_1 < i < \infty$ or $N_2 < j < \infty$, we obtain

$$y(n_1, n_2) = \sum_{i=0}^{n_1} \sum_{j=0}^{n_2} a_{ij} x(n_1 - i, n_2 - j)$$

$$+ \sum_{i=n_1+1}^{\infty} \sum_{j=0}^{n_2} a_{ij} x(n_1 - i, n_2 - j)$$

$$+ \sum_{i=0}^{n_1} \sum_{j=n_2+1}^{\infty} a_{ij} x(n_1 - i, n_2 - j)$$

$$+ \sum_{i=n_1+1}^{\infty} \sum_{j=n_2+1}^{\infty} a_{ij} x(n_1 - i, n_2 - j)$$

$$= \sum_{i=0}^{N_1} \sum_{j=0}^{N_2} a_{ij} x(n_1 - i, n_2 - j)$$

$$+ \sum_{i=N_1+1}^{n_1} \sum_{j=0}^{N_2} a_{ij}x(n_1 - i, n_2 - j)$$

$$+ \sum_{i=0}^{N_1} \sum_{j=N_2+1}^{n_2} a_{ij}x(n_1 - i, n_2 - j)$$

$$+ \sum_{i=N_1+1}^{n_1} \sum_{j=N_2+1}^{n_2} a_{ij}x(n_1 - i, n_2 - j)$$

$$= \sum_{i=0}^{N_1} \sum_{j=0}^{N_2} a_{ij}x(n_1 - i, n_2 - j) \qquad (1.7a)$$

In effect, a 2-D, linear, shift-invariant, causal, nonrecursive digital filter can be represented in terms of a linear difference equation in two variables. The ordered pair (N_1, N_2) is said to be the order of the filter.

In many applications a record of the 2-D signal to be processed is available before the start of the processing and is easily accessible. In such applications the filter need not be causal since the output for point (n_1, n_2) can be generated by using data over an *arbitrary* neighborhood of point (n_1, n_2). If the input is assumed to have a nonsymmetric half-plane support, that is, $x(n_1, n_2) = 0$ for $(n_1, n_2) \in \{(n_1, n_2): n_1 < 0, n_2 = 0\} \cup \{(n_1, n_2): -\infty < n_1 < \infty, n_2 < 0\}$ and $a_{ij} = 0$ for (i, j) outside the region $\{(i, j): 0 \le i \le N_1, j = 0\} \cup \{(i, j): -N_1 \le i \le N_1, 1 \le j \le N_2\}$, then

$$y(n_1, n_2) = \sum_{i=0}^{N_1} a_{i0}x(n_1 - i, n_2) + \sum_{i=-N_1}^{N_1} \sum_{j=1}^{N_2} a_{ij}x(n_1 - i, n_2 - j) \qquad (1.7b)$$

A filter characterized by Eq. (1.7b) is said to be a nonsymmetric half-plane nonrecursive filter.

1.4.2 Recursive Filters

A 2-D recursive digital filter differs from a nonrecursive one in that the output is a function of input as well as output values. A linear, shift-invariant, and causal recursive filter can be described in terms of the difference equation

$$y(n_1, n_2) = \sum_{i=0}^{N_1} \sum_{j=0}^{N_2} a_{ij}x(n_1 - i, n_2 - j)$$

$$- \sum_{i=0}^{N_1} \sum_{j=0}^{N_2} b_{ij}y(n_1 - i, n_2 - j) \qquad (1.8a)$$

where $b_{00} = 0$. The ordered pair (N_1, N_2) is said to be the order of the filter. Note that Eq. (1.8a) reduces to Eq. (1.7a) if $b_{ij} = 0$; in effect, the nonrecursive digital filter is a special case of the recursive one.

A nonsymmetric half-plane recursive filter is characterized by

$$y(n_1, n_2) = \left[\sum_{i=0}^{N_1} a_{i0} x(n_1 - i, n_2) + \sum_{i=-N_1}^{N_1} \sum_{j=1}^{N_2} a_{ij} x(n_1 - i, n_2 - j) \right]$$

$$- \left[\sum_{i=1}^{N_1} b_{i0} x(n_1 - i, n_2) + \sum_{i=-N_1}^{N_1} \sum_{j=1}^{N_2} b_{ij} y(n_1 - i, n_2 - j) \right]$$

(1.8b)

1.5   REPRESENTATION IN TERMS OF FLOW GRAPHS AND NETWORKS

The basic arithmetic operations inherent in the characterization of a 2-D digital filter are the shift operation with respect to each of the two independent variables, the addition, and the multiplication, as can be seen in Eq. (1.8a). Since 2-D digital filters are often implemented in terms of digital hardware, it is useful to represent them in terms of networks. The basic elements needed for the implementation of a causal 2-D digital filter are the horizontal unit shifter, the vertical unit shifter, the adder, and the multiplier. The symbols and characterizations for these elements are given in Fig. 1.4 and Table 1.1, respectively.

The analysis of digital-filter networks is usually simple and can be carried out by using the element equations as illustrated in the following example.

Example 1.1   (a) Analyze the network of Fig. 1.5a. (b) Repeat part a for the network of Fig. 1.5b.

Solution.   (a) If the signal at node A of Fig. 1.5 is assumed to be $v(n_1, n_2)$, then

$$v(n_1, n_2) = x(n_1, n_2) + e^{\alpha_1} v(n_1 - 1, n_2)$$
$$= x(n_1, n_2) + e^{\alpha_1} E_1^{-1} v(n_1, n_2) \qquad (1.9)$$

and

$$y(n_1, n_2) = v(n_1, n_2) + e^{\alpha_2} y(n_1, n_2 - 1)$$
$$= v(n_1, n_2) + e^{\alpha_2} E_2^{-1} y(n_1, n_2) \qquad (1.10)$$

$x(n_1, n_2)$ $y(n_1, n_2) = x(n_1 - 1, n_2)$

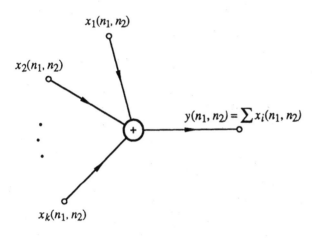

(a)

$x(n_1, n_2)$ $y(n_1, n_2) = x(n_1, n_2 - 1)$

(b)

$x_1(n_1, n_2)$

$x_2(n_1, n_2)$

$y(n_1, n_2) = \sum x_i(n_1, n_2)$

$x_k(n_1, n_2)$

(c)

m

$x(n_1, n_2)$ $y(n_1, n_2) = mx(n_1, n_2)$

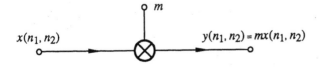

(d)

Figure 1.4   Basic elements of 2-D digital filters.

Table 1.1 Basic Elements of 2-D Digital Filters

Element	Characterization
Horizontal shifter	$y(n_1, n_2) = x(n_1 - 1, n_2)$
Vertical shifter	$y(n_1, n_2) = x(n_1, n_2 - 1)$
Adder	$y(n_1, n_2) = \sum_{i=0}^{k} x_i(n_1, n_2)$
Multiplier	$y(n_1, n_2) = mx(n_1, n_2)$

where $E_i, i = 1, 2,$ is the shift operator of numerical analysis. On eliminating $v(n_1, n_2)$, we obtain

$$y(n_1, n_2) = x(n_1, n_2) + e^{\alpha_1}y(n_1 - 1, n_2)$$
$$+ e^{\alpha_2}y(n_1, n_2 - 1) - e^{\alpha_1 + \alpha_2}y(n_1 - 1, n_2 - 1)$$

(b) As above, the output signal is obtained as

$$y(n_1, n_2) = x(n_1, n_2) + b_1 y(n_1 - 1, n_2) + b_2 y(n_1 - 1, n_2 - 2)$$

Since the elements of Fig. 1.4 can be represented in terms of simple flow graphs, 2-D digital filters can be represented in terms of flow graphs. Consequently, flow-graph methods can be applied for the analysis of 2-D digital filters. Flow graphs can readily be obtained by replacing each unit shifter in the digital-filter network by a branch with a transmittance equal to the appropriate shift operator, each multiplier by a branch with transmittance equal to the coefficient of the multiplier, and each adder by a node with one outgoing branch and as many incident branches as there are inputs in the adder. The flow graphs for the digital filters of Fig. 1.5 are illustrated in Fig. 1.6. Evidently, signal flow graphs have the advantage of simplicity and will be preferred in subsequent sections of this book.

1.6 INTRODUCTION TO SPACE-DOMAIN ANALYSIS

Space-domain analysis in 2-D digital filters is analogous to time-domain analysis in 1-D digital filters. It is the process of finding the output $y(n_1, n_2)$ of a 2-D digital filter to some arbitrary input $x(n_1, n_2)$ using some characterization or model for the filter.

1.6.1 Elementary Signals

Space-domain analysis can be simplified through the use of elementary 2-D signals such as the 2-D unit impulse, unit step, unit ramp, exponential,

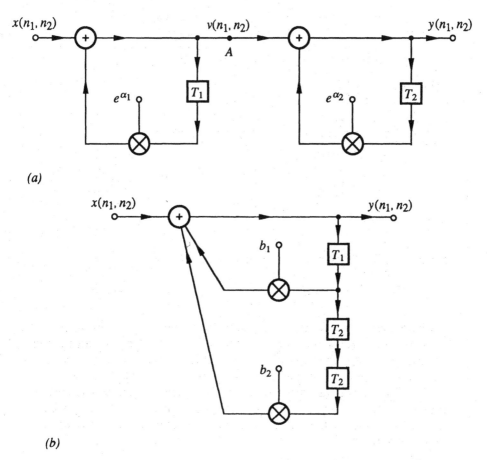

Figure 1.5 Two-dimensional digital-filter networks: (a) Example 1.1, part (a); (b) Example 1.1, part (b).

and sinusoid. The unit impulse is defined as

$$\delta(n_1, n_2) = \begin{cases} 1 & \text{for } n_1 = n_2 = 0 \\ 0 & \text{otherwise} \end{cases} \tag{1.11}$$

and has a finite support $S = (0, 0)$. The unit step, on the other hand, is defined as

$$u(n_1, n_2) = \begin{cases} 1 & \text{for } n_1, n_2 \geq 0 \\ 0 & \text{otherwise} \end{cases} \tag{1.12}$$

and has a support $S = \{(n_1, n_2) : n_1 \geq 0 \text{ and } n_2 \geq 0\}$. The unit impulse and unit step are illustrated in the discrete contour maps of Fig. 1.7a and

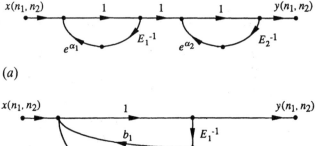

(a)

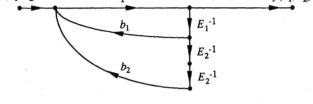

(b)

Figure 1.6   Signal flow-graph representation of 2-D digital filters.

b where each dot represents a signal value of unity. The more important of these two fundamental signals is the unit impulse since it can be used to synthesize any other 2-D discrete signal, or conversely, any 2-D signal can be decomposed into impulses, as will be demonstrated below.

Two signals that are often useful in 2-D digital filters can readily be synthesized by assuming unit impulses on the positive n_1 and n_2 axes of the (n_1, n_2) plane. They can be referred to as unit-line impulses and can be expressed as

$$\delta_1(n_1, n_2) = \begin{cases} 1 & \text{for } n_1 \geq 0, n_2 = 0 \\ 0 & \text{otherwise} \end{cases}$$

$$= \sum_{k_1=0}^{\infty} \delta(n_1 - k_1, n_2) \qquad (1.13)$$

and

$$\delta_2(n_1, n_2) = \begin{cases} 1 & \text{for } n_1 = 0, n_2 \geq 0 \\ 0 & \text{otherwise} \end{cases}$$

$$= \sum_{k_2=0}^{\infty} \delta(n_1, n_2 - k_2) \qquad (1.14)$$

The two unit-line impulses are illustrated in Fig. 1.8a and b. By repeating the unit-line impulse $\delta_2(n_1, n_2)$ along the n_1 axis or the unit-line impulse

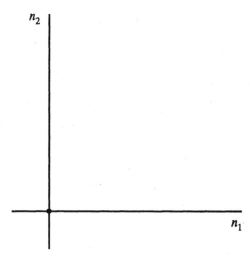

(a)

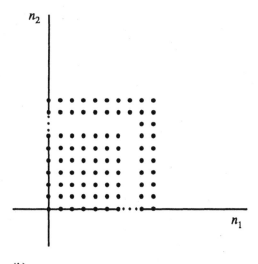

(b)

Figure 1.7 Elementary signals: (a) the unit impulse; (b) the unit step.

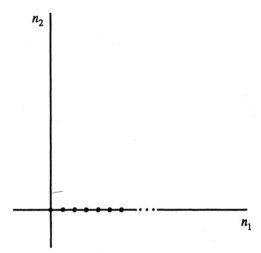

(a)

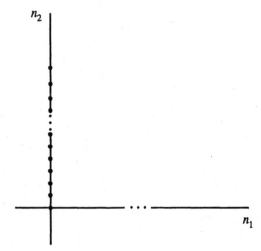

(b)

Figure 1.8 Unit-line impulses: (a) unit-line impulse along n_1 axis; (b) unit-line impulse along n_2 axis.

$\delta_1(n_1, n_2)$ along the n_2 axis, the unit step can be expressed in terms of unit impulses as

$$u(n_1, n_2) = \sum_{k_1=0}^{\infty} \delta_2(n_1 - k_1, n_2) = \sum_{k_2=0}^{\infty} \delta_1(n_1, n_2 - k_2)$$

$$= \sum_{k_1=0}^{\infty} \sum_{k_2=0}^{\infty} \delta(n_1 - k_1, n_2 - k_2) \tag{1.15}$$

A 2-D unit-line pulse is a signal comprising a series of consecutive unit-line impulses. Such signals can be synthesized by using the unit step. A unit-line pulse along the n_1 axis of width k_2 can be obtained as

$$p_{n_1 k_2}(n_1, n_2) = u(n_1, n_2) - u(n_1, n_2 - k_2) \tag{1.16}$$

Similarly, a unit-line pulse along the n_2 axis of width k_1 can be obtained as

$$p_{k_1 n_2}(n_1, n_2) = u(n_1, n_2) - u(n_1 - k_1, n_2) \tag{1.17}$$

These signals are illustrated in Fig. 1.9a and b. With unit-line pulses available, a $k_1 \times k_2$ rectangular unit pulse can now be synthesized as

$$p_{k_1 k_2}(n_1, n_2) = p_{n_1 k_2}(n_1, n_2) - p_{n_1 k_2}(n_1 - k_1, n_2)$$

$$= p_{k_1 n_2}(n_1, n_2) - p_{k_1 n_2}(n_1, n_2 - k_2) \tag{1.18}$$

The 2-D unit pulse obtained is shown in Fig. 1.10.

Other elementary signals of interest are the 2-D discrete sinusoid and exponential. The 2-D sinusoid is given by

$$x(n_1, n_2) = \sin(\omega_1 n_1 + \omega_2 n_2) \tag{1.19}$$

whereas the 2-D discrete exponential can be generated by forming the complex signal

$$x(n_1, n_2) = \cos(\omega_1 n_1 + \omega_2 n_2) + j \sin(\omega_1 n_1 + \omega_2 n_2)$$

$$= e^{j(\omega_1 n_1 + \omega_2 n_2)}$$

$$= e^{j\omega_1 n_1} e^{j\omega_2 n_2} \tag{1.20}$$

The last two signals play the same key role in frequency-domain analysis as their 1-D counterparts. Note that the exponential signal is separable while the sinusoid can be expressed as a linear combination of products of signals by using a trigonometric identity.

1.6.2 Induction Method

Space-domain analysis is normally carried out through the use of powerful state-space or transform methods. In certain simple examples, however,

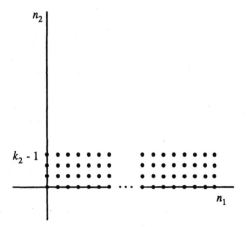

(a)

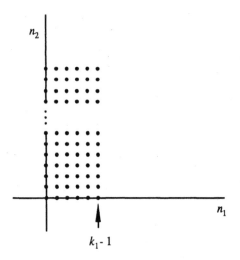

(b)

Figure 1.9 Unit-line pulses: (a) unit-line pulse along n_1 axis of width k_2; (b) unit-line pulse along n_2 axis of width k_1.

it can also be carried out through induction, as is illustrated below by some examples. Although the approach is not the most efficient, it does provide understanding of the fundamental principles involved.

Some more advanced methods for space-domain analysis will be examined in Chapter 2.

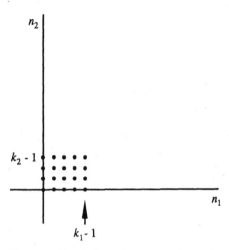

Figure 1.10   A $k_1 \times k_2$ rectangular unit pulse.

Example 1.2   (a) Find the impulse response of the filter described in Fig. 1.5a. (b) Find the impulse response of the filter characterized by the equation

$$y(n_1, n_2) = x(n_1, n_2) + by(n_1 - 1, n_2 - 1) \qquad (1.21)$$

Assume that the filter is initially relaxed, that is,

$$y(n_1, n_2) = 0 \qquad \text{for } n_1 < 0 \text{ or } n_2 < 0$$

Solution.   (a) With $x(n_1, n_2) = \delta(n_1, n_2)$, Eq. (1.9) becomes

$$v(n_1, n_2) = \delta(n_1, n_2) + e^{\alpha_1}v(n_1 - 1, n_2)$$

If $n_2 = 0$, then

$$v(0, 0) = 1 + e^{\alpha_1}v(-1, 0) = 1$$
$$v(1, 0) = 0 + e^{\alpha_1}v(0, 0) = e^{\alpha_1}$$
$$v(2, 0) = 0 + e^{\alpha_1}v(1, 0) = e^{2\alpha_1}$$
$$\cdots$$
$$v(n_1, 0) = e^{n_1\alpha_1}$$

If $n_2 > 0$, then

$$v(0, n_2) = 0 + e^{\alpha_1}v(-1, n_2) = 0$$
$$v(1, n_2) = 0 + e^{\alpha_1}v(0, n_2) = 0$$
$$\cdots$$
$$v(n_1, n_2) = 0$$

Hence

$$v(n_1, n_2) = e^{n_1\alpha_1}\delta_1(n_1, n_2)$$

With $v(n_1, n_2)$ determined as above, Eq. (1.10) becomes

$$y(n_1, n_2) = e^{n_1\alpha_1}\delta_1(n_1, n_2) + e^{\alpha_2}y(n_1, n_2 - 1)$$

Thus

$$y(n_1, 0) = e^{n_1\alpha_1}\delta_1(n_1, 0) + e^{\alpha_2}y(n_1, -1) = e^{n_1\alpha_1}\delta_1(n_1, 0)$$

$$y(n_1, 1) = e^{n_1\alpha_1}\delta_1(n_1, 1) + e^{\alpha_2}y(n_1, 0) = 0 + e^{\alpha_2}e^{n_1\alpha_1}\delta_1(n_1, 0)$$

$$y(n_1, 2) = 0 + e^{\alpha_2}y(n_1, 1) = e^{2\alpha_2}e^{n_1\alpha_1}\delta_1(n_1, 0)$$

$$\cdots$$

$$y(n_1, n_2) = e^{n_1\alpha_1}e^{n_2\alpha_2}\delta_1(n_1, 0) \qquad \text{for } n_2 \geq 0$$

Since

$$y(n_1, n_2) = 0 \qquad \text{for } n_2 < 0$$

we obtain the impulse response of the filter as

$$h(n_1, n_2) = e^{n_1\alpha_1}e^{n_2\alpha_2}\delta_1(n_1, 0)\delta_2(0, n_2)$$

$$= e^{n_1\alpha_1}e^{n_2\alpha_2}u(n_1, n_2)$$

Evidently, we can write

$$h(n_1, n_2) = [e^{n_1\alpha_1}\delta_1(n_1, 0)][e^{n_2\alpha_2}\delta_2(0, n_2)]$$

that is, the impulse response of the filter is separable. Under these circumstances, the filter is said to be separable since it can be viewed as a cascade arrangement of two distinct 1-D digital filters.

(b) On replacing $x(n_1, n_2)$ in Eq. (1.21) by $\delta(n_1, n_2)$ we obtain

$$y(n_1, n_2) = \delta(n_1, n_2) + by(n_1 - 1, n_2 - 1)$$

This equation can be solved recursively along a set of parallel straight lines. Starting with point $(0, 0)$, we can write

$$y(0, 0) = \delta(0, 0) + by(-1, -1) = 1$$

$$y(1, 0) = 0 + by(0, -1) = 0$$

$$y(0, 1) = 0 + by(-1, 0) = 0$$

$$y(2, 0) = 0 + by(1, -1) = 0$$

$$y(1, 1) = 0 + by(0, 0) = b$$

$$y(0, 2) = 0 + by(-1, 1) = 0$$

$$y(3, 0) = 0 + by(2, -1) = 0$$
$$y(2, 1) = 0 + by(1, 0) = 0$$
$$y(1, 2) = 0 + by(0, 1) = 0$$
$$y(0, 3) = 0 + by(-1, 2) = 0$$

$$\cdots$$

Proceeding in the same way, one can show that

$$y(n_1, n_2) = \begin{cases} b^n & \text{for } n_1 \geq 0 \text{ and } n_2 \geq 0, \text{ and } n_1 = n_2 = n \\ 0 & \text{otherwise} \end{cases}$$

Therefore, the impulse response is given by

$$h(n_1, n_2) = b^{n_1}\delta_1(n_1, n_1 - n_2)$$

or

$$h(n_1, n_2) = b^{n_2}\delta_2(n_1 - n_2, n_2)$$

1.6.3 Convolution Summation

By expressing an arbitrary input in terms of a weighted sum of unit impulses, it is possible to express the response of a 2-D linear, shift-invariant digital filter in terms of its impulse response.

An arbitrary input $x(n_1, n_2)$ can be expressed as

$$x(n_1, n_2) = \sum_{i=-\infty}^{\infty} \sum_{j=-\infty}^{\infty} x_{ij}(n_1, n_2)$$

where

$$x_{ij}(n_1, n_2) = \begin{cases} x(n_1, n_2) & \text{for } i = n_1 \text{ and } j = n_2 \\ 0 & \text{otherwise} \end{cases}$$

Alternatively,

$$x_{ij}(n_1, n_2) = x(i, j)\delta(n_1 - i, n_2 - j)$$

and hence

$$x(n_1, n_2) = \sum_{i=-\infty}^{\infty} \sum_{j=-\infty}^{\infty} x(i, j)\delta(n_1 - i, n_2 - j) \qquad (1.22)$$

Now let us assume that a 2-D digital filter is characterized by

$$y(n_1, n_2) = \Re x(n_1, n_2)$$

and that its impulse response is given by

$$h(n_1, n_2) = \Re\delta(n_1, n_2)$$

If the filter is linear and shift-invariant, then Eq. (1.22) yields

$$y(n_1, n_2) = \Re \sum_{i=-\infty}^{\infty} \sum_{j=-\infty}^{\infty} x(i, j)\delta(n_1 - i, n_2 - j)$$

$$= \sum_{i=-\infty}^{\infty} \sum_{j=-\infty}^{\infty} x(i, j)\Re\delta(n_1 - i, n_2 - j)$$

$$= \sum_{i=-\infty}^{\infty} \sum_{j=-\infty}^{\infty} x(i, j)h(n_1 - i, n_2 - j) \qquad (1.23a)$$

$$= \sum_{i=-\infty}^{\infty} \sum_{j=-\infty}^{\infty} h(i, j)x(n_1 - i, n_2 - j) \qquad (1.23b)$$

If the filter is causal, then any two inputs $x_1(n_1, n_2)$ and $x_2(n_1, n_2)$ such that

$$x_1(n_1, n_2) = x_2(n_1, n_2)$$

for $n_1 \leq i$ and $n_2 \leq j$, and

$$x_1(n_1, n_2) \neq x_2(n_1, n_2)$$

for some pair of values of n_1 and n_2 such that $n_1 > i$ or $n_2 > j$, will produce the same response for $n_1 \leq i$ and $n_2 \leq j$ (see Sec. 1.3.3); that is

$$y_1(n_1, n_2) - y_2(n_1, n_2) = \sum_{i=-\infty}^{\infty} \sum_{j=-\infty}^{\infty} [x_1(i,j) - x_2(i,j)]h(n_1 - i, n_2 - j) = 0$$

or

$$\sum\sum_{i\leq n_1 \text{ and } j\leq n_2} [x_1(i, j) - x_2(i, j)]h(n_1 - i, n_2 - j)$$

$$+ \sum\sum_{i>n_1 \text{or} j>n_2} [x_1(i, j) - x_2(i, j)]h(n_1 - i, n_2 - j) = 0$$

From the definition of the two input signals, the first term is zero. However, the second term is zero if and only if $h(n_1, n_2) = 0$ for $n_1 < 0$ or $n_2 < 0$ and therefore Eq. (1.23) can be expressed as

$$y(n_1, n_2) = \sum_{i=-\infty}^{n_1} \sum_{j=-\infty}^{n_2} x(i, j)h(n_1 - i, n_2 - j)$$

$$= \sum_{i=0}^{\infty} \sum_{j=0}^{\infty} h(i, j)x(n_1 - i, n_2 - j) \qquad (1.24)$$

Now if $x(n_1, n_2)$ has a first-quadrant support, we obtain

$$y(n_1, n_2) = \sum_{i=0}^{n_1} \sum_{j=0}^{n_2} x(i, j)h(n_1 - i, n_2 - j) \qquad (1.25a)$$

$$= \sum_{i=0}^{n_1} \sum_{j=0}^{n_2} h(i, j)x(n_1 - i, n_2 - j) \qquad (1.25b)$$

This relation is called the *2-D convolution summation* and is often represented in terms of the shorthand notation $x \otimes h$ or $h \otimes x$.

The convolution summation can be carried out as depicted in Fig. 1.11. The impulse response given in Fig. 1.11b is folded about the i and j axes, as illustrated in Fig. 1.11d, and is then shifted in the positive directions of the two axes, as illustrated in Fig. 1.11c. The values in Fig. 1.11c are then multiplied by the corresponding values in Fig. 1.11a to form Fig. 1.11e. The sum of all the products in Fig. 1.11e gives the response of the filter for n_1 and n_2, as illustrated in Fig. 1.11f.

1.7 STABILITY

A 2-D digital filter is said to be stable in the bounded-input, bounded-output (BIBO) sense, if a bounded input such that

$$|x(n_1, n_2)| \le M < \infty \qquad \text{for all } n_1 \text{ and } n_2 \text{ and some } M$$

yields a bounded output, that is,

$$|y(n_1, n_2)| \le N < \infty \qquad \text{for all } n_1 \text{ and } n_2 \text{ and some } N$$

For a linear, shift-invariant, and causal filter, Eq. (1.24) gives

$$|y(n_1, n_2)| \le \sum_{i=0}^{\infty} \sum_{j=0}^{\infty} |h(i, j)| \, |x(n_1 - i, n_2 - j)|$$

and if

$$|x(n_1, n_2)| \le M < \infty \qquad \text{for all } n_1 \text{ and } n_2$$

we have

$$|y(n_1, n_2)| \le M \sum_{i=0}^{\infty} \sum_{j=0}^{\infty} |h(i, j)|$$

If

$$\sum_{i=0}^{\infty} \sum_{j=0}^{\infty} |h(i, j)| \le K < \infty \qquad (1.26)$$

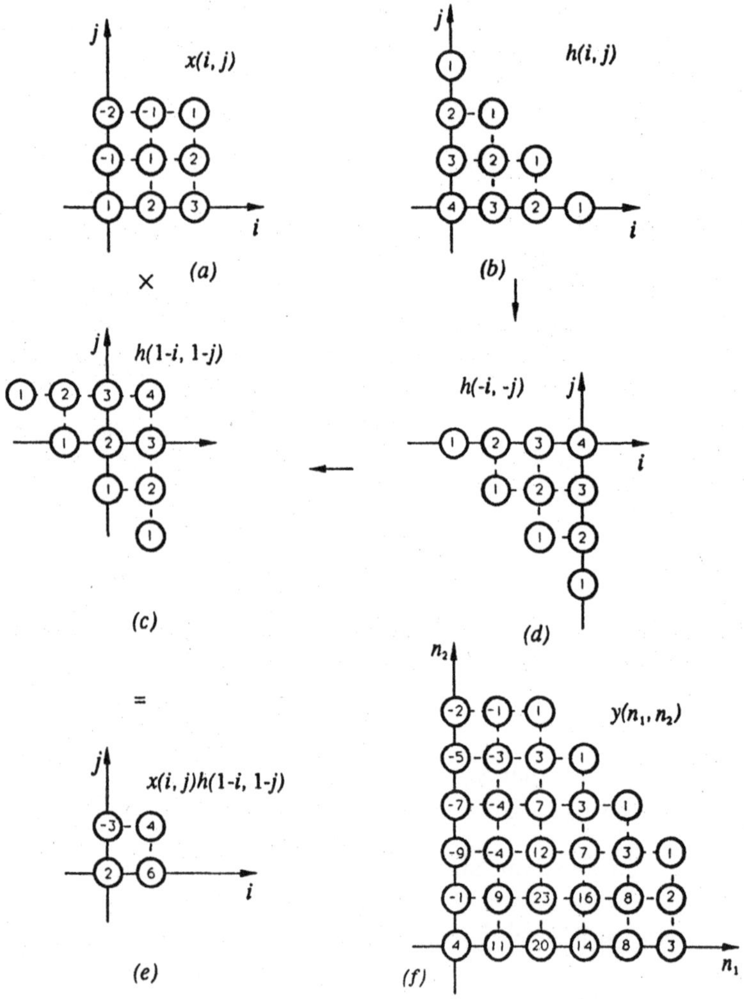

Figure 1.11 Convolution summation: (a) input; (b) impulse response; (c) shifted folded impulse response; (d) folded impulse response; (e) product of input values by the shifted folded impulse response values; (f) response of the filter at n_1 and n_2.

then

$$|y(n_1, n_2)| \le MK \le \infty \qquad \text{for all } n_1 \text{ and } n_2$$

that is, Eq. (1.26) is a sufficient condition for stability.

A 2-D filter can be considered to be stable if the output is bounded for all possible inputs. Let us consider the input

$$x(n_1 - i, n_2 - j) = \begin{cases} M & \text{if } h(i, j) \ge 0 \\ -M & \text{if } h(i, j) < 0 \end{cases}$$

where M is a positive constant. From Eq. (1.24)

$$y(n_1, n_2) = |y(n_1, n_2)| = \sum_{i=0}^{\infty} \sum_{j=0}^{\infty} M|h(i, j)|$$

or

$$|y(n_1, n_2)| = M \sum_{i=0}^{\infty} \sum_{j=0}^{\infty} |h(i, j)|$$

Therefore, the output is bounded for all n_1 and n_2 if and only if Eq. (1.26) is satisfied. In effect, the condition in Eq. (1.26) is a necessary and sufficient condition for stability.

1.8 REALIZATION

The process of converting the mathematical characterization of a 2-D digital filter into a network is said to be the realization of the filter.

For a given characterization, several digital-filter networks can usually be obtained by using different realization methods. Although the possible networks that can be obtained are all equivalent when infinite-precision arithmetic is assumed, significant differences arise in practice when the networks are implemented in terms of finite-precision hardware.

1.8.1 Direct Structure

The simplest approach to realization consists of converting the difference equation of the digital filter into a network without modifying or manipulating the difference equation. The structure obtained is said to be direct for obvious reasons.

1.8.2   Parallel Structure

Let us consider a 2-D digital filter whose impulse response can be expressed as a sum of two functions of n_1 and n_2, that is,

$$h(n_1, n_2) = h_1(n_1, n_2) + h_2(n_1, n_2)$$

From the convolution summation of Eq. (1.23), we can write

$$y(n_1, n_2) = y_1(n_1, n_2) + y_2(n_1, n_2)$$

where

$$y_1(n_1, n_2) = \sum_{i=-\infty}^{\infty} \sum_{j=-\infty}^{\infty} h_1(i, j)x(n_1 - i, n_2 - j)$$

$$y_2(n_1, n_2) = \sum_{i=-\infty}^{\infty} \sum_{j=-\infty}^{\infty} h_2(i, j)x(n_1 - i, n_2 - j)$$

Therefore, the impulse response $h(n_1, n_2)$ can be realized by connecting in parallel two filters with impulse responses $h_1(n_1, n_2)$ and $h_2(n_1, n_2)$, as depicted in Fig. 1.12a. Similarly, if the impulse response can be expressed as a sum of k impulse responses, a structure consisting of k parallel sections is obtained.

1.8.3   Cascade Structure

Let us assume that the impulse response of the filter can be expressed as a convolution of two functions of n_1, n_2 such that

$$h(n_1, n_2) = \sum_{l=-\infty}^{\infty} \sum_{m=-\infty}^{\infty} h_2(l, m)h_1(n_1 - l, n_2 - m)$$

The response of the filter to an input $x(n_1, n_2)$ can be written as

$$y(n_1, n_2) = \sum_{i=-\infty}^{\infty} \sum_{j=-\infty}^{\infty} x(i, j)h(n_1 - i, n_2 - j)$$

$$= \sum_{i=-\infty}^{\infty} \sum_{j=-\infty}^{\infty} x(i, j) \sum_{l=-\infty}^{\infty} \sum_{m=-\infty}^{\infty} h_2(l, m)h_1(n_1 - l - i, n_2 - m - j)$$

$$= \sum_{l=-\infty}^{\infty} \sum_{m=-\infty}^{\infty} h_2(l, m) \sum_{i=-\infty}^{\infty} \sum_{j=-\infty}^{\infty} x(i, j)h_1(n_1 - l - i, n_2 - m - j)$$

$$= \sum_{l=-\infty}^{\infty} \sum_{m=-\infty}^{\infty} h_2(l, m)w(n_1 - l, n_2 - m)$$

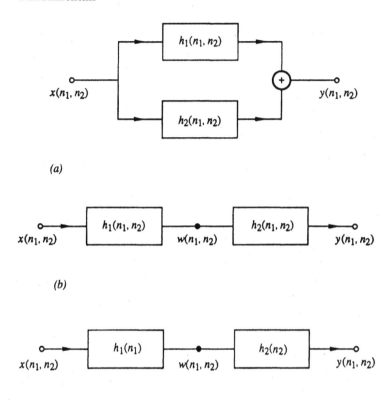

(a)

(b)

(c)

Figure 1.12 Realization: (a) parallel structure; (b) cascade structure; (c) separable structure.

where

$$w(n_1, n_2) = \sum_{i=-\infty}^{\infty} \sum_{j=-\infty}^{\infty} x(i, j) h_1(n_1 - i, n_2 - j)$$

From Eq. (1.23), it follows that $w(n_1, n_2)$ and $y(n_1, n_2)$ are the responses of filters with impulse responses $h_1(n_1, n_2)$ and $h_2(n_1, n_2)$ and inputs $x(n_1, n_2)$ and $w(n_1, n_2)$, respectively. Therefore, the impulse response $h(n_1, n_2)$ can be realized by connecting two filters with impulse responses $h_1(n_1, n_2)$ and $h_2(n_1, n_2)$ in cascade, as depicted in Fig. 1.12b. If the impulse response can be expressed in terms of k nested convolutions of the form

$$h = h_1 \otimes (h_2 \otimes (h_3 \otimes \cdots (h_k) \cdots))$$

then a structure consisting of k cascade sections is obtained.

1.8.4 Separable Structures

The case where the impulse response of a filter is separable is of particular importance. Let

$$h(n_1, n_2) = h_1(n_1)h_2(n_2)$$

where $h_1(n_1)$ and $h_2(n_2)$ are independent of n_2 and n_1, respectively. The convolution summation of Eq. (1.23b) gives the response of the filter to an input $x(n_1, n_2)$ as

$$y(n_1, n_2) = \sum_{i=-\infty}^{\infty} \sum_{j=-\infty}^{\infty} h_1(i)h_2(j)x(n_1 - i, n_2 - j)$$

$$= \sum_{i=-\infty}^{\infty} h_1(i) \sum_{j=-\infty}^{\infty} h_2(j)x(n_1 - i, n_2 - j)$$

Alternatively,

$$y(n_1, n_2) = \sum_{i=-\infty}^{\infty} h_1(i)w(n_1 - i, n_2)$$

where

$$w(n_1, n_2) = \sum_{j=-\infty}^{\infty} h_2(i)x(n_1, n_2 - j)$$

In effect, $w(n_1, n_2)$ and $y(n_1, n_2)$ can be regarded as the responses of two 1-D digital filters with impulse responses $h_2(n_1)$ and $h_1(n_2)$ and inputs $x(n_1, n_2)$ and $w(n_1, n_2)$, respectively, as depicted in Fig. 1.12c. The structure obtained is said to be separable for obvious reasons. Separability is of considerable practical importance since it allows the design of a 2-D digital filter to be carried out by designing two 1-D digital filters and then connecting them in cascade.

The preceding principles can be extended to the case where the impulse response can be expressed as a linear combination of separable impulse responses, as in Eq. (1.1). In such a case, each of the products can be realized by two 1-D filters in cascade and the sum of the products can be formed by connecting the filters obtained in parallel, as depicted in Fig. 1.13.

1.9 MULTIPLE-INPUT–MULTIPLE-OUTPUT FILTERS

In certain applications 2-D digital filters are required that have several inputs and several outputs. In the case where there are m inputs and p outputs, as depicted in Fig. 1.14, each output is a function of each and

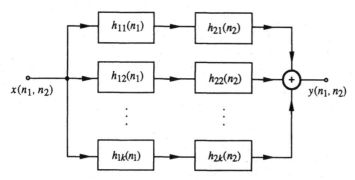

Figure 1.13 Parallel structure with separable cascade sections.

every input as well as each and every output. Consequently, the filter can be represented by p equations as

$$y_1(n_1, n_2) = \sum_{i=0}^{N_1} \sum_{j=0}^{N_2} [a_{11ij}x_1(n_1 - i, n_2 - j) + a_{12ij}x_2(n_1 - i, n_2 - j)$$

$$+ \cdots + a_{1mij}x_m(n_1 - i, n_2 - j)]$$

$$- \sum_{i=0}^{M_1} \sum_{j=0}^{M_2} [b_{11ij}y_1(n_1 - i, n_2 - j) + b_{12ij}y_2(n_1 - i, n_2 - j)$$

$$+ \cdots + b_{1pij}y_p(n_1 - i, n_2 - j)]$$

$$y_2(n_1, n_2) = \sum_{i=0}^{N_1} \sum_{j=0}^{N_2} [a_{21ij}x_1(n_1 - i, n_2 - j) + a_{22ij}x_2(n_1 - i, n_2 - j)$$

$$+ \cdots + a_{2mij}x_m(n_1 - i, n_2 - j)]$$

$$- \sum_{i=0}^{M_1} \sum_{j=0}^{M_2} [b_{21ij}y_1(n_1 - i, n_2 - j) + b_{22ij}y_2(n_1 - i, n_2 - j)$$

$$+ \cdots + b_{2pij}y_p(n_1 - i, n_2 - j)]$$

$$\cdots\cdots\cdots$$

$$y_p(n_1, n_2) = \sum_{i=0}^{N_1} \sum_{j=0}^{N_2} [a_{p1ij}x_1(n_1 - i, n_2 - j) + a_{p2ij}x_2(n_1 - i, n_2 - j)$$

$$+ \cdots + a_{pmij}x_m(n_1 - i, n_2 - j)]$$

$$- \sum_{i=0}^{M_1} \sum_{j=0}^{M_2} [b_{p1ij}y_2(n_1 - i, n_2 - j) + b_{p2ij}y_2(n_1 - i, n_2 - j)$$

$$+ \cdots + b_{ppij}y_p(n_1 - i, n_2 - j)]$$

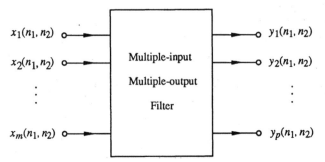

Figure 1.14 Multiple-input–multiple-output filter

This system of equations can be written in terms of matrix notation as

$$\mathbf{y}(n_1, n_2) = \sum_{i=0}^{N_1} \sum_{j=0}^{N_2} \mathbf{A}_{ij}\mathbf{x}(n_1 - i, n_2 - j)$$

$$- \sum_{i=0}^{M_1} \sum_{j=0}^{M_2} \mathbf{B}_{ij}\mathbf{y}(n_1 - i, n_2 - j)$$

(1.27)

where $\mathbf{x}(n_1, n_2)$ and $\mathbf{y}(n_1, n_2)$ are vectors of m and p dimensions, respectively, and

$$\mathbf{A}_{ij} = \begin{bmatrix} a_{11ij} & a_{12ij} & \cdots & a_{1mij} \\ a_{21ij} & a_{22ij} & \cdots & a_{2mij} \\ \cdot & \cdot & & \cdot \\ \cdot & \cdot & & \cdot \\ \cdot & \cdot & & \cdot \\ a_{p1ij} & a_{p2ij} & \cdots & a_{pmij} \end{bmatrix},$$

$$\mathbf{B}_{ij} = \begin{bmatrix} b_{11ij} & b_{12ij} & \cdots & b_{1pij} \\ b_{21ij} & b_{22ij} & \cdots & b_{2pij} \\ \cdot & \cdot & & \cdot \\ \cdot & \cdot & & \cdot \\ \cdot & \cdot & & \cdot \\ b_{p1ij} & b_{p2ij} & \cdots & b_{ppij} \end{bmatrix}, \qquad \mathbf{B}_{00} = \mathbf{0}$$

1.10 MULTIDIMENSIONAL FILTERS

In certain applications the signals to be processed are quantities that depend on several integer variables, say $n_1, n_2, \ldots, n_k$. Such signals can be processed by multidimensional digital filters that are natural extensions of their 2-D counterparts. These filters can be characterized by difference equations of the form

$$y(n_1, n_2, \ldots, n_k) = \sum_{i=0}^{N_1} \sum_{j=0}^{N_2} \cdots \sum_{m=0}^{N_k} a_{ij\cdots m} x(n_1 - i, n_2 - j, \ldots, n_k - m)$$

$$- \sum_{i=0}^{M_1} \sum_{j=0}^{M_2} \cdots \sum_{m=0}^{M_k} b_{ij\cdots m} y(n_1 - i, n_2 - j, \ldots, n_k - m)$$

with $b_{00\cdots0} = 0$ and are subject to the same underlying principles as 2-D digital filters.

REFERENCE

1. G. W. Stewart, *Introduction to Matrix Computations*, New York: Academic Press, 1973.

PROBLEMS

1.1 Let $x_1(n_1, n_2)$ and $x_2(n_1, n_2)$ be two periodic signals with periods (N_{11}, N_{12}) and (N_{21}, N_{22}), respectively. Show that $x_1(n_1, n_2) + x_2(n_1, n_2)$ and $x_1(n_1, n_2)x_2(n_1, n_2)$ are periodic signals.

1.2 Consider a 2-D signal $x_1(n_1, n_2)$ given by

$$x(n_1, n_2) = [1 \quad n_1^{-1} \quad n_1^{-2} \cdots n_1^{-N_1}]C[1 \quad n_2^{-1} \quad n_2^{-2} \cdots n_2^{-N_1}]^T$$

where C is an $N_1 \times N_2$ constant matrix. Show that $x(n_1, n_2)$ is a separable signal if and only if the rank of C is equal to one.

1.3 Check the filters characterized by the following equations for linearity, shift invariance, and causality:
 (a) $\Re x(n_1, n_2) = 2\pi x(n_1 - 1, n_2) - 3x(n_1, n_2 - 1) - 0.5n_2 x(n_1 - 1, n_2 - 1)$.
 (b) $\Re x(n_1, n_2) = 1 - x(n_1 - 1, n_2) + 0.3x(n_1, n_2 - 1)$.
 (c) $\Re x(n_1, n_2) = 0.5x(n_1 - 1, n_2) - 0.03x^2(n_1, n_2 - 1) - 0.2x (n_1 - 1, n_2 - 1)$.
 (d) $\Re x(n_1, n_2) = 0.5x(n_1 - 1, n_2) - 0.5x(n_1 - 1, n_2 + 1)$.

1.4 Repeat Problem 1.3 for the filters characterized by the following equations:
 (a) $\Re x(n_1, n_2) = e^{-x(n_1 - 1, \, n_2)} + e^{-0.5x(n_1, \, n_2 - 1)}$.
 (b) $\Re x(n_1, n_2) = 0.01(n_1 + 1)^2 x(n_1 - 1, n_2) + 0.1x(n_1, n_2 - 1) - 0.1x(n_1 - 1, n_2 - 1)$.

(c) $\mathscr{R}x(n_1, n_2) = 0.5x(n_1 - 2, 0) - 0.5x(0, n_2 - 2)$.

(d) $\mathscr{R}x(n_1, n_2) = -0.2x(n_1 - 1, n_2) + 0.3|x(n_1, n_2 - 1)| + 0.2x$ $(n_1 - 1, n_2 - 1)$.

1.5 (a) Analyze the filter network shown in Fig. P1.5a where F_i represents a 1-D filter characterized by the network depicted in Fig. P1.5b.

(b) Repeat part (a) for the network of Fig. P1.5c where F_i is the same as in part (a).

1.6 Analyze the filter network shown in Fig. P1.6.

1.7 (a) Find the impulse response of the filter illustrated in Fig. P1.5a.

(b) Find the impulse response of the filter illustrated in Fig. P1.5c.

(c) Find the impulse response of the filter characterized by the equation

$$y(n_1, n_2) = x(n_1, n_2) + 0.5y(n_1 - 1, n_2) - 0.3y(n_1 - 2, n_2)$$

(a)

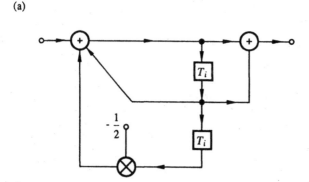

(b)

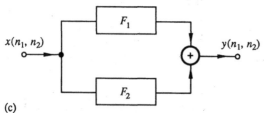

(c)

Figure P1.5

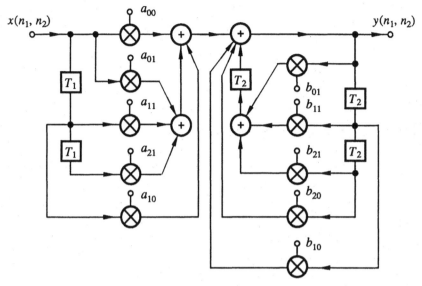

Figure P1.6

1.8 (a) Show that if the impulse response of a 2-D filter is separable, then a separable input signal to the filter always yields a separable output signal.

(b) Show that the filter represented by the network in Fig. P1.8 is stable if and only if both F_1 and F_2 represent stable 1-D filters.

1.9 Compute the output of a filter whose impulse response is given in Fig. 1.11b to the input signals given by

(a) $x(n_1, n_2) = \begin{cases} n_2 - n_1 - 1 & \text{for } 1 \leq n_1 \leq 3,\, 1 \leq n_2 \leq 3 \\ 0 & \text{otherwise} \end{cases}$

(b) $x(n_1, n_2) = \begin{cases} |n_2 - n_1| + 1 & \text{for } 1 \leq n_1 \leq 3,\, 1 \leq n_2 \leq 3 \\ 0 & \text{otherwise} \end{cases}$

1.10 Show that a 2-D filter represented by Fig. 1.12c is BIBO stable if and only if the two 1-D filters involved are BIBO stable.

1.11 Check the BIBO stability of the filters represented by Fig. P1.5a and P1.5b.

Figure P1.8

2
State-Space Methods

2.1 INTRODUCTION

Like 1-D digital filters and other types of systems, 2-D digital filters can be represented in terms of state-space models. In this approach a set of internal signals referred to as state variables is used to describe completely the operation of the filter. The approach has proved very useful in the analysis, design, and implementation of digital filters. It has the advantage that the characterization of the digital filter is in terms of matrices, which are easy to manipulate by means of array or vector processors. State-space models for 2-D digital filters have been proposed by Attasi [1], Givone and Roesser [2], and Fornasini and Marchesini [3]. The model of Givone and Roesser follows naturally from a network theoretic approach and has been used extensively in the past [4] owing to its generality and relative simplicity. In fact, the models of Attasi and Fornasini and Marchesini can actually be derived from it.

In this chapter the model of Givone and Roesser, which will be referred to as the Givone–Roesser model, is described and is then applied to the space-domain analysis and realization of 2-D digital filters. In later chapters it will be applied in the stability analysis, the study of quantization effects, and the design and implementation of 2-D digital filters.

2.2 THE GIVONE–ROESSER MODEL

As was shown in Sec. 1.4.2, a linear, shift-invariant, causal, recursive 2-D digital filter can be represented by the difference equation

$$y(n_1, n_2) = \sum_{i=0}^{N_1} \sum_{j=0}^{N_2} a_{ij} x(n_1 - i, n_2 - j)$$

$$- \sum_{i=0}^{N_1} \sum_{j=0}^{N_2} b_{ij} y(n_1 - i, n_2 - j) \tag{2.1}$$

where $b_{00} = 0$. Straightforward manipulation will show that Eq. (2.1) can be represented by the network depicted in Fig. 2.1 where N and D are nonrecursive filters characterized by the difference equations

$$y(n_1, n_2) = \sum_{i=0}^{N_1} \sum_{j=0}^{N_2} a_{ij} v(n_1 - i, n_2 - j) \tag{2.2}$$

and

$$w(n_1, n_2) = \sum_{i=0}^{N_1} \sum_{j=0}^{N_2} b_{ij} v(n_1 - i, n_2 - j) \tag{2.3}$$

respectively. Equation (2.2) can be expressed as

$$y(n_1, n_2) = \sum_{j=0}^{N_2} E_2^{-j} \sum_{i=0}^{N_1} a_{ij} v(n_1 - i, n_2)$$

$$= \sum_{j=0}^{N_2} E_2^{-j} y_j(n_1, n_2) \tag{2.4}$$

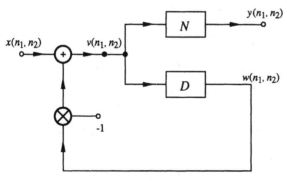

Figure 2.1 Realization of Eq. (2.1).

where

$$y_j(n_1, n_2) = \sum_{i=0}^{N_1} a_{ij} v(n_1 - i, n_2) \tag{2.5}$$

Evidently, Eqs. (2.4) and (2.5) represent the subnetworks in Fig. 2.2a and b, and by combining them, nonrecursive filter N in Fig. 2.1 can readily be obtained. Similarly, Eq. (2.3) can be expressed as

$$w(n_1, n_2) = \sum_{j=0}^{N_2} E_2^{-j} \sum_{i=0}^{N_1} b_{ij} v(n_1 - i, n_2)$$

$$= \sum_{j=0}^{N_2} E_2^{-j} w_j(n_1, n_2) \tag{2.6}$$

where

$$w_j(n_1, n_2) = \sum_{i=0}^{N_1} b_{ij} v(n_1 - i, n_2) \tag{2.7}$$

The last two equations represent the subnetworks in Fig. 2.3a and b, which can be interconnected to give nonrecursive filter D in Fig. 2.1. On interconnecting filters N and D as in Fig. 2.1 and using a common set of horizontal shifters for filters N and D, the realization in Fig. 2.4 is obtained.

By treating the outputs of shifters as state variables, a 2-D state vector $q(i, j)$ can be defined as

$$\mathbf{q}(i, j) = \begin{bmatrix} \mathbf{q}^H(i, j) \\ \mathbf{q}^V(i, j) \end{bmatrix} \tag{2.8}$$

where

$$\mathbf{q}(i, j)^H = \begin{bmatrix} q_1^H(i, j) \\ \cdot \\ \cdot \\ \cdot \\ q_{N_1}^H(i, j) \end{bmatrix} \quad \text{and} \quad \mathbf{q}(i, j)^V = \begin{bmatrix} q_1^{V_1}(i, j) \\ \cdot \\ \cdot \\ \cdot \\ q_{N_2}^{V_1}(i, j) \\ q_1^{V_2}(i, j) \\ \cdot \\ \cdot \\ q_{N_2}^{V_2}(i, j) \end{bmatrix}$$

The vector with superscript H represents the outputs of the horizontal shifters in Fig. 2.4 and the vectors with superscripts V_1 and V_2 represent the outputs of the two groups of vertical shifters.

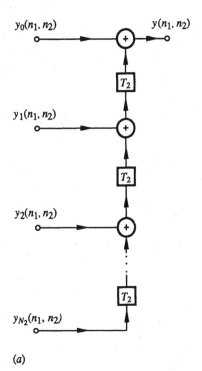

(a)

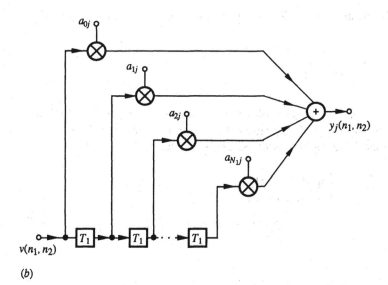

(b)

Figure 2.2 Realization of filter N: (a) realization of Eq. (2.4); (b) realization of Eq. (2.5).

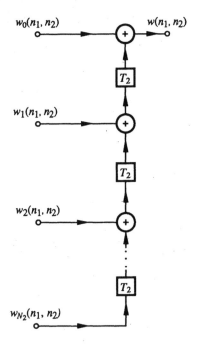

$w_0(n_1, n_2)$

$w(n_1, n_2)$

$w_1(n_1, n_2)$

$w_2(n_1, n_2)$

$w_{N_2}(n_1, n_2)$

(a)

Figure 2.3   Realization of filter D: (a) realization of Eq. (2.6); (b) realization of Eq. (2.7).

The horizontal state variables for $n_1 = i + 1$, and the vertical state variables for $n_2 = j + 1$, as well as the output of the filter $y(i, j)$, can now be expressed in terms the state variables defined in Eq. (2.8) as

$$\begin{bmatrix} \mathbf{q}^H(i + 1, j) \\ \mathbf{q}^V(i, j + 1) \end{bmatrix} = \mathbf{A}\mathbf{q}(i, j) + \mathbf{b}x(i, j) \qquad (2.9a)$$

$$y(i, j) = \mathbf{c}\mathbf{q}(i, j) + \mathbf{d}x(i, j) \qquad (2.9b)$$

where

$$\mathbf{A} = \begin{bmatrix} \mathbf{A}_1 & \mathbf{A}_2 \\ \mathbf{A}_3 & \mathbf{A}_4 \end{bmatrix}, \qquad \mathbf{b} = \begin{bmatrix} \mathbf{b}_1 \\ \mathbf{b}_2 \end{bmatrix}, \qquad \mathbf{c} = [\mathbf{c}_1 \ \mathbf{c}_2], \qquad \text{and} \quad \mathbf{d} = a_{00}$$

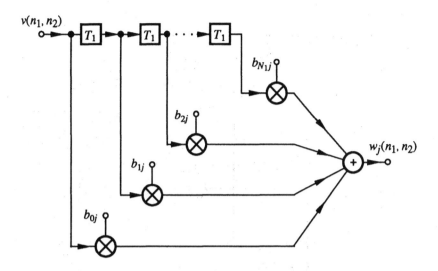

(b)

Figure 2.3 Continued.

with

$$
\mathbf{A}_1 =
\begin{bmatrix}
-b_{10} & -b_{20} & \cdots & -b_{(N_1-1)0} & -b_{N_10} \\
1 & 0 & \cdots & 0 & 0 \\
\cdot & \cdot & & \cdot & \cdot \\
\cdot & \cdot & & \cdot & \cdot \\
\cdot & \cdot & & \cdot & \cdot \\
0 & 0 & \cdots & 1 & 0
\end{bmatrix},
$$

$$
\mathbf{A}_2 =
\begin{bmatrix}
-1 & 0 & \cdots & 0 & 0 & \cdots & 0 \\
0 & 0 & \cdots & 0 & 0 & \cdots & 0 \\
\cdot & & & & & & \cdot \\
\cdot & & & & & & \\
\cdot & & & & & & \cdot \\
0 & 0 & \cdots & 0 & 0 & \cdots & 0
\end{bmatrix}
$$

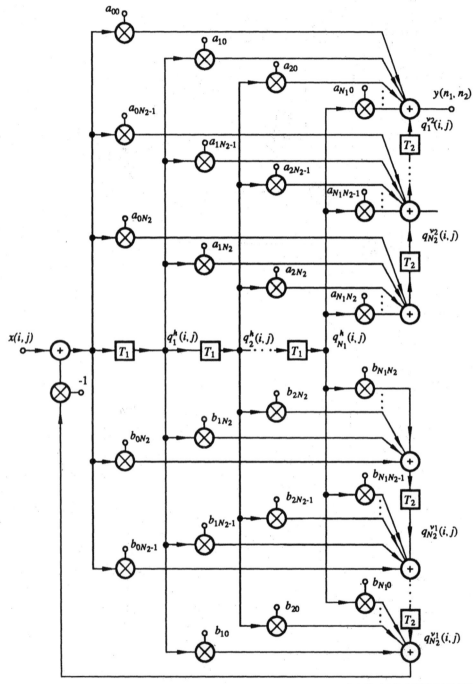

Figure 2.4 Realization for state-space model.

$$
\mathbf{A}_3 = \begin{bmatrix}
\tilde{b}_{11} & \tilde{b}_{21} & \cdots & \tilde{b}_{N_11} \\
\cdot & \cdot & & \cdot \\
\cdot & \cdot & & \cdot \\
\cdot & \cdot & & \cdot \\
\tilde{b}_{1N_2} & \tilde{b}_{2N_2} & \cdots & \tilde{b}_{N_1N_2} \\
\tilde{a}_{11} & \tilde{a}_{21} & \cdots & \tilde{a}_{N_11} \\
\cdot & \cdot & & \cdot \\
\cdot & \cdot & & \cdot \\
\cdot & \cdot & & \cdot \\
\tilde{a}_{1N_2} & \tilde{a}_{2N_2} & \cdots & \tilde{a}_{N_1N_2}
\end{bmatrix},
$$

$$
\mathbf{A}_4 = \begin{bmatrix}
-b_{01} & 1 & \cdots & 0 & 0 & 0 & \cdots & 0 \\
\cdot & \cdot & & \cdot & \cdot & \cdot & & \cdot \\
\cdot & \cdot & & \cdot & \cdot & \cdot & & \cdot \\
\cdot & \cdot & & \cdot & \cdot & \cdot & & \cdot \\
-b_{0(N_2-1)} & 0 & \cdots & 1 & 0 & 0 & \cdots & 0 \\
-b_{0N_2} & 0 & \cdots & 0 & 0 & 0 & \cdots & 0 \\
-a_{01} & 0 & \cdots & 0 & 0 & 1 & \cdots & 0 \\
\cdot & \cdot & & \cdot & \cdot & \cdot & & \cdot \\
\cdot & \cdot & & \cdot & \cdot & \cdot & & \cdot \\
\cdot & \cdot & & \cdot & \cdot & \cdot & & \cdot \\
-a_{0(N_2-1)} & 0 & & 0 & 0 & 0 & & 1 \\
-a_{0N_2} & 0 & \cdots & 0 & 0 & 0 & \cdots & 0
\end{bmatrix}.
$$

$$
\mathbf{b}_1 = \begin{bmatrix} 1 \\ 0 \\ \cdot \\ \cdot \\ \cdot \\ 0 \end{bmatrix}, \quad
\mathbf{b}_2 = \begin{bmatrix} b_{01} \\ \cdot \\ \cdot \\ \cdot \\ b_{0N_2} \\ a_{01} \\ \cdot \\ \cdot \\ \cdot \\ a_{0N_2} \end{bmatrix}
$$

$$
\mathbf{c}_1 = [\tilde{c}_{10} \cdots \tilde{a}_{N_10}], \quad \mathbf{c}_2 = [-a_{00}\ 0 \cdots 0\ 1\ 0 \cdots 0]
$$

and

$$
\tilde{a}_{ij} = a_{ij} - b_{i0}a_{0j}, \quad 1 \le i \le N_1,\ 0 \le j \le N_2
$$
$$
\tilde{b}_{ij} = b_{ij} - b_{i0}b_{0j}, \quad 1 \le i \le N_1,\ 1 \le j \le N_2
$$

If the digital filter has p distinct inputs and p distinct outputs, then it can be characterized by the difference equation

$$y(n_1, n_2) = \sum_{i=0}^{N_1} \sum_{j=0}^{N_2} \mathbf{A}_{ij}\mathbf{x}(n_1 - i, n_2 - j)$$

$$- \sum_{i=0}^{N_1} \sum_{j=0}^{N_2} \mathbf{B}_{ij}\mathbf{y}(n_1 - i, n_2 - j) \qquad (2.10)$$

(see Sec. 1.9) where

$$\mathbf{x}(i, j) = \begin{bmatrix} x_1(i, j) \\ \cdot \\ \cdot \\ \cdot \\ x_p(i, j) \end{bmatrix} \quad \text{and} \quad \mathbf{y}(i, j) = \begin{bmatrix} y_1(i, j) \\ \cdot \\ \cdot \\ \cdot \\ y_p(i, j) \end{bmatrix}$$

are the input and output signal vectors, respectively, $\mathbf{A}_{ij}$ and $\mathbf{B}_{ij}$ are matrices of dimension $p \times p$ and $\mathbf{B}_{00} = \mathbf{0}$.

Equation (2.10) can be realized as in the single-input, single-output case, as shown in Fig. 2.4, except that the multiplier constants a_{ij} and b_{ij} are replaced by constant matrices $\mathbf{A}_{ij}$ and $\mathbf{B}_{ij}$, respectively, and any constants that are unity or zero are replaced by the identity or zero matrix of dimension p. With these modifications the Givone–Roesser model for a p-input, p-output digital filter can be obtained as

$$\begin{bmatrix} \mathbf{q}^H(i + 1, j) \\ \mathbf{q}^V(i, j + 1) \end{bmatrix} = \mathbf{A}\mathbf{q}(i, j) + \mathbf{B}\mathbf{x}(i, j) \qquad (2.11a)$$

$$\mathbf{y}(i, j) = \mathbf{C}\mathbf{q}(i, j) + \mathbf{D}\mathbf{x}(i, j) \qquad (2.11b)$$

where

$$\mathbf{A} = \begin{bmatrix} \mathbf{A}_1 & \mathbf{A}_2 \\ \mathbf{A}_3 & \mathbf{A}_4 \end{bmatrix}, \quad \mathbf{B} = \begin{bmatrix} \mathbf{B}_1 \\ \mathbf{B}_2 \end{bmatrix}, \quad \mathbf{C} = [\mathbf{C}_1 \ \mathbf{C}_2], \quad \text{and} \quad \mathbf{D} = \mathbf{A}_{00}$$

are constant matrices of dimensions $(N_1 + 2N_2)p \times (N_1 + 2N_2)p$, $(N_1 + 2N_2)p \times p$, $p \times (N_1 + 2N_2)p$, and $p \times p$, respectively. Note that while $\mathbf{A}$ is partitioned into submatrices $\mathbf{A}_1, \ldots, \mathbf{A}_4$ as in the single-input, single-output case, the dimensions of the various submatrices are quite different.

The preceding generalization of the Givone–Roesser model has been derived for a p-input–p-output 2-D filter for the sake of simplicity. Nevertheless, it is also applicable for the representation of an m-input, p-output filter except that the dimensions of matrices $\mathbf{A}$, $\mathbf{B}$, $\mathbf{C}$, and $\mathbf{D}$ become $[(N_1 + N_2)p + N_2m] \times [(N_1 + N_2)p + N_2m]$, $[(N_1 + N_2)p + N_2m] \times m$, $p \times [(N_1 + N_2)p + N_2m]$, and $p \times m$, respectively.

Example 2.1 (a) Find a state-space representation for the filter described in Fig. 1.5b. (b) Repeat part (a) for the filter described by Eq. (1.21).

Solution. (a) The difference equation of the filter depicted in Fig. 1.5b is given by

$$y(n_1, n_2) = x(n_1, n_2) + b_1 y(n_1 - 1, n_2) + b_2 y(n_1 - 1, n_2 - 2)$$

Hence $N_1 = 1$, $N_2 = 2$,

$$a_{00} = 1, \qquad a_{01} = a_{02} = a_{10} = a_{11} = a_{12} = 0,$$
$$b_{01} = b_{02} = b_{11} = 0, \qquad b_{10} = -b_1, \qquad b_{12} = -b_2,$$

and

$$\tilde{a}_{10} = b_1, \qquad \tilde{a}_{11} = \tilde{a}_{12} = 0, \qquad \tilde{b}_{11} = 0, \qquad \tilde{b}_{12} = -b_2$$

From Eq. (2.1), we conclude that the filter can be represented by Eq. (2.9) with

$$\mathbf{A}_1 = [b_1], \qquad \mathbf{A}_2 = [-1 \ 0 \ 0 \ 0],$$

$$\mathbf{A}_3 = \begin{bmatrix} 0 \\ -b_2 \\ 0 \\ 0 \end{bmatrix}, \qquad \mathbf{A}_4 = \begin{bmatrix} 0 & 1 & 0 & 0 \\ 0 & 0 & 0 & 0 \\ 0 & 0 & 0 & 1 \\ 0 & 0 & 0 & 0 \end{bmatrix}$$

$$\mathbf{b}_1 = [1], \qquad \mathbf{b}_2 = \begin{bmatrix} 0 \\ 0 \\ 0 \\ 0 \end{bmatrix}$$

$$\mathbf{c}_1 = [b_1], \qquad \mathbf{c}_2 = [-1 \ 0 \ 1 \ 0], \qquad \text{and} \quad \mathbf{d} = 1$$

(b) From Eq. (1.21), we have $N_1 = N_2 = 1$,

$$a_{00} = 1, \qquad a_{01} = a_{10} = a_{11} = 0, \qquad b_{01} = b_{10} = 0,$$
$$b_{11} = -b, \tilde{a}_{10} = 0, \qquad \tilde{a}_{11} = 0, \tilde{b}_{11} = -b$$

Therefore,

$$\mathbf{A} = \left[\begin{array}{c|cc} 0 & -1 & 0 \\ \hline -b & 0 & 0 \\ 0 & 0 & 0 \end{array} \right], \qquad \mathbf{b} = \left[\begin{array}{c} 1 \\ \hline 0 \\ 0 \end{array} \right],$$

$$\mathbf{c} = [0 \ \vdots \ -1 \ 1], \qquad \text{and} \quad \mathbf{d} = [1]$$

2.3 SPACE–DOMAIN ANALYSIS

Let us assume that the input $x(i, j)$ of a single-input, single-output 2-D digital filter is zero for all pairs (i, j). From Eq. (2.9), we can write

$$
\mathbf{q}(i, j) = \begin{bmatrix} \mathbf{q}^H(i, j) \\ \mathbf{q}^V(i, j) \end{bmatrix} = \begin{bmatrix} \mathbf{A}_1\mathbf{q}^H(i - 1, j) + \mathbf{A}_2\mathbf{q}^V(i - 1, j) + \mathbf{b}_1 \cdot 0 \\ \mathbf{A}_3\mathbf{q}^H(i, j - 1) + \mathbf{A}_4\mathbf{q}^V(i, j - 1) + \mathbf{b}_2 \cdot 0 \end{bmatrix}
$$

$$
= \begin{bmatrix} \mathbf{A}_1 & \mathbf{A}_2 \\ 0 & 0 \end{bmatrix} \mathbf{q}(i - 1, j) + \begin{bmatrix} 0 & 0 \\ \mathbf{A}_3 & \mathbf{A}_4 \end{bmatrix} \mathbf{q}(i, j - 1)
$$

$$
= \mathbf{A}^{1,0}\mathbf{q}(i - 1, j) + \mathbf{A}^{0,1}\mathbf{q}(i, j - 1) \tag{2.12}
$$

where

$$
A^{1,0} = \begin{bmatrix} \mathbf{A}_1 & \mathbf{A}_2 \\ 0 & 0 \end{bmatrix} \quad \text{and} \quad \mathbf{A}^{0,1} = \begin{bmatrix} 0 & 0 \\ \mathbf{A}_3 & \mathbf{A}_4 \end{bmatrix}
$$

With this notation, $\mathbf{q}(i, j)$ can be expressed as

$$
\begin{aligned}
\mathbf{q}(i, j) &= \mathbf{A}^{1,0}[\mathbf{A}^{1,0}\mathbf{q}(i - 2, j) + \mathbf{A}^{0,1}\mathbf{q}(i - 1, j - 1)] \\
&\quad + \mathbf{A}^{0,1}[\mathbf{A}^{1,0}\mathbf{q}(i - 1, j - 1) + \mathbf{A}^{0,1}\mathbf{q}(i, j - 2)] \\
&= \mathbf{A}^{1,0} \cdot \mathbf{A}^{1,0}\mathbf{q}(i - 2, j) + (\mathbf{A}^{1,0} \cdot \mathbf{A}^{0,1} + \mathbf{A}^{0,1} \cdot \mathbf{A}^{1,0}) \\
&\quad \times \mathbf{q}(i - 1, j - 1)] + \mathbf{A}^{0,1} \cdot \mathbf{A}^{0,1}\mathbf{q}(i, j - 2) \\
&= \mathbf{A}^{2,0}q(i - 2, j) + \mathbf{A}^{1,1}\mathbf{q}(i - 1, j - 1) \\
&\quad + \mathbf{A}^{0,2}\mathbf{q}(i, j - 2)
\end{aligned} \tag{2.13}
$$

where

$$
\mathbf{A}^{2,0} = \mathbf{A}^{1,0} \cdot \mathbf{A}^{1,0}, \quad \mathbf{A}^{1,1} = \mathbf{A}^{1,0} \cdot \mathbf{A}^{0,1} + \mathbf{A}^{0,1} \cdot \mathbf{A}^{1,0},
$$

$$
\text{and} \quad \mathbf{A}^{0,2} = \mathbf{A}^{0,1} \cdot \mathbf{A}^{0,1}
$$

The state vectors at the right-hand side of Eq. (2.13) can now be eliminated using Eq. (2.12) repeatedly. Eventually, the relation

$$
\mathbf{q}(i, j) = \sum_{k=0}^{j} \mathbf{A}^{i,j-k} \begin{bmatrix} \mathbf{q}^H(0, k) \\ 0 \end{bmatrix} + \sum_{l=0}^{i} \mathbf{A}^{i-l,j} \begin{bmatrix} 0 \\ \mathbf{q}^V(l, 0) \end{bmatrix} \tag{2.14}
$$

is obtained where

$$
\mathbf{A}^{ij} = \mathbf{A}^{1,0} \cdot \mathbf{A}^{i-1,j} + \mathbf{A}^{0,1} \cdot \mathbf{A}^{i,j-1} \quad \text{for} \quad i > 0, j > 0 \tag{2.15a}
$$

$$
\mathbf{A}^{0,0} = \mathbf{I} \tag{2.15b}
$$

$$
\mathbf{A}^{-i,j} = \mathbf{A}^{i, -j} = 0 \quad \text{for} \quad i \geq 1, j \geq 1 \tag{2.15c}
$$

In the particular case where all $\mathbf{q}^H(0, k)$ and $\mathbf{q}^V(l, 0)$ are zero for $1 \le k \le j$ and $1 \le l \le i$, Eq. (2.14) becomes

$$\mathbf{q}(i, j) = \mathbf{A}^{i,j}\mathbf{q}(0, 0)$$

This equation illustrates the physical significance of matrix $\mathbf{A}^{i,j}$. It represents the effect of the boundary value of the state $\mathbf{q}(0, 0)$ on state $\mathbf{q}(i, j)$. This observation along with the assumptions that the 2-D digital filter of interest is linear and shift-invariant will allow us to generate a general expression for $\mathbf{q}(i, j)$.

Let us now take into account the effect of input $x(i, j)$ under the assumption that the boundary values of $\mathbf{q}(i, j)$, namely, $\mathbf{q}^H(0, k)$ and $\mathbf{q}^V(l, 0)$, are zero for $1 \le k \le j$ and $1 \le l \le i$, and that $x(l, k)$ for some $k \le j$ and $l \le i$ is the only nonzero input. From Eq. (2.9), we can write

$$\mathbf{q}(l + 1, k) = \begin{bmatrix} \mathbf{b}_1 \\ 0 \end{bmatrix} x(l, k) \quad \text{and} \quad \mathbf{q}(l, k + 1) = \begin{bmatrix} 0 \\ \mathbf{b}_2 \end{bmatrix} x(l, k)$$

Since the filter is shift-invariant, the effects of $\mathbf{q}(l + 1, k)$ and $\mathbf{q}(l, k + 1)$ on $\mathbf{q}(i, j)$ are given by

$$\mathbf{A}^{i-(l+1),j-k}\mathbf{q}(l + 1, k) \quad \text{and} \quad \mathbf{A}^{i-l,j-(k+1)}\mathbf{q}(l, k + 1)$$

respectively. Therefore, by the law of superposition, the effect of $x(l, k)$ is obtained as

$$\mathbf{q}(i, j) = \left(\mathbf{A}^{i-(l+1),j-k}\begin{bmatrix} \mathbf{b}_1 \\ 0 \end{bmatrix} + \mathbf{A}^{i-l,j-(k+1)}\begin{bmatrix} 0 \\ \mathbf{b}_2 \end{bmatrix} \right) x(l, k)$$

Consequently, the effect of $x(l, k)$ for $l \le i$ and $k \le j$ is obtained as

$$\mathbf{q}(i, j) = \sum\sum_{(0,0) \le (l,k) \le (i,j)} \left(\mathbf{A}^{i-(l+1),j-k}\begin{bmatrix} \mathbf{b}_1 \\ 0 \end{bmatrix} + \mathbf{A}^{i-l,j-(k+1)}\begin{bmatrix} 0 \\ \mathbf{b}_2 \end{bmatrix} \right) x(l, k)$$

Now taking into account the effects of both the boundary values of the state and the input on the state $\mathbf{q}(i, j)$, we obtain the general formula

$$\mathbf{q}(i, j) = \sum_{k=0}^{j} \mathbf{A}^{i,j-k}\begin{bmatrix} \mathbf{q}^H(0, k) \\ 0 \end{bmatrix} + \sum_{l=0}^{i} \mathbf{A}^{i-l,j}\begin{bmatrix} 0 \\ \mathbf{q}^V(l, 0) \end{bmatrix}$$

$$+ \sum\sum_{(0,0) \le (l,k) \le (i,j)} \left(\mathbf{A}^{i-(l+1),j-k}\begin{bmatrix} \mathbf{b}_1 \\ 0 \end{bmatrix} \right.$$

$$\left. + \mathbf{A}^{i-l,j-(k+1)}\begin{bmatrix} 0 \\ \mathbf{b}_2 \end{bmatrix} \right) x(l, k) \tag{2.16}$$

Therefore, from Eqs. (2.9) and (2.16) the output of the filter is obtained as

$$y(i, j) = [\mathbf{c}_1 \quad \mathbf{c}_2]\left\{\sum_{k=0}^{j} \mathbf{A}^{i,j-k}\begin{bmatrix}\mathbf{q}^H(0, k) \\ 0\end{bmatrix} + \sum_{l=0}^{i} \mathbf{A}^{i-l,j}\begin{bmatrix}0 \\ \mathbf{q}^V(l, 0)\end{bmatrix}\right.$$

$$+ \sum_{(0,0)\leq(l,k)\leq(i,j)} \mathbf{A}^{i-(l+1),j-k}\begin{bmatrix}\mathbf{b}_1 \\ 0\end{bmatrix}$$

$$\left. + \mathbf{A}^{i-l,j-(k+1)}\begin{bmatrix}0 \\ \mathbf{b}_2\end{bmatrix}\right\}x(l, k) + \mathbf{d}x(i, j) \qquad (2.17)$$

Example 2.2 Find the impulse response of the filter described by Eq. (1.21) using a state-space method.

Solution. From Example 2.1b, one of the possible state-space characterizations of the filter is given by Eq. (2.9) with

$$\mathbf{A} = \begin{bmatrix} 0 & \vdots & -1 & 0 \\ \hdashline -b & \vdots & 0 & 0 \\ 0 & \vdots & 0 & 0 \end{bmatrix}, \qquad \mathbf{b} = \begin{bmatrix} 1 \\ \hdashline 0 \\ 0 \end{bmatrix},$$

$$\mathbf{c} = [0 \quad \vdots \quad -1 \quad -1], \qquad \text{and } \mathbf{d} = 1$$

To compute the impulse response, we use the unit impulse as the input signal and assume zero boundary values of the state in Eq. (2.17) to obtain

$$h(i, j) = [\mathbf{c}_1 \quad \mathbf{c}_2]\mathbf{A}^{i-1,j}\begin{bmatrix}\mathbf{b}_1 \\ 0\end{bmatrix} + \delta(i, j)$$

Since

$$[\mathbf{c}_1 \quad \mathbf{c}_2]\mathbf{A}^{i-1,j}\begin{bmatrix}\mathbf{b}_1 \\ 0\end{bmatrix} = \begin{cases} b^i & \text{if } i = j \geq 0 \\ 0 & \text{otherwise} \end{cases}$$

we have

$$h(i, j) = \begin{cases} b^i & \text{if } i = j \geq 0 \\ 0 & \text{otherwise} \end{cases}$$

Evidently, the result obtained coincides with that obtained in Example 1.2b, as should be expected.

2.4 CONTROLLABILITY AND OBSERVABILITY

The concepts of controllability and observability have been used extensively in conjunction with the state-space approach for the characterization and analysis of 1-D dynamic systems in the past. These concepts are equally useful in the analysis of 2-D digital filters.

A 2-D digital filter characterized by Eq. (2.9) is said to be *locally controllable* if for an arbitrary vector $\hat{q}$ of dimension $N + M$, where N and M are the dimensions of A_1 and A_4 of Eq. (2.9), respectively, and the boundary values of the state assumed to be zero, that is $q^H (0, k) = 0$ for $k \geq 0$ and $q^V (0, l) = 0$ for $l \geq 0$, there exist some input signal and some pair (i, j) with $i \geq 0, j \geq 0$ such that $q(i, j) = \hat{q}$. A 2-D digital filter is said to be *locally observable* if there does not exist a nonzero boundary value of the state such that for zero input, that is, $x(i, j) = 0$ for all $i \geq 0, j \geq 0$, the output is also identically equal to zero, that is, $y(i, j) = 0$ for all $i \geq 0, j \geq 0$. More informally, a 2-D digital filter is said to be locally controllable if any state value $\hat{q}$ can be achieved by applying some appropriate signal at the input of the filter. On the other hand, a 2-D digital filter is said to be locally observable if the value of a local state can be determined by carrying out a set of measurements at the output of the filter.

To test whether a 2-D filter is controllable or not, we define the controllability matrix C_{NM} as

$$C_{NM} = [p(0, 0) \cdots p(0, M) \quad p(1, 0) \cdots p(1, M)$$
$$\cdots p(N, 0) \cdots p(N, M - 1)] \tag{2.18}$$

where

$$p(i, j) = A^{i-1,j} \begin{bmatrix} b_1 \\ 0 \end{bmatrix} + A^{i,j-1} \begin{bmatrix} 0 \\ b_2 \end{bmatrix}$$

It can be shown that a filter is locally controllable if and only if the rank of C_{NM} is equal to $N + M$ [5].

To test whether a 2-D filter is observable or not, we define the observability matrix O_{NM} as

$$O_{NM} = \begin{bmatrix} c \\ cA^{0,1} \\ \vdots \\ cA^{0,M} \\ cA^{1,0} \\ \vdots \\ cM^{1,M} \\ \vdots \\ cA^{N,0} \\ \vdots \\ cA^{N,M-1} \end{bmatrix} \tag{2.19}$$

It can be shown that a filter is locally observable if and only if the rank of O_{NM} is equal to $N + M$ [5].

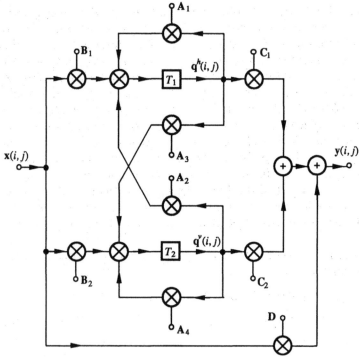

Figure 2.5   State-space realization of an m-input, p-output 2-D digital filter.

2.5 REALIZATION

The model of Eq. (2.11) leads readily to the state-space realization of an m-input, p-output 2-D digital filter shown in Fig. 2.5 where the multiplier coefficients are matrices. In the case of a one-input, one-output filter, coefficients A_1 to A_4 are matrices and coefficients B_1, B_2, C_1, and C_2 are vectors.

REFERENCES

1. S. Attasi, Systèmes lineaires homogène à deux indices, *IRIA Rapport LA-BORIA*, no. 31, 1973.
2. D. D. Givone and R. P. Roesser, Minimization of multidimensional linear iterative circuits, *IEEE Trans. Comput.*, vol. C-22, pp. 673–678, July 1973.
3. E. Fornasini and G. Marchesini, Algebraic realization theory of two-dimensional filters, in *Variable Structure Systems*, edited by A. Ruberti and R. Mohler, New York: Springer Verlag, 1975.

4. S. Y. Kung, B. C. Levy, M. Morf, and T. Kailath, New results in 2-D systems theory, Part II: 2-D state-space models—realization and the notions of controllability, observability, and minimality, *Proc. IEEE*, vol. 65, pp. 945–961, June 1977.

5. R. P. Roesser, A discrete state-space model for linear image processing, *IEEE Trans. Automat. Contr.*, vol. AC-20, pp. 1–10, Feb. 1975

PROBLEMS

2.1 Derive the Givone–Roesser model for the filters represented by the following networks:
(a) the network given in Fig. P1.5a and b.
(b) the network given in Fig. P1.5c and b.
(c) the network given in Fig. P1.6.

2.2 Derive the Givone–Roesser model for the filters represented by the following difference equations:
(a) the difference equation given in Problem 1.7c.
(b) $y(n_1, n_2) = -y(n_1 - 2, n_2 - 2) + x(n_1 - 1, n_2 - 1) + x(n_1 - 2, n_2 - 2)$.
(c) $y(n_1, n_2) = y(n_1 - 1, n_2 - 1) - x(n_1, n_2 - 1) - x(n_1 - 1, n_2)$.

2.3 Compute the impulse response $h(i, j)$ of the filters given in Problem 2.1 by using Eq. (2.17) for $0 \le i, j \le 3$.

2.4 Check the controllability and observability of the filters in Problem 2.1a–b.

2.5 The Givone–Rosser model of a 2-D filter is given by Eq. (2.9) where

$$\mathbf{A}_1 = \begin{bmatrix} 0.5 & 0.1 \\ -0.1 & 0.4 \end{bmatrix}, \quad \mathbf{A}_2 = \begin{bmatrix} 0 & 0.2 \\ 0.1 & -0.1 \end{bmatrix}, \quad \mathbf{A}_3 = \begin{bmatrix} 0 & 0 \\ 0 & 0 \end{bmatrix}$$

$$\mathbf{A}_1 = \begin{bmatrix} -0.6 & 0.2 \\ 0.0 & 0.1 \end{bmatrix}, \quad \mathbf{b}_1 = \begin{bmatrix} 1 \\ 0 \end{bmatrix}, \quad \mathbf{b}_2 = \begin{bmatrix} 0.5 \\ -0.5 \end{bmatrix}$$

$$\mathbf{c}_1 = [1 \quad -1], \quad \mathbf{c}_2 = [2 \quad 0], \quad \text{and} \quad d = 0$$

(a) Find the difference equation that characterizes the filter.
(b) Compute the impulse response $h(n_1, n_2)$ of the filter for $0 \le n_1$, $n_2 \le 3$ by using Eq. (2.17).

2.6 Consider the filter given in Problem 2.5.
(a) Compute $h(i, j)$ for $0 \le i, j \le 3$ by using the difference equation obtained in Problem 2.5a.
(b) Check the controllability and observability of the filter.

2.7 Consider the filter described in Problem 2.5.

  (a) Construct a network of the type shown in Fig. P1.5a that implements the filter, where F_1 is independent of delay element T_2 and F_2 is independent of delay element T_1.

  (b) The filter can be realized by the network shown in Fig. 2.5. Compare the numbers of multiplications and additions required to compute an output value with those required by the network obtained in part (a).

3
Transform Methods

3.1 INTRODUCTION

As in the case of 1-D linear shift-invariant digital filters, the analysis and design of 2-D linear shift-invariant digital filters can be simplified considerably by using transform methods. Through the use of the 2-D z transform, a difference equation characterizing a 2-D digital filter can be transformed into an algebraic equation that is usually much easier to manipulate than the difference equation. On the other hand, through the use of the 2-D Fourier transform, it is possible to establish the mathematical relations between discrete and continuous 2-D signals that make possible the processing of 2-D continuous signals by means of 2-D digital filters.

This chapter introduces the 2-D z transform and its inverse and discusses some of their properties. Then the complex convolution is introduced as a tool for obtaining the z transforms of products of space-domain functions and for the derivation of Parseval's formula for discrete signals.

In the second half of the chapter, the 2-D Fourier transform and its inverse are introduced and some of their properties are discussed. The 2-D sampling theorem is then introduced as the fundamental link between 2-D continuous and discrete signals. The various mathematical interrelations [1] between discrete and continuous signals are then established.

3.2   THE 2-D z TRANSFORM

3.2.1   Definition

The z transform of a 2-D discrete signal $f(n_1, n_2)$ is defined as

$$F(z_1, z_2) = \sum_{n_1=-\infty}^{\infty} \sum_{n_2=-\infty}^{\infty} f(n_1, n_2) z_1^{-n_1} z_2^{-n_2} \qquad (3.1)$$

for all (z_1, z_2) for which the preceding double summation converges. The arguments of $F(z_1, z_2)$ (i.e., z_1 and z_2) are complex variables. The z transform is usually represented in terms of the simplified notation

$$F(z_1, z_2) = \mathscr{Z}f(n_1, n_2)$$

3.2.2   Region of Convergence

The region of convergence for the series in the right-hand side of Eq. (3.1) consists of the set of points (z_1, z_2) for which

$$\sum_{n_1=-\infty}^{\infty} \sum_{n_2=-\infty}^{\infty} |f(n_1, n_2)| \, |z_1|^{-n_1} |z_2|^{-n_2} \qquad (3.2)$$

converges to a finite real number. It turns out that in its region of convergence, the 2-D z transform is an analytic function in two variables.

As can be seen, the convergence of (3.2) depends on $|f(n_1, n_2)|$, $|z_1|$, and $|z_2|$; consequently, the region of convergence of the z transform of $f(n_1, n_2)$ is an area in the first quadrant of the $(|z_1|, |z_2|)$ plane, as illustrated in the following examples.

Example 3.1   Obtain the z transform of the 2-D first-quadrant signal

$$f(n_1, n_2) = a^{n_1} b^{n_2} u(n_1, n_2)$$

and determine its region of convergence.

Solution.   The z transform of $f(n_1, n_2)$ is given by Eq. (3.1) as

$$F(z_1, z_2) = \sum_{n_1=0}^{\infty} \sum_{n_2=0}^{\infty} a^{n_1} b^{n_2} z_1^{-n_1} z_2^{-n_2}$$

$$= \sum_{n_1=0}^{\infty} \left(\frac{z_1}{a}\right)^{-n_1} \sum_{n_2=0}^{\infty} \left(\frac{z_2}{b}\right)^{-n_2}$$

$$= \frac{1}{(1 - az_1^{-1})(1 - bz_2^{-1})} = \frac{z_1 z_2}{(z_1 - a)(z_2 - b)}$$

Evidently, the region of convergence for the preceding z transform is given by

$$A = \{(z_1, z_2) : |z_1| > |a|, |z_2| > |b|\}$$

and is depicted in Fig. 3.1.

Example 3.2 Obtain the z transform of the 2-D first-quadrant signal

$$f(n_1, n_2) = \begin{cases} a^n & \text{if } n_1 = n_2 = n \geq 0 \\ 0 & \text{otherwise} \end{cases}$$

and determine its region of convergence.

Solution. The z transform of $f(n_1, n_2)$ is given by Eq. (3.1) as

$$\begin{aligned}
F(z_1, z_2) &= \sum_{n_1=-\infty}^{\infty} \sum_{n_2=-\infty}^{\infty} a^{n_1} \delta(n_1 - n_2, 0) u(n_1, n_2) z_1^{-n_1} z_2^{-n_2} \\
&= \sum_{n_1=0}^{\infty} \sum_{n_2=0}^{\infty} a^{n_1} \delta(n_1 - n_2, 0) z_1^{-n_1} z_2^{-n_2} \\
&= \sum_{n_1=0}^{\infty} (a z_1^{-1} z_2^{-1})^{n_1} \\
&= \frac{1}{1 - a z_1^{-1} z_2^{-1}} = \frac{z_1 z_2}{z_1 z_2 - a}
\end{aligned}$$

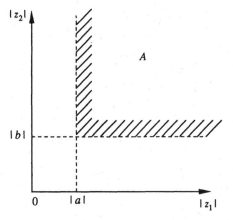

Figure 3.1 Region of convergence for Example 3.1.

The region of convergence for the preceding z transform is given by

$$A = \{(z_1, z_2) : |z_1| \, |z_2| > |a|\}$$

and is depicted in Fig. 3.2

3.2.3 *z*-Transform Theorems

The properties of the z transform can be described by means of a number of theorems that are straightforward extensions of their 1-D counterparts [2].

Theorem 3.1   Linearity   *If a and b are arbitrary constants and*

$$\mathcal{Z}f(n_1, n_2) = F(z_1, z_2)$$

$$\mathcal{Z}g(n_1, n_2) = G(z_1, z_2)$$

where $f(n_1, n_2)$ and $g(n_1, n_2)$ are arbitrary discrete 2-D signals, then

$$\mathcal{Z}[af(n_1, n_2) + bg(n_1, n_2)] = aF(z_1, z_2) + bG(z_1, z_2)$$

Theorem 3.2   Translation

$$\mathcal{Z}f(n_1 + k_1, \ n_2 + k_2) = z_1^{k_1} z_2^{k_2} F(z_1, z_2)$$

Theorem 3.3   Complex Scale Change

$$\mathcal{Z}w_1^{-n_1} w_2^{-n_2} f(n_1, n_2) = F(w_1 z_1, w_2 z_2)$$

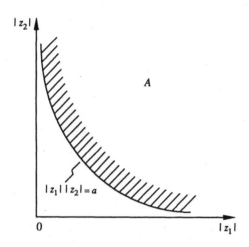

Figure 3.2 Region of convergence for Example 3.2.

In effect, the region of convergence is scaled by $|w_1|^{-1}$ in variable z_1 and by $|w_2|^{-1}$ in variable z_2.

Theorem 3.4 Complex Differentiation

$$\mathcal{Z} n_1 f(n_1, n_2) = -z_1 \frac{\partial F(z_1, z_2)}{\partial z_1}$$

$$\mathcal{Z} n_2 f(n_1, n_2) = -z_2 \frac{\partial F(z_1, z_2)}{\partial z_2}$$

$$\mathcal{Z} n_1 n_2 f(n_1, n_2) = z_1 z_2 \frac{\partial^2 F(z_1, z_2)}{\partial z_1 \, \partial z_2}$$

Theorem 3.5 Real Convolution

$$\mathcal{Z} \sum_{k_1 = -\infty}^{\infty} \sum_{k_2 = -\infty}^{\infty} f(k_1, k_2) g(n_1 - k_1, n_2 - k_2) = F(z_1, z_2) G(z_1, z_2) \quad (3.3a)$$

$$\mathcal{Z} \sum_{k_1 = -\infty}^{\infty} \sum_{k_2 = -\infty}^{\infty} f(n_1 - k_1, n_2 - k_2) g(k_1, k_2) = F(z_1, z_2) G(z_1, z_2) \quad (3.3b)$$

Theorem 3.6 Initial-Value Theorem *If $f(n_1, n_2) = 0$ for $n_1 < 0$ or $n_2 < 0$, then*

$$\lim_{z_1 \to \infty} F(z_1, z_2) = \sum_{n_2 = -\infty}^{\infty} f(0, n_2) z_2^{-n_2}$$

$$\lim_{z_2 \to \infty} F(z_1, z_2) = \sum_{n_1 = -\infty}^{\infty} f(n_1, 0) z_1^{-n_1}$$

$$\lim_{z_1 \to \infty, \; z_2 \to \infty} F(z_1, z_2) = f(0, 0)$$

3.2.4 The Inverse z Transform

Function $f(n_1, n_2)$ is said to be the inverse z transform of $F(z_1, z_2)$. Assuming that there exist two simple closed contours Γ_1 and Γ_2 that encircle the origins of the z_1 and z_2 planes, respectively, then $f(n_1, n_2)$ can be determined as

$$f(n_1, n_2) = \frac{1}{(2\pi j)^2} \oint_{\Gamma_2} \oint_{\Gamma_1} F(z_1, z_2) z_1^{n_1 - 1} z_2^{n_2 - 1} \, dz_1 \, dz_2 \quad (3.4)$$

where the two integrals are evaluated in the counterclockwise sense over contours Γ_1 and Γ_2.

The evaluation of the 2-D inverse z transform can be much more involved than its 1-D counterpart. This is due to the fact that the singularities of $F(z_1, z_2)$ need not be isolated and, consequently, the general methods used to find the coefficients in complex series are not always applicable for the evaluation of the inverse z transform.

In the case where the z transform can be expressed as a product of functions $F_1(z_1)F_2(z_2)$ where $F_1(z_1)$ and $F_2(z_2)$ are independent of z_2 and z_1, respectively, Eq. (3.4) can be expressed as

$$f(n_1, n_2) = \frac{1}{(2\pi j)^2} \oint_{\Gamma_2} \oint_{\Gamma_1} F_1(z_1)F_2(z_2)z_1^{n_1-1}z_2^{n_2-1}\, dz_1\, dz_2 \qquad (3.5a)$$

$$= \frac{1}{2\pi j} \oint_{\Gamma_1} F_1(z_1)z_1^{n_1-1}\, dz_1 \frac{1}{2\pi j} \oint_{\Gamma_2} F_2(z_2)z_2^{n_2-1}\, dz_2 \qquad (3.5b)$$

$$= f_1(n_1)f_2(n_2) \qquad (3.5c)$$

where $f_1(n_1)$ and $f_2(n_2)$ denote the inverse z transforms of $F_1(z_1)$ and $F_2(z_2)$, respectively. Therefore, if a 2-D z transform is separable, then its inverse is also separable and vice versa, as might be expected. In such a case, the numerous techniques for obtaining the 1-D inverse z transform [1] are also applicable in the case of the 2-D inverse z transform.

If $F(z_1, z_2)$ is not separable, the evaluation of the double integral in Eq. (3.4) may be necessary. This can often be carried out by expressing $F(z_1, z_2)$ as a function of one complex variable and treating the second complex variable as a parameter.

Example 3.3 Find the inverse z transform of

$$F(z_1, z_2) = \frac{1}{1 - 0.5z_1^{-1} - 0.5z_2^{-1} + 0.25z_1^{-1}z_2^{-1}}$$

Solution. $F(z_1, z_2)$ can be written as

$$F(z_1, z_2) = \frac{1}{(1 - 0.5z_1^{-1})(1 - 0.5z_2^{-1})}$$

If contours Γ_1 and Γ_2 are assumed to be the unit circles

$$U_1 = \{z_1 : |z_1| = 1\} \text{ and } U_2 = \{z_2 : |z_2| = 1\}$$

respectively, then Eq. (3.5b) yields

$$f(n_1, n_2) = \frac{1}{2\pi j} \oint_{U_1} \frac{1}{1 - 0.5z_1^{-1}} z_1^{n_1-1}\, dz_1 \frac{1}{2\pi j} \oint_{U_2} \frac{1}{1 - 0.5z_2^{-1}} z_2^{n_2-1}\, dz_2$$

$$= \frac{1}{2\pi j} \oint_{U_1} \frac{z_1^{n_1}}{z_1 - 0.5}\, dz_1 \frac{1}{2\pi j} \oint_{U_2} \frac{z_2^{n_2}}{z_2 - 0.5}\, dz_2$$

Now by applying the residue theorem to each integral, we have

$$f(n_1, n_2) = 0.5^{n_1+n_2}u(n_1, n_2)$$

Note that this result is the inverse of the z transform in Example 3.1 for the case where $a = b = 0.5$.

Example 3.4 Find the inverse z transforms of

$$(a) \quad F(z_1, z_2) = \frac{2}{2 - z_1^{-1} z_2^{-1}}$$

$$(b) \quad F(z_1, z_2) = \frac{1}{1 - 0.4 z_1^{-1} - 0.4 z_2^{-1}}$$

Solution. (a) The given 2-D z transform can be expressed as

$$F(z_1, z_2) = \frac{z_1}{z_1 - 0.5 z_2^{-1}}$$

and hence

$$f(n_1, n_2) = \frac{1}{(2\pi j)^2} \oint_{\Gamma_2} \oint_{\Gamma_1} \frac{z_1^{n_1} z_2^{n_2 - 1}}{z_1 - 0.5 z_2^{-1}} \, dz_1 \, dz_2 \qquad (3.6a)$$

$$= \frac{1}{2\pi j} \oint_{\Gamma_2} z_2^{n_2 - 1} \left(\frac{1}{2\pi j} \oint_{\Gamma_1} \frac{z_1^{n_1}}{z_1 - 0.5 z_2^{-1}} \, dz_1 \right) dz_2 \qquad (3.6b)$$

Evidently, the z transform converges in the region

$$|z_1| \, |z_2| > 0.5$$

and, consequently, the unit circles U_1 and U_2 are within the region of convergence. Further, for each fixed $z_2 \in U_2$, the function

$$\frac{z_1^{n_1}}{z_1 - 0.5 z_2^{-1}}$$

has only a single pole at $z_1 = 0.5 z_2^{-1}$ that lies within unit circle U_1 and, consequently, applying the residue theorem to the inner integral in Eq. (3.6b) yields

$$\frac{1}{2\pi j} \oint_{U_1} \frac{z_1^{n_1}}{z_1 - 0.5 z_2^{-1}} \, dz_1 = \begin{cases} (0.5 z_2^{-1})^{n_1} & \text{if } n_1 \geq 0 \\ 0 & n_1 < 0 \end{cases}$$

Therefore, for $n_1 \geq 0$, we obtain

$$f(n_1, n_2) = \frac{1}{2\pi j} \oint_{U_2} z_2^{n_2 - 1} (0.5 z_2^{-1})^{n_1} \, dz_2$$

$$= \frac{0.5^{n_1}}{2\pi j} \oint_{U_2} z_2^{n_2 - n_1 - 1} \, dz_2 = \begin{cases} 0.5^n & \text{if } n_1 = n_2 = n \geq 0 \\ 0 & \text{otherwise} \end{cases}$$

The preceding result may be verified by comparing it with the result in Example 3.2.

(b) The given z transform can be expressed as

$$F(z_1, z_2) = \frac{1}{1 - 0.4z_1^{-1} - 0.4z_2^{-1}} \tag{3.7a}$$

$$= \frac{z_1 z_2}{z_1 z_2 - 0.4z_2 - 0.4z_1} \tag{3.7b}$$

$$= \frac{z_1 G(z_2)}{z_1 - 0.4G(z_2)} \tag{3.7c}$$

where

$$G(z_2) = \frac{z_2}{z_2 - 0.4}$$

On the unit circles U_1 and U_2, we have

$$|0.4z_1^{-1} + 0.4z_2^{-1}| \le 0.4|z_1|^{-1} + 0.4|z_2|^{-1} = 0.8 < 1$$

and hence U_1 and U_2 are within the region of convergence of the z transform and can, as a consequence, be chosen as contours Γ_1 and Γ_2, respectively.

Now by substituting Eq. (3.7c) into Eq. (3.4) and applying the residue theorem for the case $n_1 \ge 0$ and $n_2 \ge 0$, we obtain

$$f(n_1, n_2) = \frac{1}{(2\pi j)^2} \oint_{U_2} \oint_{U_1} \frac{G(z_2) z_1}{z_1 - 0.4G(z_2)} z_1^{n_1 - 1} z_2^{n_2 - 1} \, dz_1 \, dz_2$$

$$= \frac{1}{2\pi j} \oint_{U_2} z_2^{n_2 - 1} G(z_2) \left(\frac{1}{2\pi j} \oint_{U_1} \frac{z_1^{n_1}}{z_1 - 0.4G(z_2)} \, dz_1 \right) dz_2$$

$$= \frac{1}{2\pi j} \oint_{U_2} z_2^{n_2 - 1} 0.4^{n_1} G(z_2)^{n_1 + 1} \, dz_2$$

$$= \frac{0.4^{n_1}}{2\pi j} \oint_{U_2} \frac{z_2^{n_1 + n_2}}{(z_2 - 0.4)^{n_1 + 1}} \, dz_2$$

$$= 0.4^{n_1 + n_2} \frac{(n_1 + n_2)!}{n_1! n_2!}$$

For the case $n_1 < 0$ or $n_2 < 0$, $f(n_1, n_2) = 0$ and therefore

$$f(n_1, n_2) = 0.4^{n_1 + n_2} \frac{(n_1 + n_2)!}{n_1! n_2!} u(n_1, n_2)$$

3.2.5 Complex Convolution

By virtue of Theorem 3.5, the inverse z transform of a product of z transforms can be formed by using the real convolution. By analogy, the z

transform of a product of two space-domain functions can be formed by using the complex convolution.

Theorem 3.7 Complex Convolution *If*

$$\mathscr{Z}f(n_1, n_2) = F(z_1, z_2) \quad and \quad \mathscr{Z}g(n_1, n_2) = G(z_1, z_2)$$

then

$$Y(z_1, z_2) = \mathscr{Z}[f(n_2, n_2)g(n_1, n_2)]$$

$$= \frac{1}{(2\pi j)^2} \oint_{\Gamma_2} \oint_{\Gamma_1} F(v_1, v_2) G\left(\frac{z_1}{v_1}, \frac{z_2}{v_2}\right) \frac{dv_1}{v_1} \frac{dv_2}{v_2}$$

$$= \frac{1}{(2\pi j)^2} \oint_{\Gamma_2'} \oint_{\Gamma_1'} F\left(\frac{z_1}{v_1}, \frac{z_2}{v_2}\right) G(v_1, v_2) \frac{dv_1}{v_1} \frac{dv_2}{v_2}$$

where Γ_1 and Γ_2 (or Γ_1' and Γ_2') are contours in the common region of convergence of $F(v_1, v_2)$ and $G(z_1/v_1, z_2/v_2)$ [or $F(z_1/v_1, z_2/v_2)$ and $G(v_1, v_2)$].

Proof.

$$Y(z_1, z_2)$$

$$= \sum_{n_1=-\infty}^{\infty} \sum_{n_2=-\infty}^{\infty} f(n_1, n_2)g(n_1, n_2)z_1^{-n_1}z_2^{-n_2}$$

$$= \sum_{n_1=-\infty}^{\infty} \sum_{n_2=-\infty}^{\infty} \left[\frac{1}{(2\pi j)^2} \oint_{\Gamma_2} \oint_{\Gamma_1} F(v_1, v_2)v_1^{n_1-1}v_2^{n_2-1} dv_1 dv_2\right] g(n_1, n_2)z_1^{-n_1}z_2^{-n_2}$$

$$= \frac{1}{(2\pi j)^2} \oint_{\Gamma_2} \oint_{\Gamma_1} F(v_1, v_2)\left[\sum_{n_1=-\infty}^{\infty} \sum_{n_2=-\infty}^{\infty} g(n_1, n_2)\left(\frac{z_1}{v_1}\right)^{-n_1}\left(\frac{z_2}{v_2}\right)^{-n_2}\right] \frac{dv_1 dv_2}{v_1 v_2}$$

$$= \frac{1}{(2\pi j)^2} \oint_{\Gamma_2} \oint_{\Gamma_1} F(v_1, v_2)G\left(\frac{z_1}{v_1}, \frac{z_2}{v_2}\right) \frac{dv_1 dv_2}{v_1 v_2} \quad \blacksquare$$

The following theorem can be readily derived from the complex convolution theorem.

Theorem 3.8 Parseval's Formula for Discrete Signals *If $f(n_1, n_2)$ and $g(n_1, n_2)$ are 2-D signals and $(z_1, z_2) = (1, 1)$ is in the region of convergence of the z transform of $f(n_1, n_2)g^*(n_1, n_2)$ where $g^*(n_1, n_2)$ is the complex* *conjugate of $g(n_1, n_2)$, then*

$$\sum_{n_1=-\infty}^{\infty} \sum_{n_2=-\infty}^{\infty} f(n_1, n_2)g^*(n_1, n_2)$$

$$= \frac{1}{(2\pi j)^2} \oint_{\Gamma_2} \oint_{\Gamma_1} F(v_1, v_2)G^*\left(\frac{1}{v_1^*}, \frac{1}{v_2^*}\right) \frac{dv_1}{v_1} \frac{dv_2}{v_2} \tag{3.8}$$

Proof. The z transform of $g^*(n_1, n_2)$ is given by

$$\sum_{n_1=-\infty}^{\infty} \sum_{n_2=-\infty}^{\infty} g^*(n_1, n_2)z_1^{-n_1}z_2^{-n_2} = G^*(z_1^*, z_2^*)$$

where $G^*(z_1^*, z_2^*)$ is the complex conjugate of $G(z_1^*, z_2^*)$. On applying Theorem 3.7 to functions $f(n_1, n_2)$ and $g^*(n_1, n_2)$, we obtain

$$\sum_{n_1=-\infty}^{\infty} \sum_{n_2=-\infty}^{\infty} f(n_1, n_2)g^*(n_1, n_2)z_1^{-n_1}z_2^{-n_2}$$

$$= \frac{1}{(2\pi j)^2} \oint_{\Gamma_2} \oint_{\Gamma_1} F(v_1, v_2)G^*\left(\frac{z_1^*}{v_1^*}, \frac{z_2^*}{v_2^*}\right) \frac{dv_1}{v_1} \frac{dv_2}{v_2}$$

Now substituting $z_1 = 1$ and $z_2 = 1$ in the preceding equation yields Eq. (3.8). In particular, if $f(n_1, n_2) = g(n_1, n_2)$, then Eq. (3.8) assumes the form

$$\sum_{n_1=-\infty}^{\infty} \sum_{n_2=-\infty}^{\infty} |f(n_1, n_2)|^2 = \frac{1}{(2\pi j)^2} \oint_{\Gamma_2} \oint_{\Gamma_1} F(v_1, v_2)$$

$$F^*\left(\frac{1}{v_1^*}, \frac{1}{v_2^*}\right) \frac{dv_1}{v_1} \frac{dv_2}{v_2} \tag{3.9}$$

Further, if we assume that the unit circles U_1 and U_2 are in the region of convergence of the z transform, then

$$\sum_{n_1=-\infty}^{\infty} \sum_{n_2=-\infty}^{\infty} |f(n_1, n_2)|^2 = \frac{1}{(2\pi j)^2} \oint_{U_2} \oint_{U_1} F(v_1, v_2)F^*\left(\frac{1}{v_1^*}, \frac{1}{v_2^*}\right) \frac{dv_1}{v_1} \frac{dv_2}{v_2} \tag{3.10a}$$

$$= \frac{1}{4\pi^2} \int_{-\pi}^{\pi} \int_{-\pi}^{\pi} |F(e^{j\omega_1}, e^{j\omega_2})|^2 \, d\omega_1 \, d\omega_2 \tag{3.10b}$$

In the case where signal $f(n_1, n_2)$ represents an appropriate physical quantity, this equation relates the energy in the space-domain signal to the energy in its spectrum. ∎

Example 3.5 Show that the z transform in Example 3.4(a) satisfies Parseval's formula.

Solution. From example 3.4a

$$\mathscr{Z}0.5^m\delta(n_1 - n_2, 0)u(n_1, n_2) = \frac{2}{2 - z_1^{-1}z_2^{-1}}$$

Hence the left-hand side of Eq. (3.10a) can be expressed as

$$\sum_{n_1 = -\infty}^{\infty} \sum_{n_2 = -\infty}^{\infty} |f(n_1, n_2)|^2 = \sum_{n_1 = -\infty}^{\infty} \sum_{n_2 = -\infty}^{\infty} 0.5^{2m}\delta^2(n_1 - n_2, 0)u^2(n_1, n_2)$$

$$= \sum_{n_1 = 0}^{\infty} \sum_{n_2 = 0}^{\infty} 0.25^m\delta(n_1 - n_2, 0)$$

$$= \sum_{n_1 = 0}^{\infty} 0.25^{n_1}$$

$$= \frac{4}{3}$$

On the other hand, the right-hand side of Eq (3.10b) can be expressed as

$$\frac{1}{4\pi^2} \int_{-\pi}^{\pi} \int_{-\pi}^{\pi} \left| \frac{2}{2 - e^{-j\omega_1}e^{-j\omega_2}} \right|^2 d\omega_1 \, d\omega_2$$

$$= \frac{1}{4\pi^2} \int_{-\pi}^{\pi} \int_{-\pi}^{\pi} \frac{1}{1.25 - \cos(\omega_1 + \omega_2)} d\omega_1 \, d\omega_2$$

$$= \frac{1}{2\pi} \int_{-\pi}^{\pi} \frac{1}{1.25 - \cos\omega} d\omega$$

$$= \frac{1}{2\pi} \frac{8\pi}{3}$$

$$= \frac{4}{3}$$

3.3 THE 2-D FOURIER TRANSFORM

The 2-D Fourier transform is a straightforward extension of its 1-D counterpart and its importance in the study of 2-D digital filters arises from the fact that 2-D discrete signals are very frequently generated by sampling corresponding 2-D continuous signals.

3.3.1 Definition

The Fourier transform of a 2-D continuous signal $f(t_1, t_2)$ is defined as

$$F(j\omega_1, j\omega_2) = \int_{-\infty}^{\infty} \int_{-\infty}^{\infty} f(t_1, t_2)e^{-j(\omega_1 t_1 + \omega_2 t_2)} \, dt_1 \, dt_2 \qquad (3.11)$$

and is often referred to as the frequency spectrum of $f(t_1, t_2)$. In general, $F(j\omega_1, j\omega_2)$ is complex and can be expressed as

$$F(j\omega_1, j\omega_2) = A(\omega_1, \omega_2)e^{j\phi(\omega_1, \omega_2)}$$

where

$$A(\omega_1, \omega_2) = |F(j\omega_1, j\omega_2)| \qquad \text{and} \qquad \phi(\omega_1, \omega_2) = \arg F(j\omega_1, j\omega_2)$$

The functions $A(\omega_1, \omega_2)$ and $\phi(\omega_1, \omega_2)$ are called the amplitude spectrum and phase spectrum of $f(t_1, t_2)$, respectively.

The function $f(t_1, t_2)$ is said to be the inverse Fourier transform of $F(j\omega_1, j\omega_2)$ and is given by

$$f(t_1, t_2) = \frac{1}{4\pi^2} \int_{-\infty}^{\infty} \int_{-\infty}^{\infty} F(j\omega_1, j\omega_2)e^{j(\omega_1 t_1 + \omega_2 t_2)} \, d\omega_1 \, d\omega_2 \qquad (3.12)$$

Equations (3.11) and (3.12) can be represented by

$$F(j\omega_1, j\omega_2) = \mathcal{F}f(t_1, t_2) \qquad \text{and} \qquad f(t_1, t_2) = \mathcal{F}^{-1}F(j\omega_1, j\omega_2)$$

respectively. An alternative shorthand notation combining the preceding two equations is

$$f(t_1, t_2) \leftrightarrow F(j\omega_1, j\omega_2)$$

3.3.2 Theorems

The properties of the 2-D Fourier transform are essentially the same as for the 1-D Fourier transform and can be summarized in terms of a number of theorems. If

$$f(t_1, t_2) \leftrightarrow F(j\omega_1, j\omega_2) \qquad \text{and} \qquad g(t_1, t_2) \leftrightarrow G(j\omega_1, j\omega_2)$$

and a, b, t_{0i}, and ω_{0i} $(i = 1, 2)$ are arbitrary constants, the following theorems hold. The proofs of these theorems are straightforward extensions of their 1-D counterparts as detailed by Papoulis [3].

Theorem 3.9 Convergence *If*

$$\lim_{T_1 \to \infty, T_2 \to \infty} \int_{-T_1}^{T_1} \int_{-T_2}^{T_2} |f(t_1, t_2)| \, dt_1 \, dt_2 < \infty$$

then the Fourier transform $F(j\omega_1, j\omega_2)$ exists and satisfies Eq. (3.12)

Theorem 3.10 Linearity

$$af(t_1, t_2) + bg(t_1, t_2) \leftrightarrow aF(j\omega_1, j\omega_2) + bG(j\omega_1, j\omega_2)$$

Theorem 3.11 Symmetry

$$F(jt_1, jt_2) \leftrightarrow 4\pi^2 f(-\omega_1, -\omega_2)$$

Theorem 3.12 Space Scaling

$$f(at_1, bt_2) \leftrightarrow \frac{1}{|ab|} F\left(\frac{j\omega_1}{a}, \frac{j\omega_2}{b}\right)$$

Theorem 3.13 Space Shifting

$$f(t_1 - t_{01}, t_2 - t_{02}) \leftrightarrow F(j\omega_1, j\omega_2)e^{-j(\omega_1 t_{01} + \omega_2 t_{02})}$$

Theorem 3.14 Frequency Shifting

$$e^{j(\omega_{01}t_1 + \omega_{02}t_2)}f(t_1, t_2) \leftrightarrow F(j\omega_1 - j\omega_{01}, j\omega_2 - j\omega_{02})$$

Theorem 3.15 Space Convolution

$$\int_{-\infty}^{\infty} \int_{-\infty}^{\infty} f(\tau_1, \tau_2)g(t_1 - \tau_1, t_2 - \tau_2) \, d\tau_1 \, d\tau_2 \leftrightarrow F(j\omega_1, j\omega_2)G(j\omega_1, j\omega_2)$$

Theorem 3.16 Frequency Convolution

$$f(t_1, t_2)g(t_1, t_2) \longleftrightarrow \frac{1}{4\pi^2} \int_{-\infty}^{\infty} \int_{-\infty}^{\infty} F(jv_1, jv_2)G(j\omega_1 - jv_1, j\omega_2 - jv_2) \, dv_1 \, dv_2$$

Theorem 3.17 Parseval's Formula for Continuous Signals

$$\int_{-\infty}^{\infty} \int_{-\infty}^{\infty} |f(t_1, t_2)|^2 \, dt_1 \, dt_2 \longleftrightarrow \frac{1}{4\pi^2} \int_{-\infty}^{\infty} \int_{-\infty}^{\infty} |F(j\omega_1, j\omega_2)|^2 \, d\omega_1 \, d\omega_2$$

Theorem 3.18 Poisson's Summation Formula (a) *If $f(t_1, t_2) \neq 0$ for $t_1 < 0$ or $t_2 < 0$, then*

$$\sum_{n_1=-\infty}^{\infty} \sum_{n_2=-\infty}^{\infty} f(n_1 T_1, n_2 T_2) = \frac{1}{T_1 T_2} \sum_{n_1=-\infty}^{\infty} \sum_{n_2=-\infty}^{\infty} F(jn_1\omega_{s1}, jn_2\omega_{s2})$$

where

$$\omega_{s1} = \frac{2\pi}{T_1} \quad and \quad \omega_{s2} = \frac{2\pi}{T_2}$$

(b) *If* $f(t_1, t_2) = 0$ *for* $t_1 < 0$ *or* $t_2 < 0$, *then*

$$\sum_{n_1=0}^{\infty} \sum_{n_2=0}^{\infty} f(n_1 T_1, n_2 T_2) = \frac{f(0+, 0+)}{2} + \frac{1}{T_1 T_2} \sum_{n_1=0}^{\infty} \sum_{n_2=0}^{\infty} F(jn_1\omega_{s1}, jn_2\omega_{s2})$$

where $f(0, 0) \equiv f(0+, 0+)$.

3.4 THE SAMPLING PROCESS

Digital filters and digital signal processing in general owe their widespread applications to the sampling theorem. This very important theorem states that a band-limited 2-D continuous signal, that is, a 2-D continuous signal whose frequency spectrum is zero for all frequencies (ω_1, ω_2) such that $|\omega_1| \geq \omega_{B1}$ or $|\omega_2| \geq \omega_{B2}$, can be completely recovered from a sampled version of the signal provided that the sampling frequencies used, namely ω_{s1} and ω_{s2}, are sufficiently high. Consequently, if a band-limited 2-D continuous signal needs to be processed, it can be converted into a corresponding 2-D discrete signal that can be processed by a 2-D digital filter. The processed signal can then be converted back into a 2-D continuous signal, if necessary. Provided that the requirements of the sampling theorem are satisfied, the discrete signal obtained during the sampling process constitutes a complete representation of the continuous signal and, further, the processed discrete signal is a processed version of the continuous signal.

In this section the sampling process is examined and relations are established between continuous and discrete signals and their frequency spectrums. The sampling theorem is then introduced as the mathematical tool that provides the means by which a 2-D continuous signal can be converted into a discrete signal and back to a continuous signal without loss of information.

3.4.1 2-D Sampled Signals

In order to establish the relations between continuous and discrete signals it is necessary to define an intermediate type of signal halfway between the continuous and discrete types of signals. The required type of signal is both continuous and sampled at the same time and has been referred to in the past as a sampled-data signal. It will be referred to here as a sampled signal, for the sake of simplicity, but it is hoped that it will not be confused with the discrete signal, which is a different mathematical entity.

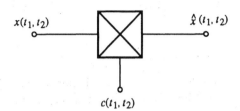

Figure 3.3   Ideal impulse sampler.

A 2-D sampled signal, denoted as $\hat{x}(t_1, t_2)$, can be generated by sampling a 2-D continuous signal by means of an ideal impulse sampler, as depicted in Fig. 3.3. This device is actually a multiplier and is characterized by

$$\hat{x}(t_1, t_2) = c(t_2, t_2)x(t_1, t_2) \tag{3.13}$$

where

$$c(t_1, t_2) = \sum_{n_1=-\infty}^{\infty} \sum_{n_2=-\infty}^{\infty} \delta(t_1 - n_1 T_1, t_2 - n_2 T_2) \tag{3.14}$$

The quantity $\delta(t_1, t_2)$ is the 2-D continuous impulse function and is assumed to have been defined as a generalized function [1,3,4] so as to ensure that

$$\delta(t_1 - t_{01}, t_2 - t_{02}) \longleftrightarrow e^{-j(\omega_1 t_{01} + \omega_2 t_{02})}$$

From Eqs. (3.13) and (3.14), we obtain

$$\hat{x}(t_1, t_2) = \sum_{n_1=-\infty}^{\infty} \sum_{n_2=-\infty}^{\infty} x(n_1 T_1, n_2 T_2)\delta(t_1 - n_1 T_1, t_2 - n_2 T_2) \tag{3.15}$$

In effect, a 2-D sampled signal is an array of continuous impulses. On comparing Eq. (3.15) with Eq. (1.22), the relation between a sampled and a discrete signal becomes evident. A discrete signal can be derived from a sampled signal and vice versa by replacing the continuous impulses by discrete impulses and vice versa.

The Fourier transform of $\hat{x}(t_1, t_2)$ is obtained as

$$\hat{X}(j\omega_1, j\omega_2)$$

$$= \mathscr{F} \sum_{n_1=-\infty}^{\infty} \sum_{n_2=-\infty}^{\infty} x(n_1 T_1, n_2 T_2)\delta(t_1 - n_1 T_1, t_2 - n_2 T_2)$$

$$= \sum_{n_1=-\infty}^{\infty} \sum_{n_2=-\infty}^{\infty} x(n_1 T_1, n_2 T_2)\mathscr{F}\delta(t_1 - n_1 T_1, t_2 - n_2 T_2)$$

$$= \sum_{n_1=-\infty}^{\infty} \sum_{n_2=-\infty}^{\infty} x(n_1 T_1, n_2 T_2)e^{-j(\omega_1 n_1 T_1 + \omega_2 n_2 T_2)}$$

$$\tag{3.16}$$

Evidently,

$$\hat{X}(j\omega_1, j\omega_2) = X_D(e^{j\omega_1 T_1}, e^{j\omega_2 T_2}) \tag{3.17}$$

where

$$X_D(z_1, z_2) = \mathscr{Z}x(n_1 T_1, n_2 T_2)$$

that is, the Fourier transform of a sampled signal $\hat{x}(t_1, t_2)$ is numerically equal to the z transform of the corresponding discrete signal $x(n_1 T_1, n_2 T_2)$ evaluated on the unit bicircle

$$T^2 = \{(z_1, z_2) : |z_1| = 1, |z_2| = 1\}$$

As may be expected, the frequency spectrum of $\hat{x}(t_1, t_2)$ is closely related to the frequency spectrum of $x(t_1, t_2)$. From Theorems 3.18a and 3.14, it can readily be shown that

$$\sum_{n_1 = -\infty}^{\infty} \sum_{n_2 = -\infty}^{\infty} x(n_1 T_1, n_2 T_2) e^{-j(\omega_1 n_1 T_1 + \omega_2 n_2 T_2)}$$

$$= \frac{1}{T_1 T_2} \sum_{n_1 = -\infty}^{\infty} \sum_{n_2 = -\infty}^{\infty} X(j\omega_1 + jn_1 \omega_{s1}, j\omega_2 + jn_2 \omega_{s2}) \tag{3.18}$$

where

$$\omega_{s1} = \frac{2\pi}{T_1} \quad \text{and} \quad \omega_{s2} = \frac{2\pi}{T_2}$$

and

$$x(t_1, t_2) \longleftrightarrow X(j\omega_1, j\omega_2)$$

Therefore, from Eqs. (3.16)–(3.18), we obtain

$$\hat{X}(j\omega_1, j\omega_2) = X_D(e^{j\omega_1 T_1}, e^{j\omega_2 T_2})$$

$$= \frac{1}{T_1 T_2} \sum_{n_1 = -\infty}^{\infty} \sum_{n_2 = -\infty}^{\infty} X(j\omega_1 + jn_1 \omega_{s1}, j\omega_2 + jn_2 \omega_{s2}) \tag{3.19}$$

that is, if the Fourier transform of $x(t_1, t_2)$ is known, that of $\hat{x}(t_1, t_2)$ can be uniquely determined. Furthermore, Eq. (3.19) shows that the Fourier transform of $\hat{x}(t_1, t_2)$ is a periodic function of (ω_1, ω_2) with period $(\omega_{s1}, \omega_{s2})$.

3.4.2 The 2-D Sampling Theorem

The 2-D sampling theorem states that a band-limited signal $x(t_1, t_2)$ for which

$$X(j\omega_1, j\omega_2) = 0 \quad \text{for} \quad |\omega_1| \geq \frac{\omega_{s1}}{2} \quad \text{or} \quad |\omega_2| \geq \frac{\omega_{s2}}{2} \tag{3.20}$$

where

$$\omega_{s1} = \frac{2\pi}{T_1} \quad \text{and} \quad \omega_{s2} = \frac{2\pi}{T_2}$$

can be uniquely determined from its values $x(n_1 T_1, n_2 T_2)$.

Like the 1-D sampling theorem, the 2-D sampling theorem is proved by deriving an interpolation formula that gives the band-limited signal $x(t_1, t_2)$ in terms of its values $x(n_1 T_1, n_2 T_2)$. With Eq. (3.20) satisfied, the quantity $T_1 T_2 \hat{X}(j\omega_1, j\omega_2)$ given by Eq. (3.19) as

$$T_1 T_2 \hat{X}(j\omega_1, j\omega_2) = \sum_{n_1 = -\infty}^{\infty} \sum_{n_2 = -\infty}^{\infty} X(j\omega_1 + jn_1\omega_{s1}, j\omega_2 + jn_2\omega_{s2})$$

is a periodic continuation of $X(j\omega_1, j\omega_2)$ and so

$$X(j\omega_1, j\omega_2) = T_1 T_2 H(j\omega_1, j\omega_2)\hat{X}(j\omega_1, j\omega_2) \qquad (3.21)$$

where

$$H(j\omega_1, j\omega_2) = \begin{cases} 1 & \text{for } |\omega_1| < \omega_{s1}/2 \text{ and } |\omega_2| < \omega_{s2}/2 \\ 0 & \text{otherwise} \end{cases} \qquad (3.22)$$

as depicted in Fig. 3.4. From Eqs. (3.16) and (3.21), we have

$X(j\omega_1, j\omega_2)$

$$= T_1 T_2 H(j\omega_1, j\omega_2) \sum_{n_1 = -\infty}^{\infty} \sum_{n_2 = -\infty}^{\infty} x(n_1 T_1, n_2 T_2)e^{-j(\omega_1 n_1 T_1 + \omega_2 n_2 T_2)} \qquad (3.23)$$

and, consequently,

$x(t_1, t_2)$

$$= \mathcal{F}^{-1}\left[T_1 T_2 H(j\omega_1, j\omega_2) \sum_{n_1 = -\infty}^{\infty} \sum_{n_2 = -\infty}^{\infty} x(n_1 T_1, n_2 T_2)e^{-j(\omega_1 n_1 T_1 + \omega_2 n_2 T_2)} \right]$$

$$= T_1 T_2 \sum_{n_1 = -\infty}^{\infty} \sum_{n_2 = -\infty}^{\infty} x(n_1 T_1, n_2 T_2)\mathcal{F}^{-1}[H(j\omega_1, j\omega_2)e^{-j(\omega_1 n_1 T_1 + \omega_2 n_2 T_2)}]$$

$$(3.24)$$

Since

$$\frac{\sin(\omega_{s1} t_1/2) \sin(\omega_{s2} t_2/2)}{\pi^2 t_1 t_2} \longleftrightarrow H(j\omega_1, j\omega_2) \qquad (3.25)$$

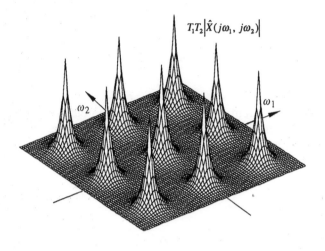

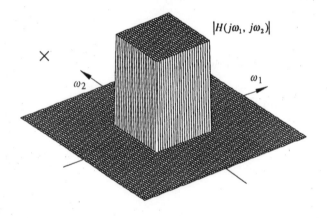

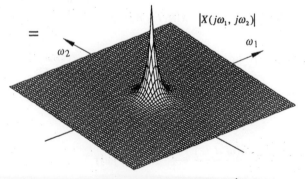

Figure 3.4 Derivation of $X(j\omega_1, j\omega_2)$ from $\hat{X}(j\omega_1, j\omega_2)$.

applying Theorem 3.13 yields

$$\frac{\sin[\omega_{s1}(t_1 - n_1T_1)/2]\,\sin[\omega_{s2}(t_2 - n_2T_2)/2]}{\pi^2(t_1 - n_1T_1)(t_2 - n_2T_2)}$$

$$\longleftrightarrow H(j\omega_1, j\omega_2)e^{-j(\omega_1 n_1 T_1 + \omega_2 n_2 T_2)} \qquad (3.26)$$

Therefore, from Eq. (3.24)

$$x(t_1, t_2) = \sum_{n_1=-\infty}^{\infty} \sum_{n_2=-\infty}^{\infty} x(n_1T_1, n_2T_2)$$

$$\times \frac{\sin[\omega_{s1}(t_1 - n_1T_1)/2]\,\sin[\omega_{s2}(t_2 - n_2T_2)/2]}{\omega_{s1}\omega_{s2}(t_1 - n_1T_1)(t_2 - n_2T_2)/4} \qquad (3.27)$$

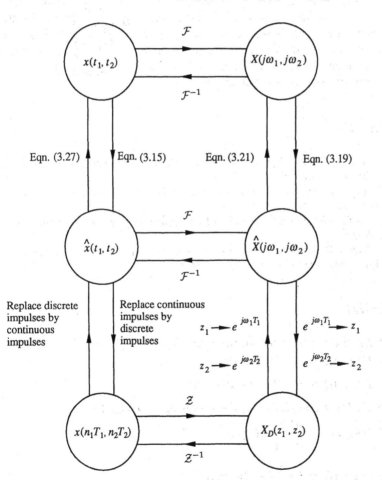

Figure 3.5   Interrelations between 2-D continuous, sampled, and discrete signals.

Equation (3.21) can be regarded as the characterization of an ideal 2-D continuous lowpass filter with input $\hat{x}(t_1, t_2)$, output $x(t_1, t_2)$, and transfer function $T_1 T_2 H(j\omega_1, j\omega_2)$. This provides a means by which a sampled signal can be converted back to the original continuous signal.

If Eq. (3.20) is not satisfied, then $T_1 T_2 \hat{X}(j\omega_1, j\omega_2)$ is not a periodic continuation of $X(j\omega_1, j\omega_2)$ and the use of a 2-D lowpass filter will yield an aliased or distorted version of $x(t_1, t_2)$.

3.4.3 Interrelations

In the preceding two sections it has been shown that a given band-limited 2-D continuous signal can be converted into a corresponding sampled signal, which can, in turn, be converted into a corresponding discrete signal. Furthermore, the discrete signal obtained can be converted back into the original sampled signal, which can, in turn, be used to obtain the original continuous signal. The space-domain and frequency-domain interrelations involved are illustrated in the diagram of Fig. 3.5.

REFERENCES

1. A. Antoniou, *Digital Filters: Analysis and Design,* New York: McGraw-Hill, 1979 (2nd ed. in press).
2. E. I. Jury, *Theory and Application of the z-Transform Method,* New York: Wiley, 1968.
3. A. Papoulis, *The Fourier Integral and Its Applications,* New York: McGraw-Hill, 1962.
4. M. J. Lighthill, *Introduction to Fourier Analysis and Generalized Functions,* London: Cambridge University Press, 1958.

PROBLEMS

3.1 Obtain the z transforms of the following 2-D signals:

(a) $f(n_1, n_2) = a^{n_1 + n_2} u(n_1, n_2), \quad |a| < 1.$

(b) $f(n_1, n_2) = a^{|n_1 - n_2|} u(n_1, n_2), \quad |a| < 1.$

3.2 Obtain the z transforms of the following 2-D signals:

(a) $f(n_1, n_2) = \dfrac{a^{n_1} b^{n_2}}{n_1! n_2!} u(n_1, n_2).$

(b) $f(n_1, n_2) = \dfrac{(-n_2)^{n_1}}{n_1! n_2!} u(n_1, n_2).$

3.3 Obtain the z transforms of the following 2-D signals:

(a) $f(n_1, n_2) = \begin{cases} 1 & -1 \le n_2 \le 3 \quad \text{and} \quad |n_1| \le n_2 + 1 \\ 0 & \text{otherwise.} \end{cases}$

(b) $f(n_1, n_2) = \begin{cases} 1 & |n_1| \le 3, |n_2| \le 3, n_1 n_2 \ge 0 \\ 0 & \text{otherwise} \end{cases}$

(c) $f(n_1, n_2) = \begin{cases} 1 & |n_1| \le 3, |n_2| \le 3 \\ 0 & \text{otherwise.} \end{cases}$

3.4 Prove Theorems 3.1–3.3.

3.5 Prove Theorems 3.4–3.5.

3.6 Prove Theorem 3.6.

3.7 Find the inverse z transform of

(a) $F(z_1, z_2) = \dfrac{z_1^{-1} z_2^{-1}}{1 + 0.2 z_1^{-1} + 0.5 z_1^{-1} z_2^{-1}}.$

(b) $F(z_1, z_2) = \dfrac{1 + z_1^{-1} - 0.5 z_1^{-1} z_2^{-1}}{1 + 0.8 z_1^{-1} - 0.5 z_2^{-1} - 0.4 z_1^{-1} z_2^{-1}}.$

3.8 Assuming that the 1-D inverse z transforms of $1/F_1(z_1)$ and $1/F_2(z_2)$ are known as $\tilde{f}_1(n_1)$ and $\tilde{f}_2(n_2)$, respectively, show that the inverse z transform of

$$F(z_1, z_2) = \frac{\sum_{i=0}^{M_1} \sum_{j=0}^{M_2} a_{ij} z_1^{-i} z_2^{-j}}{F_1(z_1) F_2(z_2)}$$

is given by

$$f(n_1, n_2) = \sum_{i=0}^{M_1} \sum_{j=0}^{M_2} a_{ij} \tilde{f}_1(n_1 - i) \tilde{f}_2(n_2 - j)$$

3.9 Apply the formula obtained in Problem 3.8 to $F(z_1, z_2)$ of Problem 3.7b.

3.10 Prove Theorems 3.10–3.14.

3.11 Prove Theorems 3.15–3.16.

3.12 Prove Theorem 3.17.

3.13 Obtain the Fourier transforms of the following 2-D signals:
(a) $f(t_1, t_2) = \delta(t_1, t_2)$ where $\delta(t_1, t_2)$ is the generalized impulse function defined by

$$\int_{-\infty}^{\infty} \int_{-\infty}^{\infty} \delta(t_1, t_2) \phi(t_1, t_2) \, dt_1 \, dt_2 = \phi(0, 0)$$

which is the 2-D extension of the 1-D generalized impulse function. (See Chapter 6 of [1].)

(b) $f(t_1, t_2) = 1$.

(c) $f(t_1, t_2) = \delta(t_1 - t_{10}, t_2 - t_{20})$ where t_{10} and t_{20} are constants.

3.14 Obtain the Fourier transforms of the following 2-D signals:

(a) $f(t_1, t_2) = e^{j(\omega_{10}t_1 + \omega_{20}t_2)}$.

(b) $f(t_1, t_2) = e^{-(\omega_{10}t_1 + \omega_{20}t_2)}$.

(c) $f(t_1, t_2) = e^{-(\omega_{10}t_1^2 + \omega_{20}t_2^2)}$.

3.15 Obtain the Fourier transforms of the following 2-D signals:

(a) $f(t_1, t_2) = \begin{cases} 1 & |t_1| \leq 1, \ -0.5 \leq t_2 \leq 1 \\ 0 & \text{otherwise.} \end{cases}$

(b) $f(t_1, t_2) = \begin{cases} 1 & |t_1| \leq 1, |t_2| \leq 1 \quad \text{and} \quad t_1 t_2 \geq 0 \\ 0 & \text{otherwise.} \end{cases}$

3.16 Using the functions in Problems 3.14c and 3.15a, verify Theorem 3.15.

3.17 Find the inverse Fourier transform of

$$F(j\omega_1, j\omega_2) = \begin{cases} e^{-(|\omega_1| + |\omega_2|)} & |\omega_1| \leq \pi, |\omega_2| \leq \pi \\ 0 & \text{otherwise.} \end{cases}$$

3.18 Using function $f(t_1, t_2)$ obtained in Problem 3.17, verify the formula in Eq. (3.27).

4

The Application of the *z* Transform

4.1 INTRODUCTION

Like a 1-D digital filter, a 2-D digital filter can be represented by a discrete transfer function that can be used to find the space-domain response of the filter for an arbitrary input or to carry out a frequency-domain analysis. The discrete transfer function contains, in addition, all the necessary information to determine whether the filter is stable, marginally stable, or unstable or to study the effects of using finite arithmetic in the implementation of the filter.

In this chapter the 2-D discrete transfer function is defined through the use of the convolution summation, as in the 1-D case, and methods for its derivation from a difference equation, a digital-filter network, or a state-space characterization are then described. Some of the properties of the discrete transfer function and its application to space-domain analysis and frequency-domain analysis are then considered.

4.2 THE TRANSFER FUNCTION

The transfer function of a 2-D digital filter is defined as the ratio of the *z* transform of the output to the *z* transform of the input.

Consider a linear, shift-invariant, 2-D digital filter, and let $x(n_1, n_2)$, $y(n_1, n_2)$, and $h(n_1, n_2)$ be the input, output, and impulse response of the filter, respectively. By using the convolution summation described in Sec. 1.6.3 (see Eq. (1.23b)), we have

$$y(n_1, n_2) = \sum_{i=-\infty}^{\infty} \sum_{j=-\infty}^{\infty} h(i, j)x(n_1 - i, n_2 - j) \qquad (4.1)$$

and, consequently, Theorem 3.5 yields

$$\mathscr{L}y(n_1, n_2) = \mathscr{L}h(n_1, n_2)\mathscr{L}x(n_1, n_2)$$

Therefore

$$H(z_1, z_2) = \frac{Y(z_1, z_2)}{X(z_1, z_2)} \qquad (4.2)$$

Evidently, the transfer function of a digital filter turns out to be the z transform of its impulse response, as in other types of systems.

The transfer function of a 2-D digital filter can readily be derived from a difference equation characterizing the filter, from a filter network, or from a state-space characterization.

4.2.1 Derivation of $H(z_1, z_2)$ from a Difference Equation

For causal recursive filters

$$y(n_1, n_2) = \sum_{i=0}^{N_1} \sum_{j=0}^{N_2} a_{ij}x(n_1 - i, n_2 - j) - \sum_{i=0}^{N_1} \sum_{j=0}^{N_2} b_{ij}y(n_1 - i, n_2 - j)$$

where $b_{00} = 0$, and on applying the z transform we obtain

$$\mathscr{L}y(n_1, n_2) = \sum_{i=0}^{N_1} \sum_{j=0}^{N_2} a_{ij}z_1^{-i}z_2^{-j}\mathscr{L}x(n_1 - i, n_2 - j)$$

$$- \sum_{i=0}^{N_1} \sum_{j=0}^{N_2} b_{ij}z_1^{-i}z_2^{-j}\mathscr{L}y(n_1 - i, n_2 - j)$$

or

$$\frac{Y(z_1, z_2)}{X(z_1, z_2)} = H(z_1, z_2) = \frac{\overline{N}(z_1^{-1}, z_2^{-1})}{\overline{D}(z_1^{-1}, z_2^{-1})}$$

$$= \frac{\sum_{i=0}^{N_1} \sum_{j=0}^{N_2} a_{ij}z_1^{-i}z_2^{-j}}{1 + \sum_{i=0}^{N_1} \sum_{j=0}^{N_2} b_{ij}z_1^{-i}z_2^{-j}} \qquad (4.3a)$$

$$= \frac{N(z_1, z_2)}{D(z_1, z_2)}$$

$$= \frac{\sum_{i=0}^{N_1} \sum_{j=0}^{N_2} a_{ij} z_1^{N_1-i} z_2^{N_2-j}}{z_1^{N_1} z_2^{N_2} + \sum_{i=0}^{N_1} \sum_{j=0}^{N_2} b_{ij} z_1^{N_1-i} z_2^{N_2-j}} \quad (4.3b)$$

Similarly, for nonsymmetric half-plane recursive filters, Eq. (1.8b) gives

$$H(z_1, z_2) = \frac{\overline{N}(z_1^{-1}, z_2^{-1})}{\overline{D}(z_1^{-1}, z_2^{-1})}$$

$$= \frac{\sum_{i=0}^{N_1} a_{i0} z_1^{-i} + \sum_{i=-N_1}^{N_1} \sum_{j=1}^{N_2} a_{ij} z_1^{-i} z_2^{-j}}{1 + \sum_{i=1}^{N_1} b_{i0} z_1^{-i} + \sum_{i=-N_1}^{N_1} \sum_{j=1}^{N_2} b_{ij} z_1^{-i} z_2^{-j}} \quad (4.3c)$$

$$= \frac{N(z_1, z_2)}{D(z_1, z_2)}$$

$$= \frac{\sum_{i=0}^{N_1} a_{i0} z_1^{N_1-i} z_2^{N_2} + \sum_{i=-N_1}^{N_1} \sum_{j=1}^{N_2} a_{ij} z_1^{N_1-i} z_2 N_2^{N_2-j}}{z_1^{N_1} z_2^{N_2} + \sum_{i=1}^{N_1} b_{i0} z_1^{N_1-i} z_2^{N_2} + \sum_{i=-N_1}^{N_1} \sum_{j=1}^{N_2} b_{ij} z_1^{N_1-i} z_2 N_2^{N_2-j}} \quad (4.3d)$$

For causal noncursive filters, $b_{ij} = 0$, for $i = 0, 1, \ldots, N_1$ and $j = 0, 1, \ldots, N_2$ and so the transfer function assumes the form

$$H(z_1, z_2) = \overline{N}(z_1^{-1}, z_2^{-1}) = \sum_{i=0}^{N_1} \sum_{j=0}^{N_2} a_{ij} z_1^{-i} z_2^{-j} \quad (4.4a)$$

$$= \frac{N(z_1, z_2)}{D(z_1, z_2)} = \frac{\sum_{i=0}^{N_1} \sum_{j=0}^{N_2} a_{ij} z_1^{N_1-i} z_2^{N_2-j}}{z_1^{N_1} z_2^{N_2}} \quad (4.4b)$$

Similarly, for nonsymmetric half-plane nonrecursive filters, Eqs. (4.3c) and (4.3d) assume the form

$$H(z_1, z_2) = \overline{N}(z_1^{-1}, z_2^{-1}) = \sum_{i=0}^{N_1} a_{i0} z_1^{-i} + \sum_{i=-N_1}^{N_1} \sum_{j=1}^{N_2} a_{ij} z_1^{-i} z_2^{-j} \quad (4.4c)$$

$$= \frac{N(z_1, z_2)}{D(z_1, z_2)}$$

$$= \frac{\sum_{i=0}^{N_1} a_{i0} z_1^{N_1-i} z_2^{N_2} + \sum_{i=-N_1}^{N_1} \sum_{j=1}^{N_2} a_{ij} z_1^{N_1-i} z_2^{N_2-j}}{z_1^{N_1} z_2^{N_2}} \quad (4.4d)$$

Two major differences between 1-D and 2-D discrete transfer functions should be mentioned at this point. First, polynomials in two variables cannot in general be factorized into products of lower-order polynomials due to the fact that the Fundamental theorem of algebra does not extend to polynomials in two variables. Second, the zeros of polynomials in two complex variables are not in general isolated points, and consequently, the singularities of the transfer function, which are the zeros of the denominator polynomial in $H(z_1, z_2)$, are not in general isolated points.

Example 4.1   Find the zeros and singularities of

$$(a) \quad H(z_1, z_2) = \frac{z_1 z_2}{z_1 z_2 - 0.5}$$

$$(b) \quad H(z_1, z_2) = \frac{1 - z_1 - z_2 + z_1 z_2}{2 - z_1 - z_2}$$

Solution.   (a) The zeros of $H(z_1, z_2)$ are the points in the sets

$$Z_1 = \{(z_1, z_2) : z_1 = 0, z_2 \text{ arbitrary}\}$$

$$Z_2 = \{(z_1, z_2) : z_1 \text{ arbitrary}, z_2 = 0\}$$

and the singularities of $H(z_1, z_2)$ are the points in the set

$$S = \{(z_1, z_2) : z_1 z_2 = 0.5\}$$

Evidently, the numbers of zeros and singularities are both infinite.

(b) In this case the zeros and singularities of the transfer function are given by

$$Z = \{(z_1, z_2) : z_1 + z_2 - z_1 z_2 = 1\}$$

and

$$S = \{(z_1, z_2) : z_1 + z_2 = 2\}$$

As in part (a) there are an infinite number of zeros or singularities.

The point $(z_1, z_2) = (1, 1)$ in Example 4.1b is of particular interest. As can be readily verified, this point is simultaneously a zero and a singularity of the transfer function since it belongs to both sets Z and S. Such points are of interest in the study of stability analysis and are referred to as nonessential singularities of the second kind.

4.2.2   Derivation of $H(z_1, z_2)$ from a Digital-Filter Network

The z-domain characterizations of the unit shifters, the adder, and the multiplier can readily be obtained from Table 1.1 (see p. 16) as

$$Y(z_1, z_2) = z_1^{-1} X(z_1, z_2) \quad \text{and} \quad Y(z_1, z_2) = z_2^{-1} X(z_1, z_2)$$

$$Y(z_1, z_2) = \sum_{i=1}^{k} X_i(z_1, z_2)$$

and

$$Y(z_1, z_2) = m X(z_1, z_2)$$

respectively. By using these relations, the transfer function of a 2-D digitial filter can be obtained directly from a network representation of the filter.

Example 4.2 Find the transfer function of the filter shown in Fig. 4.1.

Solution. From Fig. 4.1

$$Y(z_1, z_2) = X(z_1, z_2) + 0.5z_1^{-1}Y(z_1, z_2) - 0.4z_1^{-1}z_2^{-2}Y(z_1, z_2)$$

Hence

$$H(z_1, z_2) = \frac{Y(z_1, z_2)}{X(z_1, z_2)}$$

$$= \frac{1}{1 - 0.5z_1^{-1} + 0.4z_1^{-1}z_2^{-2}}$$

$$= \frac{z_1 z_2^2}{z_1 z_2^2 - 0.5z_2^2 + 0.4}$$

4.2.3 Derivation of $H(z_1, z_2)$ from a State-Space Characterization

By applying the z transform to Eq. (2.9), we have

$$\mathscr{L}\begin{bmatrix} \mathbf{q}^H(i + 1, j) \\ \mathbf{q}^V(i, j + 1) \end{bmatrix} = \mathbf{A}\mathscr{L}\begin{bmatrix} \mathbf{q}^H(i, j) \\ \mathbf{q}^V(i, j) \end{bmatrix} + \mathbf{b}\mathscr{L}x(i, j)$$

$$\mathscr{L}y(i, j) = \mathbf{c}\mathscr{L}\begin{bmatrix} \mathbf{q}^H(i, j) \\ \mathbf{q}^V(i, j) \end{bmatrix} + \mathbf{d}\mathscr{L}x(i, j)$$

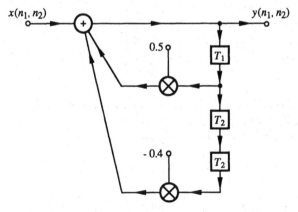

Figure 4.1 The digital filter for Example 4.2.

If we write

$$\mathscr{L} \begin{bmatrix} \mathbf{q}^H(i, j) \\ \mathbf{q}^V(i, j) \end{bmatrix} = \mathbf{Q}(z_1, z_2)$$

and note that

$$\mathscr{L} \begin{bmatrix} \mathbf{q}^H(i + 1, j) \\ \mathbf{q}^V(i, j + 1) \end{bmatrix} = \begin{bmatrix} z_1 \mathbf{I}_N & \mathbf{0} \\ \mathbf{0} & z_2 \mathbf{I}_M \end{bmatrix} \mathbf{Q}(z_1, z_2)$$

where $\mathbf{I}_N$ and $\mathbf{I}_M$ denote the identity matrices of dimensions N and M, respectively, we obtain

$$\begin{bmatrix} z_1 \mathbf{I}_N & \mathbf{0} \\ \mathbf{0} & z_2 \mathbf{I}_M \end{bmatrix} \mathbf{Q}(z_1, z_2) = \mathbf{A}\mathbf{Q}(z_1, z_2) + \mathbf{b}X(z_1, z_2)$$

$$Y(z_1, z_2) = \mathbf{c}\mathbf{Q}(z_1, z_2) + \mathbf{d}X(z_1, z_2)$$

Therefore, on eliminating $\mathbf{Q}(z_1, z_2)$, we have

$$H(z_1, z_2) = \mathbf{c} \begin{bmatrix} z_1 \mathbf{I}_N - \mathbf{A}_1 & -\mathbf{A}_2 \\ -\mathbf{A}_3 & z_2 \mathbf{I}_M - \mathbf{A}_4 \end{bmatrix}^{-1} \mathbf{b} + \mathbf{d} \tag{4.5}$$

4.3 STABILITY

As in other systems, the stability of a 2-D digital filter is closely linked with the singularities of the transfer function. Since these are not in general isolated points as in 1-D digital filters, stability analysis is much more involved in 2-D digital filters, as will be demonstrated in Chapter 5. In this section, it is shown that the necessary and sufficient space-domain condition for BIBO stability, as defined in Sec. 1.7, translates readily into a z-domain condition that is imposed on the singularities of the transfer function. This condition is due to Shanks [1–4] and is stated in terms of the following theorem.

Theorem 4.1 Shanks' Theorem *A 2-D digital filter characterized by a transfer function $H(z_1, z_2)$, as in Eq. (4.3a), is BIBO stable if $H(z_1, z_2)$ has no singularities on the unit bidisc defined by the set*

$$\overline{U}^2 = \{(z_1^{-1}, z_2^{-1}) : |z_1^{-1}| \le 1, |z_2^{-1}| \le 1\}$$

that is,

$$\overline{D}(z_1^{-1}, z_2^{-1}) \ne 0 \qquad \text{for } (z_1^{-1}, z_2^{-1}) \in \overline{U}^2 \tag{4.6}$$

where

$$\overline{D}(z_1^{-1}, z_2^{-1}) = 1 + \sum_{i=0}^{N_1} \sum_{j=0}^{N_2} b_{ij} z_1^{-i} z_2^{-j}$$

with $b_{00} = 0$.

Proof.   By the continuity of $\overline{D}(z_1^{-1}, z_2^{-1})$, there exists another set given by

$$\overline{U}_{1+\varepsilon}^2 = \{(z_1^{-1}, z_2^{-1}): |z_1^{-1}| \le 1 + \varepsilon, |z_2^{-1}| \le 1 + \varepsilon\}$$

such that

$$\overline{D}(z_1^{-1}, z_2^{-1}) \ne 0 \qquad \text{for } (z_1^{-1}, z_2^{-1}) \in \overline{U}_{1+\varepsilon}^2$$

At any point in the set $\overline{U}_{1+\varepsilon}^2$, the transfer function $H(z_1, z_2)$ is finite; that is, the series at the right-hand side of

$$H(z_1, z_2) = \sum_{n1=0}^{\infty} \sum_{n2=0}^{\infty} h(n_1, n_2) z_1^{-n_1} z_2^{-n_2}$$

is absolutely summable and so

$$\sum_{n1=-\infty}^{\infty} \sum_{n2=-\infty}^{\infty} |h(n_1, n_2)| \, |z_1^{-n_1}| \, |z_2^{-n_2}| \le K \qquad \text{for } (z_1^{-1}, z_2^{-1}) \in \overline{U}_{1+\varepsilon}^2$$

Now with the substitution of $(z_1^{-1}, z_2^{-1}) = (1, 1)$, we obtain

$$\sum_{n1=0}^{\infty} \sum_{n2=0}^{\infty} |h(n_1, n_2)| \le K$$

Therefore, from Sec. 1.7, we conclude that the 2-D filter is BIBO stable. ∎

In the preceding proof, the condition in Eq. (4.6) was shown to be sufficient for stability. Nevertheless, under certain circumstances it is also a necessary condition for stability.

Example 4.3   Check the filters characterized by the following transfer functions for BIBO stability.

$$(a) \quad H(z_1, z_2) = \frac{1}{1 - 0.4 z_1^{-1} - 0.4 z_2^{-1}}$$

$$(b) \quad H(z_1, z_2) = \frac{1}{1 - b z_1^{-1} z_2^{-1}}$$

$$(c) \quad H(z_1, z_2) = \frac{(1 - z_1^{-1})^8 (1 - z_2^{-1})^8}{2 - z_1^{-1} - z_2^{-1}}$$

Solution. (a) For $(z_1^{-1}, z_2^{-1}) \in \bar{U}^2$

$$\begin{aligned} |\bar{D}(z_1^{-1}, z_2^{-1})| &= |1 - 0.4z_1^{-1} - 0.4z_2^{-1}| \\ &\geq 1 - 0.4 |z_1^{-1} + z_2^{-1}| \\ &\geq 1 - 0.4 \times 2 \\ &> 0 \end{aligned}$$

Therefore, the filter is BIBO stable.

(b) For $(z_1^{-1}, z_2^{-1}) \in \bar{U}^2$

$$\begin{aligned} |\bar{D}(z_1^{-1}, z_2^{-1})| &= |1 - bz_1^{-1}z_2^{-1}| \\ &\geq 1 - |b| \end{aligned}$$

If $|b| < 1$, then $|\bar{D}(z_1^{-1}, z_2^{-1})| \neq 0$ for $(z_1^{-1}, z_2^{-1}) \in \bar{U}^2$ and the filter is BIBO stable. It turns out that this condition is also necessary for BIBO stability, as will be demonstrated in Chapter 5.

(c) We note that

$$|\bar{D}(z_1^{-1}, z_2^{-1})| = 2 - z_1^{-1}z_2^{-1}$$

becomes zero at point $(z_1^{-1}, z_2^{-1}) = (1, 1)$. In addition, we note that at point $(1, 1)$, the numerator of the transfer function becomes zero and that no common factor exists between the numerator and denominator. In this case we have a nonessential singularity of the second kind and, further, it is not certain whether the filter is stable or not. The role of nonessential singularities of the second kind in stability analysis will be examined in Chapter 5.

4.4 SPACE-DOMAIN ANALYSIS

The space-domain response of a 2-D digital filter to any input $x(n_1, n_2)$ can be readily obtained from Eq. (4.2) as

$$y(n_1, n_2) = \mathscr{Z}^{-1}[H(z_1, z_2)X(z_1, z_2)] \tag{4.7}$$

The inverse z transform of the product of the transfer function and the z transform of the input can be obtained by using the inversion techniques described in Sec. 3.2.4.

Example 4.4 Find the unit-step response of the filter described in Example 1.2, assuming that $b = 0.5$.

Solution.   The z transform of the unit step is given by

$$X(z_1, z_2) = \sum_{n_1 = -\infty}^{\infty} \sum_{n_2 - \infty}^{\infty} u(n_1, n_2) z_1^{-n_1} z_2^{-n_2}$$

$$= \sum_{n_1 = 0}^{\infty} \sum_{n_2 = 0}^{\infty} z_1^{-n_1} z_2^{-n_2}$$

$$= \sum_{n_1 = 0}^{\infty} z_1^{-n_1} \sum_{n_2 = 0}^{\infty} z_2^{-n_2}$$

$$= \frac{1}{(1 - z_1^{-1})(1 - z_2^{-1})}$$

$$= \frac{z_1 z_2}{(z_1 - 1)(z_2 - 1)}$$

where z_1 and z_2 satisfy the inequalities

$$|z_1| > 1 \quad \text{and} \quad |z_2| > 1$$

in order to guarantee the convergence of the z transform. From Eqs. (1.21) and (4.3b), the transfer function of the filter is obtained as

$$H(z_1, z_2) = \frac{z_1 z_2}{z_1 z_2 - 0.5}$$

and hence Eq. (4.7) gives the unit-step response of the filter as

$$y(n_1, n_2) = \frac{1}{(2\pi j)^2} \oint_{\Gamma_2} \oint_{\Gamma_1} \frac{z_1 z_2}{(z_1 z_2 - 0.5)}$$

$$\times \frac{z_1 z_2}{(z_1 - 1)(z_2 - 1)} z_1^{n_1 - 1} z_2^{n_2 - 1} \, dz_1 \, dz_2$$

where contours Γ_1 and Γ_2 are chosen as

$$\Gamma_1 = \{z_1 : |z_1| = 2\} \quad \text{and} \quad \Gamma_2 = \{z_2 : |z_2| = 2\}$$

in order to guarantee convergence. The preceding integrand can be written as

$$\frac{z_1^{n_1 + 1} z_2^{n_2 + 1}}{(z_1 z_2 - 0.5)(z_1 - 1)(z_2 - 1)} = \frac{z_2^{n_2 + 1}}{(z_2 - 1)(z_2 - 0.5)}$$

$$\times \left(\frac{1}{z_1 - 1} - \frac{1}{z_1 - 0.5 z_2^{-1}} \right) z_1^{n_1 + 1}$$

and hence for $n_1 \geq 0$ and $n_2 \geq 0$, we have

$$
\begin{aligned}
y(n_1, n_2) &= \frac{1}{2\pi j} \oint_{\Gamma_2} \frac{z_2^{n_2+1}}{(z_2 - 1)(z_2 - 0.5)} \\
&\quad \times \left[\frac{1}{2\pi j} \oint_{\Gamma_1} \left(\frac{z_1^{n_1+1}}{z_1 - 1} - \frac{z_1^{n_1+1}}{z_1 - 0.5 z_2^{-1}} \right) dz_1 \right] dz_2 \\
&= \frac{1}{2\pi j} \oint_{\Gamma_2} \frac{z_2^{n_2+1} - 0.5^{n_1+1} z_2^{n_2-n_1}}{(z_2 - 1)(z_2 - 0.5)} dz_2 \\
&= \frac{1}{2\pi j} \oint_{\Gamma_2} (2 z_2^{n_2+1} - 0.5^{n_1} z_2^{n_2-n_1}) \\
&\quad \times \left(\frac{1}{z_2 - 1} - \frac{1}{z_2 - 0.5} \right) dz_2 \\
&= \begin{cases}
2 - 0.5^{n_2} & \text{if } n_1 \geq 0, n_2 \geq 0, n_1 \geq n_2 \\
2 - 0.5^{n_1} & \text{if } n_1 \geq 0, n_2 \geq 0, n_2 \geq n_1 \\
0 & \text{otherwise}
\end{cases}
\end{aligned}
$$

4.5 FREQUENCY-DOMAIN ANALYSIS

The most fundamental application of 2-D digital filters involves the manipulation of the frequency spectrum of a 2-D discrete signal. To determine the effect of a 2-D digital filter on a 2-D signal, a frequency-domain analysis must be carried out, which is concerned with the response of the filter to a sinusoidal input.

4.5.1 Sinusoidal Response

Consider a 2-D digital filter characterized by a transfer function

$$
H(z_1, z_2) = \sum_{n_1 = -\infty}^{\infty} \sum_{n_2 = -\infty}^{\infty} h(n_1, n_2) z_1^{-n_1} z_2^{-n_2} \tag{4.8}
$$

and assume that the input of the filter is given by

$$
x(n_1, n_2) = \sin(n_1 \omega_1 T_1 + n_2 \omega_2 T_2)
$$

The response of the filter is given by the convolution summation of Eq. (4.1) as

$$y(n_1, n_2) = \sum_{k_1=-\infty}^{\infty} \sum_{k_2=-\infty}^{\infty} h(k_1, k_2)$$

$$\times \sin[(n_1 - k_1)\omega_1 T_1 + (n_2 - k_2)\omega_2 T_2]$$

$$= \frac{1}{2j} \sum_{k_1=-\infty}^{\infty} \sum_{k_2=-\infty}^{\infty} h(k_1, k_2)$$

$$\times \left[e^{j[(n_1-k_1)\omega_1 T_1 + (n_2-k_2)\omega_2 T_2]} \right.$$

$$\left. - e^{-j[(n_1-k_1)\omega_1 T_1 + (n_2+k_2)\omega_2 T_2]} \right]$$

$$= \frac{1}{2j} \left[e^{j(n_1\omega_1 T_1 + n_2\omega_2 T_2)} \right.$$

$$\times \sum_{k_1=-\infty}^{\infty} \sum_{k_2=-\infty}^{\infty} h(k_1, k_2) e^{-j\omega_1 T_1 k_1} e^{-j\omega_2 T_2 k_2}$$

$$- e^{-j(n_1\omega_1 T_1 + n_2\omega_2 T_2)}$$

$$\left. \times \sum_{k_1=-\infty}^{\infty} \sum_{k_2=-\infty}^{\infty} h(k_1, k_2) e^{j\omega_1 T_1 k_1} e^{j\omega_2 T_2 k_2} \right] \quad (4.9)$$

From Eq. (4.8), we can write

$$H(e^{j\omega_1 T_1}, e^{j\omega_2 T_2}) = M(\omega_1, \omega_2) e^{j\theta(\omega_1, \omega_2)}$$

$$= \sum_{n_1=-\infty}^{\infty} \sum_{n_2=-\infty}^{\infty} h(n_1, n_2) e^{-j\omega_1 T_1 n_1} e^{-j\omega_2 T_2 n_2} \quad (4.10)$$

where

$$M(\omega_1, \omega_2) = |H(e^{j\omega_1 T_1}, e^{j\omega_2 T_2})| \quad \text{and} \quad \theta(\omega_1, \omega_2) = \arg H(e^{j\omega_1 T_1}, e^{j\omega_2 T_2}) \quad (4.11)$$

and hence Eqs. (4.9) and (4.10) yield

$$y(n_1, n_2) = \frac{1}{2j} \left[e^{j(n_1\omega_1 T_1 + n_2\omega_2 T_1)} M(\omega_1, \omega_2) e^{j\theta(\omega_1, \omega_2)} \right.$$

$$\left. - e^{-j(n_1\omega_1 T_1 + n_2\omega_2 T_2)} M(\omega_1, \omega_2) e^{-j\theta(\omega_1, \omega_2)} \right]$$

Therefore

$$y(n_1, n_2) = M(\omega_1, \omega_2) \sin[(n_1\omega_1 T_1 + n_2\omega_2 T_2) + \theta(\omega_1, \omega_2)] \quad (4.12)$$

This result shows that the effect of a 2-D digital filter on a sinusoidal input is to introduce a gain $M(\omega_1, \omega_2)$ and a phase shift $\theta(\omega_1, \omega_2)$ as in the 1-D case.

In the preceding analysis, the input $x(n_1, n_2)$ was assumed to be sinusoidal over the entire (n_1, n_2) plane. Nevertheless, the result obtained holds for the case where the input is sinusoidal in the first quadrant of the (n_1, n_2) plane and zero elsewhere; that is, for

$$x(n_1, n_2) = u(n_1, n_2) \sin(n_1 \omega_1 T_1 + n_2 \omega_2 T_2)$$

provided that the filter is allowed to reach steady state. To demonstrate this fact, let us consider a 2-D filter whose transfer function can be expressed as

$$H(z_1, z_2) = \sum_{l=1}^{L} H_{1l}(z_1) H_{2l}(z_2) \tag{4.13}$$

Let us assume that the poles of $H_{1l}(z_1)$ and $H_{2l}(z_2)$ are located inside the unit bicircle

$$U^2 = \{(z_1, z_2) : |z_1| = 1, |z_2| = 1\}$$

to ensure that the filter is stable, and that none of the poles are multiple for the sake of simplicity. The z transform of $x(n_1, n_2)$ is given by

$$
\begin{aligned}
X(z_1, z_2) &= \sum_{n1=0}^{\infty} \sum_{n2=0}^{\infty} \sin(n_1 \omega_1 T_1 + n_2 \omega_2 T_2) z_1^{-n_1} z_2^{-n_2} \\
&= \frac{z_1 z_2 [z_1 \sin \omega_2 T_2 + z_2 \sin \omega_1 T_1 - \sin(\omega_1 T_1 + \omega_2 T_2)]}{(z_1 - e^{j\omega_1 T_1})(z_2 - e^{j\omega_2 T_2})(z_1 - e^{-j\omega_1 T_1})(z_2 - e^{-j\omega_2 T_2})} \\
&= \frac{z_1^2 z_2 \sin \omega_2 T_2 + z_1 z_2^2 \sin \omega_1 T_1 - z_1 z_2 \sin(\omega_1 T_1 + \omega_2 T_2)}{D_1(z_1) D_2(z_2)}
\end{aligned}
$$

where

$$\frac{1}{D_1(z_1)} = \frac{1}{2j \sin \omega_1 T_1} \left(\frac{1}{z_1 - e^{j\omega_1 T_1}} - \frac{1}{z_1 - e^{-j\omega_1 T_1}} \right)$$

and

$$\frac{1}{D_2(z_2)} = \frac{1}{2j \sin \omega_2 T_2} \left(\frac{1}{z_2 - e^{j\omega_2 T_2}} - \frac{1}{z_2 - e^{-j\omega_2 T_2}} \right)$$

The response of the filter is given by

$$y(n_1, n_2) = \frac{1}{(2\pi j)^2} \oint_{\Gamma_2} \oint_{\Gamma_1} H(z_1, z_2) X(z_1, z_2) z_1^{n_1-1} z_2^{n_2-1} \, dz_1 \, dz_2$$

and if contours Γ_1 and Γ_2 are taken to be in $\overline{U}^2$ (see Sec. 4.3), we obtain

$$y(n_1, n_2) = y_t(n_1, n_2) + y_s(n_1, n_2) \tag{4.14}$$

where $y_t(n_1, n_2)$ and $y_s(n_1, n_2)$ are output components due to the residues at the poles of $H(z_1, z_2)$ and $X(z_1, z_2)$, respectively. After some manipulation, we obtain

$$
\begin{aligned}
y_s(n_1, n_2) &= \frac{1}{2j} [e^{j(n_1\omega_1 T_1 + n_2\omega_2 T_2)} M(\omega_1, \omega_2) e^{j\theta(\omega_1, \omega_2)} \\
&\quad - e^{-j(n_1\omega_1 T_1 + n_2\omega_2 T_2)} M(\omega_1, \omega_2) e^{-j\theta(\omega_1, \omega_2)}] \\
&= M(\omega_1, \omega_2) \sin[(n_1\omega_1 T_1 + n_2\omega_2 T_2) + \theta(\omega_1, \omega_2)] \quad (4.15)
\end{aligned}
$$

and

$$
y_t(n_1, n_2) = \sum_{l=1}^{L} \sum_{j=1}^{M_l} \sum_{i=1}^{N_l} \operatorname*{res}_{z_1 = p_{li}} [H_{1l}(z_1)] \operatorname*{res}_{z_2 = p_{lj}} [H_{2l}(z_2)] X(p_{li}, q_{lj}) p_{li}^{n_1 - 1} q_{lj}^{n_2 - 1}
$$

where N_l and M_l denote the orders and p_{li} and q_{lj} denote the ith and jth poles of $H_{1l}(z_1)$ and $H_{2l}(z_2)$, respectively. Since $|p_{li}| < 1$ and $|q_{lj}| < 1$, we have

$$
\lim_{n_1 \to \infty, n_2 \to \infty} y_t(n_1, n_2) = 0
$$

that is, $y_t(n_1, n_2)$ is a transient component and, therefore, Eqs. (4.14) and (4.15) yield

$$
\lim_{n_1 \to \infty, n_2 \to \infty} y(n_1, n_2) = y_s(n_1, n_2)
$$

$$
= M(\omega_1, \omega_2) \sin[(n_1\omega_1 T_1 + n_2\omega_2 T_2) + \theta(\omega_1, \omega_2)]
$$

4.5.2 Frequency Response

The transfer function evaluated on the unit bicircle as a function of (ω_1, ω_2) is said to be the frequency response of the 2-D digital filter. The quantities $M(\omega_1, \omega_2)$ and $\theta(\omega_1, \omega_2)$ as functions of (ω_1, ω_2) are said to be the amplitude response and phase response, respectively. They can be represented by 3-D plots or by contour plots. As an example, the amplitude and phase responses of the 2-D filter characterized by the transfer function [5]

$$
H(z_1, z_2) = \frac{0.286(z_1 + 1)^5(z_2 + 1)^5}{\sum_{i=0}^{5} \sum_{j=0}^{5} b_{ij} z_1^{5-i} z_2^{5-j}}
$$

where coefficients b_{ij} are the elements of matrix **B** given by

$$
\mathbf{B} = \begin{bmatrix}
78.4280 & 140.2000 & 101.4700 & -36.0290 & 6.0739 & -0.4134 \\
-128.4300 & 136.7600 & -33.3130 & -7.4214 & 1.4820 & 0.7907 \\
83.0110 & -22.7050 & -7.2275 & -33.4870 & 28.7070 & -6.3054 \\
-25.5190 & -13.2490 & -29.3020 & 61.5510 & -28.7340 & 4.2941 \\
3.4667 & 3.4048 & 23.8380 & -25.8710 & 7.8851 & -0.7930 \\
-0.1645 & 0.3218 & -4.8317 & 3.3632 & -0.6450 & 0.0652
\end{bmatrix}
$$

can be represented by the 3-D plots shown in Fig. 4.2a and b or by the contour plots of Fig. 4.3a and b.

A pair of parameters that are sometimes of interest are the "group delays" of the 2-D digital filter. These are defined as

$$
\tau_1 = -\frac{\partial \theta(\omega_1, \omega_2)}{\partial \omega_1} \quad \text{and} \quad \tau_2 = -\frac{\partial \theta(\omega_1, \omega_2)}{\partial \omega_2}
$$

where the units of τ_1 and τ_2 are seconds only if $n_1 T_1$ or $n_2 T_2$ represents time. In many applications, in particular in image processing, the phase response is required to be linear in order to prevent signal distortion. In such applications the group delays must be constant. The group delays corresponding to the phase response shown in Fig 4.2b are plotted with respect to the (ω_1, ω_2) plane in Fig. 4.4.

It is easy to verify that the frequency response of a 2-D filter and, consequently, its amplitude and phase responses are periodic with periods equal to the sampling frequencies $\omega_{s1} = 2\pi/T_1$ and $\omega_{s2} = 2\pi/T_2$, that is,

$$
H(e^{j(\omega_1 + k_1 \omega_{s1}) T_1}, e^{j(\omega_2 + k_2 \omega_{s2}) T_1}) = H(e^{j\omega_1 T_1}, e^{j\omega_2 T_2})
$$

for all integers k_1 and k_2. The rectangular area defined by the fundamental period, namely,

$$
B = \left\{ (\omega_1, \omega_2) : -\frac{\omega_{s1}}{2} \leq \omega_1 \leq \frac{\omega_{s1}}{2}, -\frac{\omega_{s2}}{2} \leq \omega_2 \leq \frac{\omega_{s2}}{2} \right\}
$$

is often referred to as the baseband. The frequencies $\omega_{s1}/2$ and $\omega_{s2}/2$ are referred to as the Nyquist frequencies.

4.5.3 Symmetries

If the coefficients of the transfer function are real, the amplitude response of the filter is symmetrical with respect to the origin, that is,

$$
M(\omega_1, \omega_2) = M(-\omega_1, -\omega_2) \quad \text{and} \quad M(\omega_1, -\omega_2) = M(-\omega_1, \omega_2)
$$

Sometimes, certain symmetries are imposed by the application at hand. In

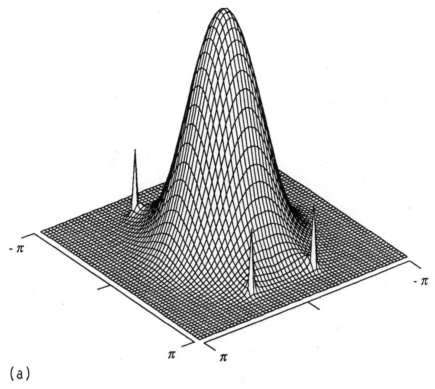

(a)

Figure 4.2   Three-D plots: (a) amplitude response; (b) phase response.

certain applications the contours of the amplitude response can be rectangular, that is,

$$M(\omega_1, \omega_2) = \text{constant} \quad \text{for } |\omega_1| = a \text{ and } |\omega_2| \leq b, |\omega_2| = b \text{ and } |\omega_1| \leq a$$

In other applications, in particular if the frequency spectrum of the 2-D signal is homogeneous with respect to the two frequency variables, a 2-D filter with circularly symmetric amplitude response may be required. In such an application the contours should be circular, that is,

$$M(\omega_1, \omega_2) = \text{constant} \quad \text{for } \sqrt{\omega_1^2 + \omega_2^2} = \text{constant}$$

Another type of symmetry that has been considered in the literature is the case where the contours of the amplitude response are ellipses whose axes are aligned in the direction of the ω_1 and ω_2 axes [6,7] or whose orientation is arbitrary [8].

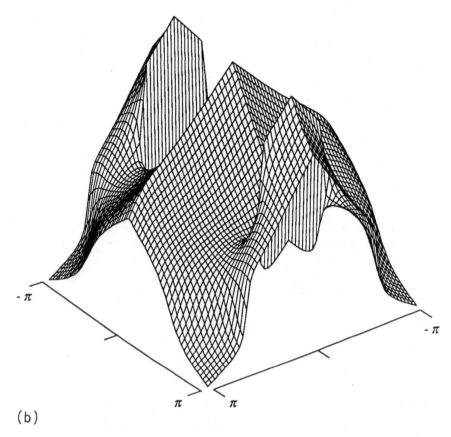

(b)

Figure 4.2 Continued.

A broader type of symmetry is the case where the amplitude response is symmetrical with respect to both frequency axes, that is

$$M(\omega_1, \omega_2) = M(\omega_1, -\omega_2) = M(-\omega_1, \omega_2) = M(-\omega_1, -\omega_2)$$

This type of symmetry is said to be quadrantal. Evidently, every filter whose amplitude response has circular or rectangular symmetry also has quadrantal symmetry. On the other hand, if the contours are ellipses that are not aligned in the direction of the ω_1 or ω_2 axis, the amplitude response does not have quadrantal symmetry.

4.5.4   Idealized Filters

The design of 2-D digital filters can often be simplified by identifying a number of idealized types of filters on the basis of the shape of the amplitude (e.g., lowpass, highpass, etc.).

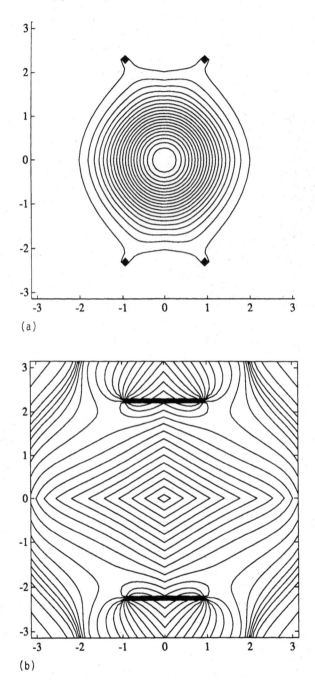

Figure 4.3 Contour plots: (a) amplitude response; (b) phase response.

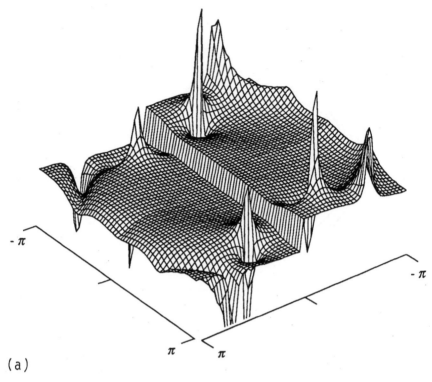

(a)

Figure 4.4 Plots of group delays for the phase response in Fig. 4.2b: (a) group delay τ_1; (b) group delay τ_2.

A filter with an amplitude response

$$M(\omega_1, \omega_2) = \begin{cases} 1 & \text{for } (\omega_1, \omega_2) \in R_1 \\ 0 & \text{otherwise} \end{cases}$$

where

$$R_1 = \{(\omega_1, \omega_2) : |\omega_1| \leq \omega_{c1} \text{ and } |\omega_2| \leq \omega_{c2}\}$$

is said to be a rectangularly symmetric lowpass filter since it will pass low-frequency components located in the rectangular area R_1 and reject high-frequency components not in R_1. The area R_1 is called the passband, the area outside R_1 is called the stopband, and the boundary of R_1 is called the passband boundary. The frequencies ω_{c1} and ω_{c2} are said to be the cutoff frequencies of the filter.

A filter with an amplitude response

$$M(\omega_1, \omega_2) = \begin{cases} 0 & \text{for } (\omega_1, \omega_2) \in R_1 \\ 1 & \text{otherwise} \end{cases}$$

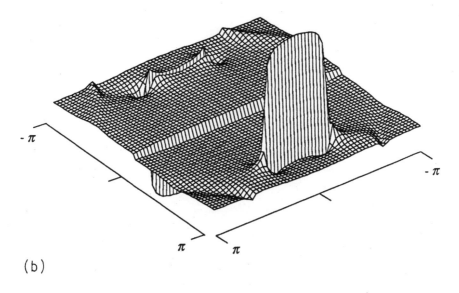

(b)

is said to be a rectangularly symmetric highpass filter.

A rectangularly symmetric bandpass filter is one that will pass frequency components located between two rectangles, that is,

$$M(\omega_1, \omega_2) = \begin{cases} 1 & \text{for } (\omega_1, \omega_2) \in R_1 \cap R_2 \\ 0 & \text{otherwise} \end{cases}$$

where

$$R_2 = \{(\omega_1, \omega_2) : |\omega_1| \leq \omega_{c3} \text{ and } |\omega_2| \leq \omega_{c4}\}$$

where $\omega_{c3} > \omega_{c1}$ and $\omega_{c4} > \omega_{c2}$. Similarly, a rectangularly symmetric band-stop filter is one that will reject frequency components located between two rectangles, that is,

$$M(\omega_1, \omega_2) = \begin{cases} 0 & \text{for } (\omega_1, \omega_2) \in R_1 \cap R_2 \\ 1 & \text{otherwise} \end{cases}$$

The amplitude responses of the four types of rectangularly symmetric filters defined above are illustrated in Fig. 4.5.

Corresponding circularly symmetric ideal filters can readily be identified by replacing rectangular areas R_1 and R_2 by discs C_1 and C_2 defined as

$$C_1 = \{(\omega_1, \omega_2) : \sqrt{\omega_1^2 + \omega_2^2} \leq \omega_{c1}\}$$

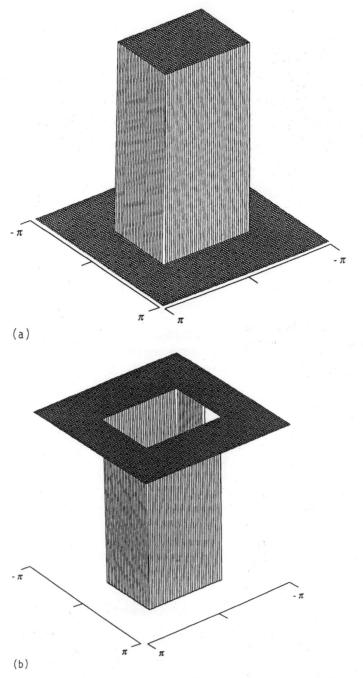

(a)

(b)

Figure 4.5 Amplitude responses of rectangularly symmetric filters: (a) lowpass; (b) highpass; (c) bandpass; (d) bandstop.

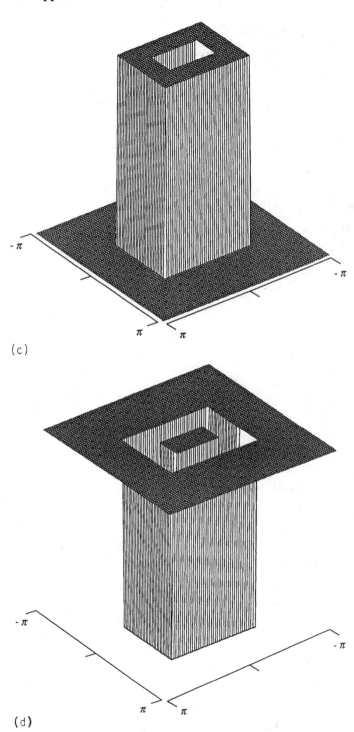

(c)

(d)

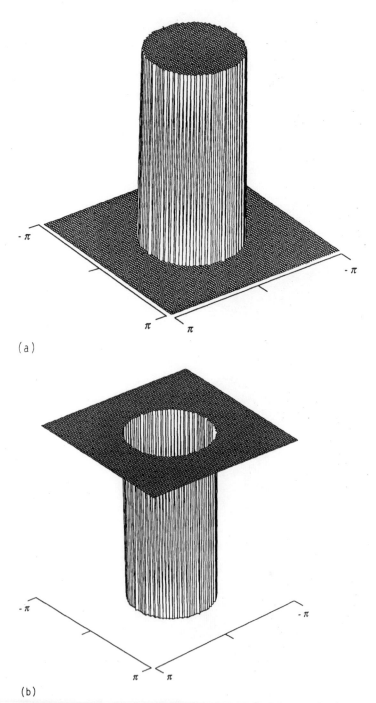

(a)

(b)

Figure 4.6 Amplitude responses of circularly symmetric filters: (a) lowpass; (b) highpass; (c) bandpass; (d) bandstop.

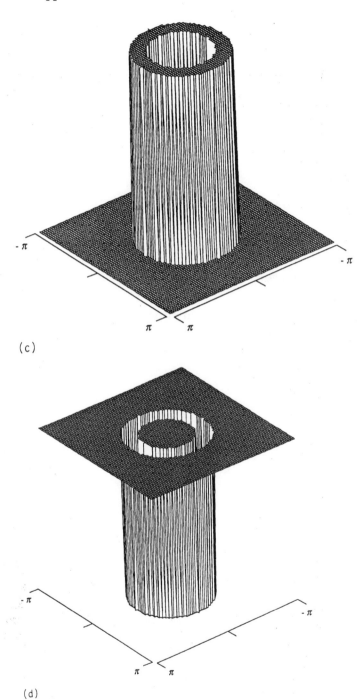

(c)

(d)

Figure 4.6 Continued.

and

$$C_2 = \{(\omega_1, \omega_2) : \sqrt{\omega_1^2 + \omega_2^2} \le \omega_{c2}\}$$

where $\omega_{c2} > \omega_{c1}$. The amplitude responses of the four types of circularly symmetric filters defined above are illustrated in Fig. 4.6.

It is worth noting at this point that the amplitude response of a rectangularly symmetric lowpass filter can be expressed as

$$M(\omega_1, \omega_2) = M_1(\omega_1)M_2(\omega_2)$$

where

$$M_1(\omega_1, \omega_2) = \begin{cases} 1 & \text{for } |\omega_1| \le \omega_{c1} \\ 0 & \text{otherwise} \end{cases}$$

and

$$M_2(\omega_1, \omega_2) = \begin{cases} 1 & \text{for } |\omega_2| \le \omega_{c2} \\ 0 & \text{otherwise} \end{cases}$$

Consequently, such a filter can be designed by designing two 1-D lowpass filters and then connecting them in cascade [9].

Similarly, in the case of a rectangularly symmetric bandpass filter, we can write

$$M(\omega_1, \omega_2) = M_1(\omega_1)M_2(\omega_2)$$

where

$$M_1(\omega_1) = \begin{cases} 0 & \text{for } (\omega_1, \omega_2) \in R_1 \\ 1 & \text{otherwise} \end{cases}$$

and

$$M_2(\omega_2) = \begin{cases} 1 & \text{for } (\omega_1, \omega_2) \in R_2 \\ 0 & \text{otherwise} \end{cases}$$

As a result a bandpass filter can be obtained by connecting in cascade a highpass filter with cutoff frequencies ω_{c1} and ω_{c2} and a lowpass filter with cutoff frequencies ω_{c3} and ω_{c4} where $\omega_{c3} > \omega_{c1}$ and $\omega_{c4} > \omega_{c2}$, Hence, like rectangularly symmetric lowpass filters, rectangularly symmetric bandpass filters can be designed in terms of 1-D filters. In effect, these filters are separable and are, as a consequence, easier to design than corresponding circularly symmetric filters.

Another type of filter that has been found useful in the processing of geophysical data is the so-called fan filter. An ideal fan filter may have an

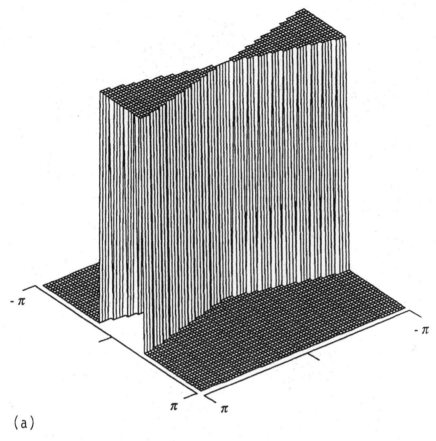

(a)

Figure 4.7 Amplitude response of fan filters: (a) bandpass fan filter; (b) bandstop fan filter.

amplitude response.

$$M_1(\omega_1, \omega_2) = \begin{cases} 1 & \text{for } (\omega_1, \omega_2) \in S_\theta \\ 0 & \text{otherwise} \end{cases} \qquad (4.16)$$

or

$$M_2(\omega_1, \omega_2) = \begin{cases} 0 & \text{for } (\omega_1, \omega_2) \in S_\theta \\ 1 & \text{otherwise} \end{cases} \qquad (4.17)$$

where S_θ is a sector given by

$$S_\theta = \{(\omega_1, \omega_2) : \left| \tan^{-1} \frac{\omega_2}{\omega_1} \right| \le \theta \quad \text{or} \quad \left| \tan^{-1} \frac{\omega_2}{\omega_1} \right| \ge \pi - \theta \}$$

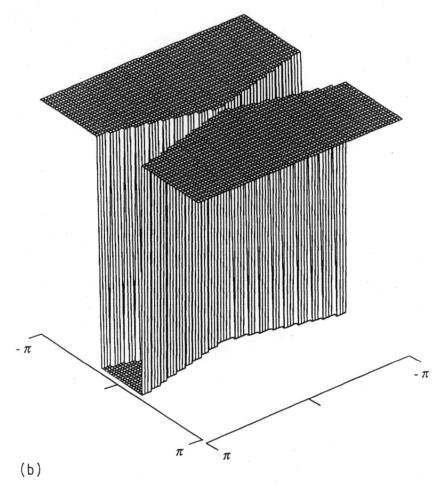

(b)

Figure 4.7 Continued.

as depicted in Fig. 4.7a and b. The amplitude responses in Eq. (4.16) and
Eq. (4.17) describe a bandpass and a bandstop fan filter, respectively.
Note, however, that a filter which is bandpass with respect to the ω_1 axis
is a bandstop filter with respect to the ω_2 axis and vice versa.

Practical 2-D filters differ from ideal ones in that transitions between
passbands and stopbands are gradual, the gain in a passband is never exactly
equal to unity, and the gain in a stopband is never exactly equal to zero.
Furthermore, the contours of the amplitude response in a rectangularly or
circularly symmetric filter are only approximately rectangular or circular.

REFERENCES

1. J. L. Shanks, Two-dimensional recursive filters, *SWIEECO Rec.*, pp. 19E1–19E8, 1969.
2. J. L. Shanks, The design of stable two-dimensional recursive filters, *Proc. UMR-M. J. Kelly Commun. Conf., Univ. Missouri*, pp. 15-2-1–15-2-2, Oct. 1970.
3. J. L. Shanks, S. Treitel, and J. H. Justice, Stability and synthesis of two-dimensional recursive filters, *IEEE Trans. Audio Electroacoust.*, vol. AU-20, pp. 115–128, June 1972.
4. T. S. Huang, Stability of two-dimensional recursive filters, *IEEE Trans. Audio Electroacoust.*, vol. AU-20, pp. 158–163, June 1972.
5. P. A. Ramamoorthy and L. T. Burton, Design of two-dimensional recursive filters, in *Topics in Applied Physics*, edited by T. S. Huang, vol. 42, pp. 41–83, New York: Springer Verlag, 1981.
6. R. M. Mersereau, W. F. G. Mecklenbräuker, and T. F. Quatieri, Jr., McClellan transformation for two-dimensional digital filtering: I—Design, *IEEE Trans. Circuits Syst.*, vol. CAS-23, pp. 405–413, July 1976.
7. M. S. Reddy and S. N. Hazra, Design of elliptically symmetric two-dimensional FIR filters using the McClellan transformation, *IEEE Trans. Circuit Syst.*, vol. CAS-34, pp. 196–198, Feb. 1987.
8. D. T. Nguyen and M. N. S. Swamy, Approximation design of 2-D digital filters with elliptical magnitude response of arbitrary orientation, *IEEE Trans. Circuits Syst.*, vol. CAS-33, pp. 597–603, June 1976.
9. A. Antoniou, M. Ahmadi, and C. Charalambous, Design for factorable low-pass 2-dimensional filters satisfying prescribed specifications, *IEE Proc.*, vol. 128, pp. 53–60, April 1981.

PROBLEMS

4.1 Obtain the transfer function of the filters represented by
    (a) the network in Fig. P1.5a and b.
    (b) the network in Fig. P1.5c and b.

4.2 Obtain the transfer function of the filters represented by
    (a) the network in Fig. P1.6.
    (b) the difference equation given in Problem 1.7c.

4.3 Obtain the transfer function of the filters represented by
    (a) the difference equation given in Problem 2.2b.
    (b) the difference equation given in Problem 2.2c.

4.4 Obtain the transfer function of the filter described by the state-space equation given in Problem 2.5.

4.5 Check the BIBO stability of the filters characterized by
(a) the state-space equation in Problem 2.5
(b) the transfer function

$$H(z_1, z_2) = \frac{1}{z_1 z_2 + 0.2 z_2 + 0.5}$$

(c) the transfer function

$$H(z_1, z_2) = \frac{z_1 z_2 + z_2 - 0.5}{z_1 z_2 + 0.8 z_2 - 0.5 z_1 - 0.4}$$

4.6 Find the unit-step response of the filter described by
(a) the transfer function in Problem 4.5b.
(b) the transfer function in Problem 4.5c.

4.7 The transfer function given by Eq. (4.3a) can be written as

$$H(z_1, z_2) = \frac{\mathbf{z}_1^T \mathbf{A} \mathbf{z}_2}{\overline{B}(z_1, z_2)}$$

where

$$\mathbf{z}_1 = [1 \ z_1^{-1} \cdots z_1^{-N_1}]^T, \qquad \mathbf{z}_2 = [1 \ z_2^{-1} \cdots z_2^{-N_2}]^T$$

and

$$\mathbf{A} = \begin{bmatrix} a_{00} & a_{01} & \cdots & a_{0N_2} \\ a_{10} & a_{11} & \cdots & a_{1N_2} \\ \vdots & \vdots & & \vdots \\ a_{N_10} & a_{N_11} & \cdots & a_{N_1N_2} \end{bmatrix}$$

By using the singular-value decomposition technique, show that the transfer function can be expressed as

$$H(z_1, z_2) = \sum_{k=1}^{r} H_k(z_1, z_2)$$

where $r = \text{rank}(\mathbf{A})$ and

$$H_k(z_1, z_2) = \frac{a_k^{(1)}(z_1) a_k^{(2)}(z_2)}{\overline{B}(z_1, z_2)}$$

4.8 Consider the same transfer function as in Problem 4.7. Show that if $\overline{B}(z_1, z_2)$ is separable, then $H(z_1, z_2)$ can be expressed as a sum of r separable subfilters, that is,

$$H(z_1, z_2) = \sum_{k=1}^{r} F_k(z_1) G_k(z_2)$$

where $F_k(z_1)$ and $G_k(z_2)$ for $1 \le k \le r$ are 1-D causal recursive filters.

4.9 Find the steady-state sinusoidal response of the filters represented by the transfer functions

(a) $H(z_1, z_2) = \dfrac{z_1 z_2}{z_1 z_2 - 0.5}$.

(b) $H(z_1, z_2) = \dfrac{z_1 z_2}{z_1 z_2 - 0.4 z_1 - 0.4 z_2}$.

4.10 Write MATLAB programs that generate the amplitude responses of the following idealized filters:*
  (a) a circularly symmetric lowpass filter with normalized cutoff frequency $\omega_c = 0.5\pi$.
  (b) a circularly symmetric highpass filter with normalized cutoff frequency $\omega_c = 0.5\pi$.

4.11 Write MATLAB programs that generate the amplitude responses of the following idealized filters:*
  (a) a circularly symmetric bandpass filter with $\omega_{c1} = 0.4\pi$ and $\omega_{c2} = 0.6\pi$.
  (b) a circularly symmetric bandstop filter with $\omega_{c1} = 0.4\pi$ and $\omega_{c2} = 0.6\pi$.

*Each of the programs in Problems 4.10 and 4.11 may be written as a MATLAB function with the cutoff frequency and the size of the sampled amplitude matrix as parameters. The recommended size of the sampled amplitude matrix is 91×91 for workstations and 71×71 for personal computers.

5
Stability Analysis

5.1 INTRODUCTION

A digital filter must be stable in order to ensure that any bounded input will produce a bounded output. Otherwise, the filter is not useful for any practical application. When a filter is being designed, it is possible to obtain a transfer function that satisfies the required amplitude- or phase-response specifications while the filter is unstable. On the other hand, a filter that has been designed to satisfy the necessary conditions for stability on the basis of infinite-precision arithmetic may become unstable when the transfer-function coefficients are represented in terms of finite-precision arithmetic. Consequently, it is of paramount importance in both the design and application of digital filters to have simple, robust, and efficient stability tests and criteria that can be used to determine with certainty whether a given filter is stable or not.

In this chapter the stability properties of 2-D digital filters are examined and the necessary and sufficient conditions for stability are identified. Then frequency-domain tests and criteria are described, which can be used to check whether a 2-D digital filter is stable or not without actually finding its singularities. Subsequently, the stability properties of 2-D digital filters are reexamined in the state-space domain, and corresponding stability conditions and tests are obtained. These results lead to the generalization of the classical Lyapunov stability test for 1-D systems to the case of 2-D digital filters. The chapter concludes with some simple stability tests for low-order filters.

5.2 STABILITY ANALYSIS IN FREQUENCY DOMAIN

Bounded-input, bounded-output (BIBO) stability can be studied in the frequency domain by examining the constraints imposed on the singularities of the transfer function by the requirement that the output of the filter be bounded when the input is bounded.

Consider a quarter-plane (causal) 2-D digital filter characterized by the transfer function

$$H(z_1, z_2) = \frac{\overline{N}(z_1^{-1}, z_2^{-1})}{\overline{D}(z_1^{-1}, z_2^{-1})} \tag{5.1}$$

where

$$\overline{N}(z_1^{-1}, z_2^{-1}) = \sum_{i=0}^{N_1} \sum_{j=0}^{N_2} a_{ij} z_1^{-i} z_2^{-j} \tag{5.2}$$

and

$$\overline{D}(z_1^{-1}, z_2^{-1}) = 1 + \sum_{i=0}^{N_1} \sum_{j=0}^{N_2} b_{ij} z_1^{-i} z_2^{-j}, \qquad b_{00} = 0 \tag{5.3}$$

If two polynomials $\overline{N}(z_1^{-1}, z_2^{-1})$ and $\overline{D}(z_1^{-1}, z_2^{-1})$ have no common factors, they are said to be factor coprime. On the other hand, if there are no points (z_1^{-1}, z_2^{-1}) at which the two polynomials assume the value of zero, the two polynomials are said to be zero coprime. Two polynomials in one variable are factor coprime if and only if they are zero coprime, as can be easily verified. Nevertheless, two polynomials in two variables can be zero coprime without being factor coprime, as can be seen in Example 4.1b. In the following analysis, $\overline{N}(z_1^{-1}, z_2^{-1})$ and $\overline{D}(z_1^{-1}, z_2^{-1})$, given by Eqs. (5.2) and (5.3), are assumed to be factor coprime but may or may not be zero coprime.

5.2.1 Stability Properties

The 2-D filter characterized by the transfer function in Eq. (5.1) is stable if and only if its impulse response $h(n_1, n_2)$ is absolutely summable, as was demonstrated in Sec. 1.7. This important result is restated as Theorem 5.1 for the sake of convenience.

Theorem 5.1

$$BIBO\ stability \rightleftarrows \{h(n_1, n_2)\} \in l_1$$

where $\{h(n_1, n_2)\}$ is said to belong to l_p if

$$\sum_{i=0}^{\infty} \sum_{j=0}^{\infty} |h(i, j)|^p < \infty \qquad (5.4)$$

The arrows $\rightarrow$ and $\leftarrow$ stand for *"implies"* and *"is implied by,"* respectively.

While Eq. (5.4) with $p = 1$ is a necessary and sufficient condition for BIBO stability, the evaluation of the double summation is highly impractical since the impulse response is not in general easy to obtain from the transfer function. An alternative approach to establish whether a 2-D digital filter is stable or not is to use Shanks' stability theorem, which was stated as Theorem 4.1 in Sec. 4.3. This theorem is repeated below for the sake of convenience.

Theorem 5.2 (Shanks) *A 2-D digital filter represented by the transfer function in Eq. (5.1) is BIBO stable if*

$$\overline{D}(z_1^{-1}, z_2^{-1}) \neq 0 \quad for \quad (z_1^{-1}, z_2^{-1}) \in \overline{U}^2 \qquad (5.5)$$

where $\overline{U}^2$ is the unit bidisc defined as

$$\overline{U}^2 = \{(z_1^{-1}, z_2^{-1}) : |z_1^{-1}| \leq 1, |z_2^{-1}| \leq 1\}$$

Shanks' condition, while sufficient for BIBO stability, is not always necessary and the question, therefore, arises as to how close it is to being necessary.

It turns out that the stability of 2-D digital filters depends on the location of nonessential singularities of the second kind. These singularities have been defined in Sec. 4.3 as points in the (z_1^{-1}, z_2^{-1}) plane at which the numerator and denominator polynomials of the transfer function become zero while being factor coprime. To illustrate the role of singularities of this type, let us consider the transfer functions

$$H(z_1, z_2) = \frac{(1 - z_1^{-1})^8(1 - z_2^{-1})^8}{2 - z_1^{-1} - z_2^{-1}}$$

and

$$H(z_1, z_2) = \frac{(1 - z_1^{-1})(1 - z_2^{-1})}{2 - z_1^{-1} - z_2^{-1}}$$

Both of these transfer functions satisfy the following conditions: (i) the numerator and denominator polynomials are factor coprime, (ii)

$\overline{D}(z_1^{-1}, z_2^{-1}) \neq 0$ for $(z_1^{-1}, z_2^{-1}) \in \overline{U}^2$ except at $(z_1^{-1}, z_2^{-1}) = (1, 1)$, and (iii) each transfer function has a nonessential singularity of the second kind at point $(z_1^{-1}, z_2^{-1}) = (1, 1)$. The first filter can be shown to be stable by demonstrating that its impulse response is absolutely summable [1], whereas the second filter can be shown to be unstable by demonstrating the opposite. Evidently, if a filter satisfies Shanks' condition except that it has nonessential singularities of the second kind on the unit bicircle U^2 given by

$$U^2 = \{(z_1^{-1}, z_2^{-1}) : |z_1^{-1}| = 1, |z_2^{-1}| = 1\}$$

a more detailed analysis is necessary.

Some additional stability properties which are sometimes useful are summarized in terms of the following three theorems. Their proofs are given by Goodman [1].

Theorem 5.3 $H(z_1, z_2)$ *has no nonessential singularities of the first kind on* $\overline{U}^2 \rightarrow$ *BIBO stability.*

A nonessential singularity of the first kind is a point (z_1^{-1}, z_2^{-1}) such that $\overline{D}(z_1^{-1}, z_2^{-1}) = 0$ and $\overline{N}(z_1^{-1}, z_2^{-1}) \neq 0$. The symbol $\rightarrow$ stands for *"does not imply."* The *"does not imply"* part of this theorem follows from the preceding two examples.

Theorem 5.4

$$\overline{D}(z_1^{-1}, z_2^{-1}) \neq 0 \quad \text{on} \quad \overline{U}^2 - U^2 \rightarrow \text{BIBO stability}$$

where

$$\overline{U}^2 - U^2 = \{(z_1^{-1}, z_2^{-1}) : (z_1^{-1}, z_2^{-1}) \in \overline{U}^2 \text{ and } (z_1^{-1}, z_2^{-1}) \notin U^2\}$$

The "does not imply" part of the theorem follows from the preceding examples.

From Theorems 5.3 and 5.4, it follows that if a 2-D digital filter is BIBO stable, then $\overline{D}(z_1^{-1}, z_2^{-1}) \neq 0$ on $\overline{U}^2$ except on U^2 where all the zeros of $D(z_1^{-1}, z_2^{-1})$ must be nonessential singularities of the second kind of $H(z_1, z_2)$.

Theorem 5.5

$$\{h(n_1, n_2)\} \in l_2 \rightleftarrows \text{BIBO stability}$$

This theorem follows from Theorem 5.1 if it is noted that

$$\{h(n_1, n_2)\} \in l_1 \rightleftarrows \{h(n_1, n_2)\} \in l_2$$

In the rest of this chapter it is assumed that the transfer function has no nonessential singularities of the second kind on U^2, for the sake of simplicity. For this class of filters, Shanks' condition is both necessary and sufficient for stability; that is,

$$\overline{D}(z_1^{-1}, z_2^{-1}) \neq 0 \quad \text{on} \quad \overline{U}^2 \rightleftarrows BIBO \; stability \qquad (5.6)$$

The stability of such filters can, in principle, be verified by demonstrating that $\overline{D}(z_1^{-1}, z_2^{-1})$ has no zeros in $\overline{U}^2$. Unfortunately, however, this type of analysis is not easy in general and researchers have spent considerable effort during the past 20 years in developing alternative necessary and sufficient conditions for BIBO stability that are simpler to check. Some of these results are stated in terms of the following three theorems and form the foundation of a number of stability criteria, as will be demonstrated in Sec. 5.2.2.

Theorem 5.6 (Huang [2]) *A 2-D digital filter is BIBO stable if and only if*

1. $\overline{D}(z_1^{-1}, 0) \neq 0 \quad for \; |z_1^{-1}| \leq 1.$ (5.7a)

2. $\overline{D}(z_1^{-1}, z_2^{-1}) \neq 0 \quad for \; |z_1^{-1}| = 1 \; and \; |z_2^{-1}| \leq 1.$ (5.7b)

Proof. The condition in Eq. (5.5) implies the conditions in Eqs. (5.7a) and (5.7b). To demonstrate the converse, let

$$n_z(z_2^{-1}) = \frac{1}{2\pi j} \oint_{|z_1^{-1}|=1} \frac{\partial \overline{D}(z_1^{-1}, z_2^{-1})}{\partial z_1^{-1}} \frac{dz_1^{-1}}{\overline{D}(z_1^{-1}, z_2^{-1})}$$

and note that for fixed z_2^{-1} with $|z_2^{-1}| \leq 1$, the condition in Eq. (5.7b) implies that the preceding integral is well defined and, therefore, continuous, and represents the number of zeros of $\overline{D}(z_1^{-1}, z_2^{-1})$ in the unit disc $|z_1^{-1}| \leq 1$ (see Silverman [3]). On the other hand, the condition in Eq. (5.7a) implies that $n_z(0) = 0$. Since $n_z(z_2^{-1})$ is simultaneously continuous and an integer-valued function, we have $n_z(z_2^{-1}) = 0$ for $|z_2^{-1}| \leq 1$, which implies that the condition in Eq. (5.5) holds. ∎

Theorem 5.7 (Strintzis [4]) *A 2-D digital filter is BIBO stable if and only if*

1. $\overline{D}(\alpha, z_2^{-1}) \neq 0 \quad for \; |z_2^{-1}| \leq 1 \; and \; some \; \alpha \; with \; |\alpha| \leq 1.$ (5.8a)

2. $\overline{D}(z_1^{-1}, \beta) \neq 0 \quad for \; |z_1^{-1}| \leq 1 \; and \; some \; \beta \; with \; |\beta| \leq 1.$ (5.8b)

3. $\overline{D}(z_1^{-1}, z_2^{-1}) \neq 0 \quad for \; (z_1^{-1}, z_2^{-1}) \in U^2.$ (5.8c)

This theorem can be proved by using an argument similar to that used to prove Theorem 5.6. The special case where $\alpha = \beta = 1$ is of particular interest and is stated in terms of the following corollary.

Corollary 5.1 *A 2-D digital filter is BIBO stable if and only if*

1. $\overline{D}(1, z_2^{-1}) \neq 0$ *for* $|z_2^{-1}| \leq 1$. $\hspace{5cm}$ (5.9a)

2. $\overline{D}(z_1^{-1}, 1) \neq 0$ *for* $|z_1^{-1}| \leq 1$. $\hspace{5cm}$ (5.9b)

3. $\overline{D}(z_1^{-1}, z_2^{-1}) \neq 0$ *for* $(z_1^{-1}, z_2^{-1}) \in U^2$. $\hspace{3.5cm}$ (5.9c)

Theorem 5.8 (DeCarlo, Murray, and Saeks [5]) *A 2-D digital filter is BIBO stable if and only if*

1. $\overline{D}(z^{-1}, z^{-1}) \neq 0$ *for* $|z^{-1}| \leq 1$. $\hspace{4.5cm}$ (5.10a)

2. $\overline{D}(z_1^{-1}, z_2^{-1}) \neq 0$ *for* $(z_1^{-1}, z_2^{-1}) \in U^2$. $\hspace{3cm}$ (5.10b)

Proof. Let k_1 and k_2 be the numbers of zeros of $\overline{D}(z_1^{-1}, 1)$ and $\overline{D}(1, z_2^{-1})$ in $|z_1^{-1}| < 1$ and $|z_2^{-1}| < 1$, respectively, and assume that the conditions in Eqs. (5.10a) and (5.10b) hold. From Rudin [6], the condition in Eq. (5.10b) implies that $\overline{D}(z^{-1}, z^{-1})$ has $k_1 + k_2$ zeros in $|z^{-1}| < 1$. Hence the conditions in Eqs. (5.10a) and (5.10b) imply that

$$\overline{D}(1, z_2^{-1}) \neq 0 \qquad \text{for } |z_2^{-1}| \leq 1$$

and

$$\overline{D}(z_1^{-1}, 1) \neq 0 \qquad \text{for } |z_1^{-1}| \leq 1$$

By Corollary 5.1, it follows that the filter is BIBO stable. To prove the *"only if"* part of the theorem, we note that the filter is unstable if either of the two conditions is violated. ∎

Some additional stability properties of 2-D digital filters and multidimensional discrete systems in general can be found in a paper published by Jury [7] and in the work of Bose [8].

5.2.2 Stability Criteria

Stability tests or criteria are efficient algorithms that can be used to determine whether a system is stable or not without actually finding the singularities of the transfer function. Classical stability criteria that have been used extensively in 1-D continuous systems in the past are the Routh–Hurwitz and Nyquist stability criteria [9]. A related and equally useful stability criterion that has been used extensively in 1-D discrete systems and digital filters is the Jury–Marden criterion [10].

Theorems 5.2 and 5.6 to 5.8 have shown that there are two types of conditions that need to be satisfied to assure the stability of a 2-D digital filter, namely, conditions where the polynomial involved is a function of one variable, e.g., the condition in Eq. (5.8a), and conditions where the polynomial involved is a function of two variables, e.g., the condition in

Eq. (5.5). Conditions of the first type can be checked by using the Jury–Marden stability criterion, which will now be described.

Given a real polynomial $p(z)$ of the form

$$p(z) = p_0 z^n + p_1 z^{n-1} + \cdots + p_n, \qquad p_0 > 0 \qquad (5.11)$$

the so-called Jury–Marden table can be constructed by first forming two initial rows as

$$\{c_{11}\, c_{12} \cdots c_{1,n+1}\} = \{p_0\, p_1 \cdots p_n\}$$

$$\{d_{11}\, d_{12} \cdots d_{1,n+1}\} = \{p_n\, p_{n-1} \cdots p_0\}$$

and then computing a number of pairs of subsequent rows as

$$c_{ij} = \det \begin{bmatrix} c_{i-1,1} & c_{i-1,n-j-i+3} \\ d_{i-1,1} & d_{i-1,n-j-i+3} \end{bmatrix} \qquad \text{for } i = 2, 3, \ldots, n-1$$

and

$$d_{ij} = c_{i,n-j-i+3}$$

Once the Jury–Marden table is constructed, stability conditions involving polynomials in one variable can be readily checked by applying the following theorem.

Theorem 5.9 *The zeros of polynomial $p(z)$, designated as z_i for $i = 1, 2, \ldots, n$, are located on the open unit disc (i.e., $|z_i| < 1$) if and only if*

1. $p(1) > 0$.
2. $(-1)^n p(-1) > 0$.
3. $|c_{i1}| > |c_{i,n-i+2}|$ *for $i = 1, 2, \ldots, n-1$.*

The preceding table along with Theorem 5.9 constitute the Jury–Marden stability criterion.

An alternative 1-D stability criterion can be obtained by using the Schur–Cohn matrix given by

$$\mathbf{K} = \{k_{ij}\} \qquad (5.12)$$

where $1 \le i, j \le n$ and

$$k_{ij} = \begin{cases} \displaystyle\sum_{r=0}^{i-1} (p_{i-1-r}\, p_{j-1-r} - p_{n+r-i+1}\, p_{n+r-j+1}) & \text{if } i \le j \\ k_{ji} & \text{otherwise} \end{cases}$$

In this case, stability conditions involving polynomials in one variable can be checked by noting the following theorem.

Theorem 5.10 *The zeros of polynomial $p(z)$ are located on the open unit disc if and only if its Schur–Cohn matrix $\mathbf{K}$ in Eq. (5.12) is positive definite.*

The construction of matrix $\mathbf{K}$ and the application of Theorem 5.10 constitute the Schur–Cohn 1-D stability criterion. Matrix $\mathbf{K}$ can be checked for positive definiteness (a) by diagonalizing it and checking its diagonal elements, (b) by computing the determinants of its principal minors, or (c) by finding its eigenvalues. If the diagonal elements in (a) or the determinants in (b) or the eigenvalues in (c) are all positive, then $\mathbf{K}$ is positive definite (see Noble [11] for details).

The Jury–Marden and Schur–Cohn stability criteria necessitate that polynomials in z_1^{-1} or z_2^{-1} be changed into polynomials in z_1 or z_2 by multiplying them by $z_1^{N_1}$ or $z_2^{N_2}$. For example, the polynomial in Eq. (5.9b) can be put in the appropriate form by multiplying it by $z_1^{N_1}$; that is,

$$p(z_1) = z_i^{N_1} \overline{D}(z_1^{-1}, 1)$$

With this modification, it is evident that the condition in Eq. (5.9b) is satisfied if and only if the zeros of $p(z_1)$ are located on the open unit disc.

If the stability of a 2-D digital filter is to be checked by using Theorem 5.6, the condition in Eq. (5.7a) can be readily checked by using the Jury–Marden or Schur–Cohn criterion since it involves a polynomial in one variable. The condition in Eq. (5.7b), however, involves a polynomial in two variables and a more advanced method is, therefore, needed. It turns out that a generalization of the Schur–Cohn criterion is applicable, as will now be demonstrated.

Let us multiply the polynomial in Eq. (5.7b) by $z_2^{N_2}$ to obtain a polynomial in z_2 as

$$c(z_2) = z_2^{N_2} \overline{D}(z_1^{-1}, z_2^{-1}) \qquad (5.13)$$

We note that if the condition in Eq. (5.7a) holds, then the condition in Eq. (5.7b) is the same as

$$c(z_2) \neq 0 \quad \text{for } |z_1^{-1}| = 1 \text{ and } |z_2| \geq 1 \qquad (5.14)$$

By writing $c(z_2)$ in the form

$$c(z_2) = \sum_{k=0}^{N_2} c_k z_2^k$$

where coefficients c_k are polynomials of z_1^{-1}, we can construct the $N_2 \times N_2$ Schur–Cohn Hermitian matrix

$$\mathbf{C}(z_1^{-1}) = \{c_{ij}\} \qquad (5.15)$$

where $1 \leq i, j \leq N_2$, and

$$
c_{ij} = \begin{cases} \displaystyle\sum_{k=1}^{i} (c_{N_2-i+k}c^*_{N_2-j+k} - c^*_{i-k}c_{j-k}) & \text{if } i \leq j \\ c^*_{ji} & \text{otherwise} \end{cases}
$$

The condition in Eq. (5.14), and therefore the condition in Eq. (5.7b), can now be checked by applying the following theorem.

Theorem 5.11   *The condition in Eq. (5.14) holds if and only if the Schur–Cohn matrix* $\mathbf{C}(z_1^{-1})$ *in Eq. (5.15) is positive definite for* $|z_1^{-1}| = 1$.

Further progress in this analysis can be made by noting the following results, which are due to Siljak [12]. Matrix $\mathbf{C}(z_1^{-1})$ is positive definite for $|z_1^{-1}| = 1$ if and only if

$$\mathbf{C}(1) \text{ is a positive-definite matrix} \tag{5.16a}$$

and

$$f(z_1^{-1}) = \det \mathbf{C}(z_1^{-1}) > 0 \qquad \text{for } |z_1^{-1}| = 1 \tag{5.16b}$$

where polynomial $f(z_1^{-1})$ always has the form

$$f(z_1^{-1}) = \sum_{k=0}^{K} a_k(z_1^k + z_1^{-k})$$

on the unit circle $|z_1^{-1}| = 1$. By defining a polynomial $g(z_1)$ as

$$g(z_1) = z_1^k f(z_1^{-1}) = \sum_{k=0}^{K} a_k(z_1^{K+k} + z_1^{K-k}) \tag{5.17}$$

the condition in Eq. (5.16b) can be transformed into the simpler but equivalent conditions

$$g(1) > 0 \tag{5.18a}$$

and

$$g(z_1) \text{ has no roots on the unit circle } |z_1^{-1}| = 1 \tag{5.18b}$$

Evidently, the condition in Eq. (5.18b) involves a polynomial in one variable and can, therefore, be easily checked by finding the roots of $g(z_1)$ using a reliable software package like MATLAB [13] or by applying a recursive procedure described by Siljak [14].

The preceding results can be summarized as

$$\text{BIBO stability} \rightleftarrows \begin{cases} \text{Eq. (5.7a)} \\ \text{Eq. (5.7b)} \end{cases}$$

$$\text{Eq. (5.7b)} \rightleftarrows \begin{cases} \text{Eq. (5.16a)} \\ \text{Eq. (5.16b)} \end{cases}$$

$$\text{Eq. (5.16b)} \rightleftarrows \begin{cases} \text{Eq. (5.18a)} \\ \text{Eq. (5.18b)} \end{cases}$$

and, therefore, a 2-D digital filter can be checked for stability by checking the one-variable conditions in Eqs. (5.7a), (5.16a), (5.18a) and (5.18b) in this order.

Example 5.1 A 2-D digital filter is characterized by the transfer function

$$H(z_1, z_2) = \frac{\overline{N}(z_1^{-1}, z_2^{-1})}{\overline{D}(z_1^{-1}, z_2^{-1})}$$

where

$$\overline{D}(z_1^{-1}, z_2^{-1}) = 1 - z_1^{-1} - 0.7z_2^{-1} + 0.25z_1^{-2}$$
$$+ 0.67z_1^{-1}z_2^{-1} - 0.16z_1^{-2}z_2^{-1}$$

Check the filter for stability.

Solution. The polynomial

$$\overline{D}(z_1^{-1}, 0) = 1 - z_1^{-1} + 0.25z_1^{-2}$$

has two zeros at $z_1^{-1} = 2$ and hence the condition in Eq. (5.7a) is satisfied. Polynomial $c(z_2)$ given by Eq. (5.13) can be formed as

$$c(z_2) = z_2\overline{D}(z_1^{-1}, z_2^{-1})$$
$$= (1 - z_1^{-1} + 0.25z_1^{-2})z_2 - (0.7 - 0.67z_1^{-1} + 0.16z_1^{-2})$$

Since the highest power of z_2 is one, the Schur–Cohn matrix for $|z_1^{-1}| = 1$ given by Eq. (5.15) is a 1×1 matrix and can be expressed as

$$\mathbf{C}(z_1^{-1}) = (1 - z_1^{-1} + 0.25z_1^{-2})(1 - z_1 + 0.25z_1^2)$$
$$- (0.7 - 0.67z_1^{-1} + 0.16z_1^{-2})(0.7 - 0.67z_1 + 0.16z_1^2)$$
$$= 1.0980 - 0.6738(z_1 + z_1^{-1}) + 0.1380(z_1^2 + z_1^{-2})$$

Hence

$$\mathbf{C}(1) = 0.0254 > 0$$

and the condition in Eq. (5.16a) is satisfied. From Eqs. (5.16b) and (5.17) the polynomial $g(z_1)$ can be obtained as

$$g(z_1) = z_1^2 \det \mathbf{C}(z_1^{-1})$$

$$= 0.1380 - 0.6738z_1 + 1.0980z_1^2 - 0.6738z_1^3 + 0.1380z_1^4$$

and so

$$g(1) = 0.0264 > 0$$

that is, the condition in Eq. (5.18a) is satisfied. Finally, the zeros of $g(z_1)$ can be obtained as

$$z_1 = 2.0, \qquad z_1 = 1.8388, \qquad z_1 = 0.5438, \qquad z_1 = 0.50$$

and, as can be seen, $g(z_1)$ has no roots on the unit circle $|z_1^{-1}| = 1$, that is, the condition in Eq. (5.18b) is satisfied. Therefore, the filter is stable.

The preceding procedure can also be used to check the two-variable condition

$$\overline{D}(z_1^{-1}, z_2^{-1}) \neq 0 \qquad \text{for } (z_1^{-1}, z_2^{-1}) \in U^2 \tag{5.19}$$

which appeared in Corollary 5.1 and Theorems 5.7 and 5.8. For this condition an alternative approach proposed by Chiasson, Brierley, and Lee [15] is also applicable. This approach is based on the resultant method for checking whether two polynomials have common zeros (see Walker [16]) and it is sometimes easier to apply.

If $p(z)$ and $q(z)$ are two polynomials given by

$$p(z) = p_n z^n + p_{n-1} z^{n-1} + \cdots + p_0, \qquad p_n \neq 0$$

and

$$q(z) = q_m z^m + q_{m-1} z^{m-1} + \cdots + q_0, \qquad q_m \neq 0$$

then $p(z)$ and $q(z)$ have no common zeros if and only if

$$\det \mathbf{R} \neq 0$$

where $\mathbf{R}$ denotes the so-called resultant matrix of polynomials $p(z)$ and $q(z)$ given by

$$\mathbf{R} = \left[\begin{array}{cccccccc} p_0 & p_1 & \cdots & p_n & 0 & \cdots & 0 \\ 0 & p_0 & p_1 & \cdots & p_n & \cdots & 0 \\ \vdots & \vdots & \vdots & \vdots & \vdots & \vdots & \vdots \\ 0 & \cdots & 0 & p_0 & p_1 & \cdots & p_n \\ q_0 & q_1 & \cdots & q_m & 0 & \cdots & 0 \\ 0 & q_0 & q_1 & \cdots & q_m & \cdots & 0 \\ \vdots & \vdots & \vdots & \vdots & \vdots & \vdots & \vdots \\ 0 & \cdots & 0 & q_0 & q_1 & \cdots & q_m \end{array} \right] \begin{array}{l} \left.\vphantom{\begin{array}{c}1\\1\\1\\1\end{array}}\right\} m \text{ rows} \\ \left.\vphantom{\begin{array}{c}1\\1\\1\\1\end{array}}\right\} n \text{ rows} \end{array}$$

To illustrate the application of the resultant method for the checking of the condition in Eq. (5.19), let us define the reciprocal polynomial of $\bar{D}(z_1^{-1}, z_2^{-1})$ as

$$\tilde{D}(z_1^{-1}, z_2^{-1}) = z_1^{-N_1} z_2^{-N_2} \bar{D}(z_1, z_2)$$

Evidently, if

$$\tilde{D}(z_1^{-1}, z_2^{-1}) = 0 \qquad \text{at } (z_1^{-1}, z_2^{-1}) \in U^2$$

then

$$\bar{D}(z_1, z_2) = 0 \qquad \text{at } (z_1, z_2) \in U^2$$

and in addition

$$\bar{D}^*(z_1, z_2) = \bar{D}(z_1^*, z_2^*) = \bar{D}(z_1^{-1}, z_2^{-1}) = 0 \qquad \text{at } (z_1^{-1}, z_2^{-1}) \in U^2$$

In other words, the condition in Eq. (5.19) will be violated if and only if $\bar{D}(z_1^{-1}, z_2^{-1})$ and $\tilde{D}(z_1^{-1}, z_2^{-1})$ have one or more common zeros on U^2. On the basis of this observation, the following step-by-step procedure [16] can be used to check the condition in Eq. (5.19).

STEP 1: Write polynomials $\bar{D}(z_1^{-1}, z_2^{-1})$ and $\tilde{D}(z_1^{-1}, z_2^{-1})$ in the form

$$\bar{D}(z_1^{-1}, z_2^{-1}) = \sum_{i=0}^{N_2} \bar{b}_i(z_1^{-1}) z_2^{-i} \quad \text{and} \quad \tilde{D}(z_1^{-1}, z_2^{-1}) = \sum_{i=0}^{N_2} \tilde{b}_i(z_1^{-1}) z_2^{-i}$$

respectively and obtain the determinant of the resultant matrix $\mathbf{R}$ as

$$r(z_1^{-1}) = \det \begin{bmatrix} \bar{b}_0(z_1^{-1}) & \bar{b}_1(z_1^{-1}) & \cdots & \bar{b}_{N_2}(z_1^{-1}) & & & 0 & \cdots & 0 \\ 0 & \bar{b}_0(z_1^{-1}) & \bar{b}_1(z_1^{-1}) & \cdots & & & \bar{b}_{N_2}(z_1^{-1}) & \cdots & 0 \\ \vdots & \vdots & \vdots & & \vdots & & \vdots & & \vdots \\ 0 & \cdots & 0 & \bar{b}_0(z_1^{-1}) & \bar{b}_1(z_1^{-1}) & \cdots & \bar{b}_{N_2}(z_1^{-1}) \\ \tilde{b}_0(z_1^{-1}) & \tilde{b}_1(z_1^{-1}) & \cdots & \tilde{b}_{N_2}(z_1^{-1}) & & & 0 & \cdots & 0 \\ 0 & \tilde{b}_0(z_1^{-1}) & \tilde{b}_0(z_1^{-1}) & \cdots & & & \tilde{b}_{N_2}(z_1^{-1}) & \cdots & 0 \\ \vdots & \vdots & \vdots & & \vdots & & \vdots & & \vdots \\ 0 & \cdots & 0 & \tilde{b}_1(z_1^{-1}) & \cdots & & \tilde{b}_{N_2}(z_1^{-1}) \end{bmatrix}$$

STEP 2: Find the roots of the 1-D algebraic equation $r(z_1^{-1}) = 0$ and represent them by $z_{1,1}^{-1}, \ldots, z_{1,K}^{-1}$. If

$$|z_{1,i}^{-1}| \neq 1 \qquad \text{for all } i \text{ in the range } 1 \le i \le K$$

then the condition in Eq. (5.19) holds. Otherwise, proceed to Step 3.

STEP 3: If $|z_{1,i}^{-1}| = 1$ for some i, then substitute root $z_{1,i}^{-1}$ in $\overline{D}(z_1^{-1}\, z_2^{-1})$ and find the roots of $\overline{D}(z_{1,i}^{-1}, z_2^{-1}) = 0$. If one of the roots obtained is on the unit circle, then the condition in Eq. (5.19) is violated. Repeat this step for each root $z_{1,i}^{-1}$ whose magnitude is one. If no roots of $\overline{D}(z_{1,i}^{-1}, z_2^{-1}) = 0$ can be found on the unit circle, then the condition in Eq. (5.19) holds.

Example 5.2 Use Theorem 5.8 to check the stability of the filter described in Example 5.1.

Solution. To check the condition in Eq. (5.10a), replace both z_1^{-1} and z_2^{-1} by z^{-1} in $\overline{D}(z_1^{-1}, z_2^{-1})$ to obtain

$$\overline{D}(z^{-1}, z^{-1}) = 1 - 1.7z + 0.92z^{-2} - 0.16z^{-3}$$

The roots of $\overline{D}(z^{-1}, z^{-1}) = 0$ are $z^{-1} = 2.5$, $z^{-1} = 2.0$, and $z^{-1} = 1.25$. Hence the condition in Eq. (5.10a) holds. To check the condition in Eq. (5.10b), we apply the steps of the resultant method. The reciprocal polynomial of $\overline{D}(z_1^{-1}, z_2^{-1})$ is given by

$$\hat{D}(z_1^{-1}, z_2^{-1}) = -0.16 + 0.25z_2^{-1} + 0.67z_1^{-1} - 0.7z_1^{-2} - z_1^{-1}z_2^{-1} + z_1^{-2}z_2^{-1}$$

The determinant of the resultant matrix $\mathbf{R}$ with respect to z_2^{-1} is given by

$$r(z_1) = 1 - 4.855z_1^{-1} + 7.957z_1^{-2} - 4.855z_1^{-3} + z_1^{-4}$$

and has zeros at $z_1^{-1} = 0.5038 \pm j0.0899$ and $z_1^{-1} = 1.9237 \pm j0.3434$. Since none of the zeros are on the unit circle, the condition in Eq. (5.10b) holds. Therefore, the filter is stable.

5.3 STABILITY ANALYSIS IN STATE-SPACE DOMAIN

The stability properties of 2-D digital filters can also be examined in terms of the state-space representation developed in Chapter 2. While the results that can be obtained are equivalent to those already presented in the preceding section, certain advantages are gained through the use of the state-space representation. First, all stability conditions are stated in terms of the spectral properties of certain matrices and as a result standard matrix analysis methods can be used to check the stability of filters. Second, the Lyapunov theory, which was found to be very useful in 1-D stability analysis, can be generalized to include the 2-D case.

A digital filter represented by the transfer function of Eq. (5.1) can also be represented by the state-space representation of Eq. (4.5) and, consequently, we can write

$$H(z_1, z_2) = \frac{\overline{N}(z_1^{-1}, z_2^{-1})}{\overline{D}(z_1^{-1}, z_2^{-1})} = \mathbf{c} \begin{bmatrix} z_1\mathbf{I}_N - \mathbf{A}_1 & -\mathbf{A}_2 \\ -\mathbf{A}_3 & z_2\mathbf{I}_M - \mathbf{A}_4 \end{bmatrix}^{-1} \mathbf{b} + \mathbf{d} \quad (5.20)$$

where

$$\overline{D}(z_1^{-1}, z_2^{-1}) = \det \begin{bmatrix} \mathbf{I}_N - z_1^{-1}\mathbf{A}_1 & -z_1^{-1}\mathbf{A}_2 \\ -z_2^{-1}\mathbf{A}_3 & \mathbf{I}_M - z_2^{-1}\mathbf{A}_4 \end{bmatrix} \quad (5.21)$$

As in the previous section, it is assumed that the numerator and denominator polynomials in Eq. (5.20) are factor coprime and that the transfer function has no nonessential singularities of the second kind on the unit bicircle U^2, for the sake of simplicity.

Straightforward analysis yields

$$\overline{D}(z_1^{-1}, z_2^{-1}) = \det(\mathbf{I}_N - z_1^{-1}\mathbf{A}_1)$$
$$\times \det\{\mathbf{I}_M - z_2^{-1}[\mathbf{A}_4 + \mathbf{A}_3(z_1\mathbf{I}_N - \mathbf{A}_1)^{-1}\mathbf{A}_2]\} \quad (5.22a)$$
$$= \det(\mathbf{I}_M - z_2^{-1}\mathbf{A}_4)$$
$$\times \det\{\mathbf{I}_N - z_1^{-1}[\mathbf{A}_1 + \mathbf{A}_2(z_2\mathbf{I}_M - \mathbf{A}_4)^{-1}\mathbf{A}_3]\} \quad (5.22b)$$

On the basis of Eqs. (5.22a)–(5.22b), state-space versions of the stability conditions stated in Theorems 5.6 and 5.8 can readily be obtained [17]. These conditions are stated in the following theorem.

Theorem 5.12 *The transfer function of Eq. (5.20) represents a BIBO-stable digital filter if and only if one of the following conditions holds:*

1. (i) *All the eigenvalues of matrix* $\mathbf{A}_1$ *are located on the open unit disc.*
   (ii) *All the eigenvalues of matrix* $\mathbf{A}_4 + \mathbf{A}_3(z_1\mathbf{I}_N - \mathbf{A}_1)^{-1}\mathbf{A}_2$ *are located on the open unit disc for each* $z_1 \in U_1 = \{z_1: |z_1| = 1\}$.
2. (i) *All the eigenvalues of matrix* $\mathbf{A}_4$ *are located on the open unit disc.*
   (ii) *All the eigenvalues of matrix* $\mathbf{A}_1 + \mathbf{A}_2(z_2\mathbf{I}_M - \mathbf{A}_4)^{-1}\mathbf{A}_3$ *are located on the open unit disc for each* $z_2 \in U_2 = \{z_2: |z_2| = 1\}$.
3. (i) *All the eigenvalues of matrix* $\mathbf{A}$ *[see Eq. (2.9a)] are located on the open unit disc.*
   (ii) *Matrix* $\mathbf{A}_1$ *has no eigenvalues on the unit circle.*
   (iii) *Matrix* $\mathbf{A}_4 + \mathbf{A}_3(z_1\mathbf{I}_N - \mathbf{A}_1)^{-1}\mathbf{A}_2$ *has no eigenvalues on the unit circle for each* $z_1 \in U_1$.
4. (i) *All the eigenvalues of matrix* $\mathbf{A}$ *are located on the open unit disc.*
   (ii) *Matrix* $\mathbf{A}_4$ *has no eigenvalues on the unit circle.*
   (iii) *Matrix* $\mathbf{A}_1 + \mathbf{A}_2(z_2\mathbf{I}_M - \mathbf{A}_4)^{-1}\mathbf{A}_3$ *has no eigenvalues on the unit circle for each* $z_2 \in U_2$.

Conditions 1(i), 2(i), and 1(ii), 2(ii) follow from conditions (1) and (2) of Theorem 5.6, respectively, conditions 3(i) and 4(i) follow from condition (1) of Theorem 5.8 and conditions 3(ii)–3(iii) and 4(ii)–4(iii) follow from condition (2) of Theorem 5.8.

An immediate consequence of the preceding theorem can be stated in terms of the following corollary.

Corollary 5.2   *If the filter represented by the transfer function in Eq. (5.20) is BIBO stable, then all eigenvalues of each of the matrices* A_1, A_4, *and* A *must lie on the open unit disc.*

In other words, if any one of the eigenvalues of A_1 or A_4 or A is not located on the open unit disc, then the filter is unstable. However, the condition that all eigenvalues of A_1, A_4, and A lie on the open unit disc is not a sufficient condition for BIBO stability, as can be seen in part (b) of the following example.

Example 5.3   Check the filters characterized by the following transfer functions for stability.

$$(a) \quad H(z_1, z_2) = \frac{z_1^{-1} + z_2^{-1} + 1.75z_1^{-1}z_2^{-1}}{1 + 0.5z_1^{-1} - 0.5z_2^{-1} - z_1^{-1}z_2^{-1}}$$

$$(b) \quad H(z_1, z_2) = \frac{z_1^{-1} + z_2^{-1} - 0.4z_1^{-1}z_2^{-1}}{1 + 0.4z_1^{-1} - 0.8z_2^{-1} - 0.16z_1^{-1}z_2^{-1}}$$

Solution.   (a) It is easy to show that the filter can be represented in the state-space domain by Eq. (2.9) with

$$A = \begin{bmatrix} A_1 & A_2 \\ A_3 & A_4 \end{bmatrix} = \begin{bmatrix} -0.5 & 0.75 \\ 1 & 0.5 \end{bmatrix}, \quad b = \begin{bmatrix} 1 \\ 1 \end{bmatrix}, \quad c = [1 \ 1], \quad d = 0$$

Since the eigenvalues of A are -1 and 1, it follows from Corollary 5.2 that the filter is unstable.

(b) This filter can be represented by the matrices

$$A = \begin{bmatrix} A_1 & A_2 \\ A_3 & A_4 \end{bmatrix} = \begin{bmatrix} -0.4 & -0.4 \\ 0.4 & 0.8 \end{bmatrix}, \quad b \begin{bmatrix} -1 \\ 1 \end{bmatrix},$$

$$c = [1 \ 1], \quad d = 0$$

Hence all eigenvalues of A_1, A_4, and A lie on the open unit disc. Nevertheless

$$A_4 + A_3(z_1I_N - A_1)^{-1}A_2 = 0.80 - \frac{0.16}{z_1 + 0.40} = 1.07 \qquad \text{at } z_1 = -1$$

In effect, condition 2(ii) of Theorem 5.12 is violated and, therefore, the filter is unstable.

5.4 2-D LYAPUNOV STABILITY THEORY

Another important approach to the study of stability is based on the classical results of Lyapunov. While these results were originally applied for the study of stability of 1-D continuous systems, they have also been applied in the case of 1-D discrete systems in recent years [18]. In this section it is shown that these results can also be applied for the study of stability in the case of 2-D digital filters.

Consider a 1-D digital filter characterized by the state-space equations

$$q(i + 1) = Aq(i) + bx(i) \tag{5.23a}$$

$$y(i) = cq(i) + dx(i) \tag{5.23b}$$

The transfer function of the filter can be easily obtained as

$$H(z) = c(zI - A)^{-1}b + d$$

and the stability of the filter can be assured if and only if all the eigenvalues of A are located on the open unit disc. The necessary and sufficient conditions for the stability of the filter can be stated in terms of the positive definiteness of the solution of a matrix equation often referred to as the discrete Lyapunov equation. This result is stated in terms of the following theorem.

Theorem 5.13 (1-D Lyapunov) *A 1-D digital filter represented by Eq. (5.23) is stable if and only if for any positive-definite matrix* Q *there exists a unique positive-definite matrix* P *that satisfies the equation*

$$A^T PA - P = -Q \tag{5.24}$$

Proof. To demonstrate the sufficiency of the state condition for stability, let λ be any one of the eigenvalues of A and let v be the associated eigenvector. We can write

$$v^*A^*PAv - v^*Pv = v^*Pv(|\lambda|^2 - 1) = -v^*Qv$$

where * denotes the complex-conjugate transpose operation. Hence

$$|\lambda|^2 - 1 = -\frac{\mathbf{v}^*\mathbf{Q}\mathbf{v}}{\mathbf{v}^*\mathbf{P}\mathbf{v}} < 0$$

that is, λ is on the open unit disc and, therefore, the filter is stable.

To demonstrate the necessity of the stated condition for stability, we note that since the filter is assumed to be stable, all the eigenvalues of $\mathbf{A}$ are on the open unit disc. Consequently,

$$\mathbf{P} = \sum_{k=0}^{\infty} (\mathbf{A}^T)^k \mathbf{Q} \mathbf{A}^k \tag{5.25}$$

is well defined and positive definite. Moreover, matrix $\mathbf{P}$ satisfies the Lyapunov equation stated as Eq. (5.24) since

$$\mathbf{A}^T\mathbf{P}\mathbf{A} - \mathbf{P} = \sum_{k=0}^{\infty} (\mathbf{A}^T)^{k+1} \mathbf{Q} \mathbf{A}^{k+1} - \sum_{k=0}^{\infty} (\mathbf{A}^T)^k \mathbf{Q} \mathbf{A}^k = -\mathbf{Q}$$

To demonstrate that the solution of Eq. (5.24) is unique, assume that $\mathbf{P}_1$ is another solution. We can write

$$\mathbf{P} = \sum_{k=1}^{\infty} (\mathbf{A}^T)^k \mathbf{Q} \mathbf{A}^k = \sum_{k=0}^{\infty} (\mathbf{A}^T)^k (\mathbf{P}_1 - \mathbf{A}^T\mathbf{P}_1\mathbf{A}) \mathbf{A}^k$$

$$= \sum_{k=0}^{\infty} (\mathbf{A}^T)^k \mathbf{P}_1 \mathbf{A}^k - \sum_{k=0}^{\infty} (\mathbf{A}^T)^{k+1} \mathbf{P}_1 \mathbf{A}^{k+1}$$

$$= \mathbf{P}_1$$

that is, $\mathbf{P}$ is a unique solution. ∎

From Theorems 5.12 and 5.13, the 2-D version of the Lyapunov theorem can now be stated as follows [19].

Theorem 5.14 (Lu and Lee)   *A 2-D digital filter represented by the transfer function of Eq. (4.5) is BIBO stable if and only if*

1.   *Matrix $\mathbf{A}_1$ has all its eigenvalues on the open unit disc.*
2.   *The Lyapunov equation*
$$\mathbf{P}^*(z)\mathbf{G}(z)\mathbf{P}(z) - \mathbf{G}(z) = -\mathbf{W}(z), \qquad z \in U \tag{5.26}$$
     where
$$\mathbf{G}(z) = \mathbf{A}_4 + \mathbf{A}_3(z\mathbf{I}_N - \mathbf{A}_1)^{-1}\mathbf{A}_2 \tag{5.27}$$
has a unique positive-definite Hermitian solution $\mathbf{P}(z)$ for any given matrix $\mathbf{W}(z)$ which is a positive-definite Hermitian matrix for $z \in U$.

The 2-D Lyapunov equation in Eq. (5.26), while similar to its 1-D counterpart in Eq. (5.24), differs in one significant way. It depends on complex variable z, which can assume values on the unit circle of the z plane. An alternative 2-D Lyapunov equation which involves only constant matrices is stated in the following theorem [20].

Theorem 5.15 (El-Agizi and Fahmy)   *A 2-D digital filter represented by the transfer function of Eq. (5.20) is BIBO stable if there exists a block-diagonal positive-definite matrix*

$$\mathbf{W} = \begin{bmatrix} \mathbf{W}_1 & 0 \\ 0 & \mathbf{W}_2 \end{bmatrix} \tag{5.28}$$

where $\mathbf{W}_1$ *and* $\mathbf{W}_2$ *are positive definite matrices of dimensions N and M, respectively, such that matrix* $\mathbf{Q}$ *given by*

$$\mathbf{Q} = \mathbf{W} - \mathbf{A}^T \mathbf{W} \mathbf{A} \tag{5.29}$$

is positive definite, where $\mathbf{A}$ *is given by Eq. (2.9a).*

Proof.   Assume that there is a point $(z_1^{-1}, z_2^{-1}) \in \bar{U}^2$, where $\bar{U}^2$ is the unit bidisc, such that

$$\bar{D}(z_1^{-1}, z_2^{-1}) = \det \begin{bmatrix} \mathbf{I}_N - z_1^{-1}\mathbf{A}_1 & -z_1^{-1}\mathbf{A}_2 \\ -z_2^{-1}\mathbf{A}_3 & \mathbf{I}_M - z_2^{-1}\mathbf{A}_4 \end{bmatrix} = 0 \tag{5.30}$$

In such a circumstance there exists a nonzero vector $\mathbf{p}$ of dimension $N + M$ such that

$$\begin{bmatrix} \mathbf{I}_N - z_1^{-1}\mathbf{A}_1 & -z_1^{-1}\mathbf{A}_2 \\ -z_2^{-1}\mathbf{A}_3 & \mathbf{I}_M - z_2^{-1}\mathbf{A}_4 \end{bmatrix} \mathbf{p} = 0$$

that is,

$$\mathbf{p} - \begin{bmatrix} z_1^{-1}\mathbf{I}_N & 0 \\ 0 & z_2^{-1}\mathbf{I}_M \end{bmatrix} \mathbf{Ap} = 0$$

and from Eq. (5.29) we have

$$\mathbf{p^*Qp} = \mathbf{p^*Wp} - (\mathbf{Ap})^*\mathbf{WAp} = \mathbf{p^*} \begin{bmatrix} (1 - |z_1|^2)\mathbf{W}_1 & 0 \\ 0 & (1 - |z_2|^2)\mathbf{W}_2 \end{bmatrix} \mathbf{p}$$

Evidently, $\mathbf{p^*Qp}$ is positive, whereas the right-hand side of the preceding equation is nonpositive. This contradicts the assumption in Eq. (5.30) and so

$$\overline{D}(z_1^{-1}, z_2^{-1}) \neq 0 \qquad \text{for } (z^{-1}, z_2^{-1}) \in \overline{U}^2$$

Therefore, the 2-D filter is stable. ■

It should be mentioned here that the existence of a block-diagonal positive-definite matrix **W** that makes matrix **Q** of Eq. (5.29) positive definite is a sufficient but not necessary condition for BIBO stability in a 2-D digital filter. For example, a 2-D digital filter with the **A** matrix.

$$\mathbf{A} = \begin{bmatrix} 0.50 & 0.007 & 0.012 & 0.008 & 0.028 & 0 & 0 & 0 \\ -0.007 & 0.05 & 0 & 0 & 0 & 0.012 & 0.008 & 0.012 \\ 0 & 0 & 0 & 1 & 0 & 0 & 0 & 0 \\ 0 & 0 & 0 & 0 & 1 & 0 & 0 & 0 \\ 0 & 1 & 0.845 & -2.657 & 2.810 & 0 & 0 & 0 \\ 0 & 0 & 0 & 0 & 0 & 0 & 1 & 0 \\ 0 & 0 & 0 & 0 & 0 & 0 & 0 & 1 \\ 1 & 0 & 0 & 0 & 0 & -0.845 & -2.657 & -2.810 \end{bmatrix}$$

is BIBO stable but there is no block-diagonal positive-definite matrix **W** that makes matrix **Q** positive definite [21].

5.5 STABILITY OF LOW-ORDER FILTERS

Frequently, the design of 2-D digital filters is carried out by using low-order filters in cascade or in parallel or an appropriate combination of cascade and parallel filters. It is, therefore, of interest to obtain general conditions that can be applied to check the stability of arbitrary low-order filters. In this section, conditions are presented that guarantee the stability of filters up to order (2, 2).

Consider the general 2-D digital filter of order (2, 2) which can be represented by the transfer function

$$H\,(z_1, z_2) = \frac{\overline{N}(z_1^{-1}, z_2^{-1})}{\overline{D}(z_1^{-1}, z_2^{-1})} = \frac{\overline{N}(z_1^{-1}, z_2^{-1})}{\sum_{i=0}^{N_1} \sum_{j=0}^{N_2} b_{ij} z_1^{-i} z_2^{-j}}, \qquad b_{00} = 1$$

If we assume that the numerator and denominator polynomials are factor coprime and that the transfer function has no nonessential singularities of the second kind on the unit bicircle, the stability properties of the filter are independent of the numerator polynomial. Consequently, the stability analysis of the preceding filter can be carried out by considering the simpler transfer function

$$H(z_1, z_2) \doteq \frac{1}{\overline{D}(z_1^{-1}, z_2^{-1})} = \frac{1}{\sum_{i=0}^{N1} \sum_{j=0}^{N2} b_{ij} z_1^{-i} z_2^{-j}}, \qquad b_{00} = 1$$

Several cases can be identified depending on the order of the denominator polynomial.

CASE 1: If

$$\overline{D}(z_1^{-1}, z_2^{-1}) = 1 + b_{10} z_1^{-1} \qquad (5.31)$$

then the filter is of order $(1, 0)$. Clearly, a 1-D filter is obtained that is stable if and only if

$$|b_{10}| < 1 \qquad (5.32)$$

CASE 2: If

$$\overline{D}(z_1^{-1}, z_2^{-1}) = 1 + b_{01} z_2^{-1} \qquad (5.33)$$

a filter of order $(1, 0)$ is again obtained that is stable if and only if

$$|b_{01}| < 1 \qquad (5.34)$$

CASE 3: If

$$\overline{D}(z_1^{-1}, z_2^{-1}) = 1 + b_{10} z_1^{-1} + b_{01} z_2^{-1} \qquad (5.35)$$

the filter is of order $(1, 1)$ and can be represented by Eq. (2.9) where

$$\mathbf{A} = \begin{bmatrix} A_1 & A_2 \\ A^3 & A^4 \end{bmatrix} = \begin{bmatrix} -b_{10} & -b_{10}b_{01} \\ -1 & -b_{01} \end{bmatrix},$$

$$\mathbf{b} = \begin{bmatrix} b_{10} \\ 1 \end{bmatrix}, \qquad \mathbf{c} = [-1 \quad -b_{01}], \qquad \mathbf{d} = 1$$

From conditions 1(i) and 1(ii) of Theorem 5.12, the filter is BIBO stable if and only if

1. $|b_{10}| < 1$.

2. $\max_{z_1 \in U_1} \left| b_{01} - \frac{b_{10}b_{01}}{z_1 + b_{10}} \right| < 1$.

Since the maximum value in condition (2) can be achieved only at $z_1 = 1$

or $z_1 = -1$, we conclude that the filter is stable if and only if

1. $|b_{10}| < 1$.
2. $\max\left\{\left|\dfrac{b_{01}}{1 + b_{10}}\right|, \left|\dfrac{b_{01}}{1 - b_{10}}\right|\right\} < 1$.

These relations can be combined into the more compact but equivalent form

$$|b_{01}| + |b_{10}| < 1 \tag{5.36}$$

CASE 4: If

$$\overline{D}(z_1^{-1}, z_2^{-1}) = 1 + b_{10}z_1^{-1} + b_{01}z_2^{-1} + b_{11}z_1^{-1}z_2^{-1} \tag{5.37}$$

we obtain, as in Case 3, a filter of order $(1, 1)$, which can be represented by the matrices

$$\mathbf{A} = \begin{bmatrix} \mathbf{A}_1 & \mathbf{A}_2 \\ \mathbf{A}_3 & \mathbf{A}_4 \end{bmatrix} = \begin{bmatrix} -b_{10} & -b_{10}b_{01} + b_{11} \\ -1 & -b_{01} \end{bmatrix},$$

$$\mathbf{b} = \begin{bmatrix} b_{10} \\ 1 \end{bmatrix}, \qquad \mathbf{c} = [-1 \ -b_{01}], \qquad \mathbf{d} = 1$$

From conditions 1(i) and 1(ii) of Theorem 5.12, the filter is BIBO stable if and only if

1. $|b_{10}| < 1$.
2. $\max\limits_{z_1 \in U_1} \left| b_{01} + \dfrac{b_{11} - b_{10}b_{01}}{z_1 + b_{10}} \right| < 1$.

and, as in Case 3, we can show that the filter is stable if and only if

1. $|b_{10}| < 1$.
2. $\max\left\{\left|\dfrac{b_{01} + b_{11}}{1 + b_{10}}\right|, \left|\dfrac{b_{01} - b_{11}}{1 - b_{10}}\right|\right\} < 1$.

These relations can be combined into the more compact but equivalent form

$$|b_{10} + b_{01}| - 1 < b_{11} < 1 - |b_{10} - b_{01}| \tag{5.38}$$

CASE 5: If

$$\overline{D}(z_1^{-1}, z_2^{-1}) = \sum_{i=0}^{N_1} \sum_{j=0}^{N_2} b_{ij}z_1^{-i}z_2^{-j} \tag{5.39}$$

the filter is of order $(2, 2)$ and the necessary and sufficient conditions for stability become quite complicated. Nevertheless, an algorithm due to O'Connor and Huang [22] is available that can be used to check the stability of an arbitrary filter of order $(2, 2)$. This algorithm involves a number of computations and is given below.

1. Compute coefficients s_i, t_j, q_j, and r_j for $0 \le i \le 2, 0 \le j \le 4$, as follows:

$$s_0 = b_{02}^2 + b_{12}^2 + b_{22}^2 - b_{00}^2 - b_{10}^2 - b_{20}^2 - 2b_{02}b_{22} + 2b_{20}b_{00}$$
$$s_1 = 2(b_{02}b_{12} + b_{12}b_{22} - b_{00}b_{10} - b_{20}b_{10})$$
$$s_2 = 4(b_{02}b_{22} - b_{20}b_{00})$$
$$t_0 = b_{02}b_{01} + b_{12}b_{11} + b_{22}b_{21} - b_{01}b_{00} - b_{11}b_{10} - b_{21}b_{20}$$
$$t_1 = b_{12}b_{01} + b_{22}b_{11} - b_{00}b_{11} - b_{21}b_{10}$$
$$t_2 = b_{02}b_{11} + b_{12}b_{21} - b_{10}b_{01} - b_{20}b_{11}$$
$$t_3 = b_{01}b_{22} - b_{21}b_{00}$$
$$t_4 = b_{02}b_{21} - b_{20}b_{01}$$
$$q_0 = t_0^2 + t_1^2 + t_2^2 + t_3^2 + t_4^2$$
$$q_1 = t_0t_1 + t_0t_2 + t_1t_3 + t_2t_4$$
$$q_2 = t_0t_3 + t_1t_2 + t_0t_4$$
$$q_3 = t_1t_4 + t_2t_3$$
$$q_4 = t_3t_4$$
$$r_0 = s_0^2 - q_0 + 2q_2 - 2q_4$$
$$r_1 = 2s_0s_1 - 2q_1 + 6q_3$$
$$r_2 = 2s_0s_2 + s_1^2 + 16q_4 - 4q_2$$
$$r_3 = 2s_1s_2 - 8q_3$$
$$r_4 = s_2^2 - 16q_4$$

2. Define polynomials $\alpha(z)$, $\beta(x)$, and $\gamma(x)$ as
$$\alpha(z) = b_{20}z^2 + b_{10}z + b_{00}$$
$$\beta(x) = s_2x^2 + s_1 x + s_0$$
$$\gamma(x) = r_4x^4 + r_3x^3 + r_2x^2 + r_1x + r_0$$

3. The filter of order $(2, 2)$ is stable if and only if
   (i) Polynomial $\alpha(z)$ has no zeros with magnitude less than equal to unity.
   (ii) $\beta(x) < 0$ for $|x| \le 1$, which reduces to
       (a) $s_0 < 0$.
       (b) $\beta(x)$ has no real zeros in the range $|x| \le 1$.
   (iii) $\gamma(x) > 0$ for $|x| \le 1$ which reduces to
       (a) $r_0 > 0$.
       (b) $\gamma(x)$ has no real zeros in the range $|x| \le 1$.

REFERENCES

1. D. Goodman, Some stability properties of two-dimensional linear shift-invariant digital filters, *IEEE Trans. Circuits Syst.*, vol. CAS-24, pp. 201–208, April 1977.

2. T. S. Huang, Stability of two-dimensional recursive filters, *IEEE Trans. Audio Electroacoust.*, vol. AU-20, pp. 158–163, June 1972.

3. H. Silverman, *Complex Variables*, Boston: Houghton Mifflin, 1975, pp. 220–222.

4. M. G. Strintzis, Tests of stability of multidimensional filters, *IEEE Trans. Circuits Syst.*, vol. CAS-24, pp. 432–437, Aug. 1977.

5. R. A. DeCarlo, J. Murray, and R. Saeks, Multivariable Nyquist theory, *Int. J. Control*, vol. 25, pp. 657–675, 1977.

6. W. Rudin, *Function Theory in Polydiscs*, New York: Benjamin, 1969, p. 90.

7. E. I. Jury, Stability of multidimensional scalar and matrix polynomials, *Proc. IEEE*, vol. 66, pp. 1018–1047, Sept. 1978.

8. N. K. Bose, *Applied Multidimensional Systems Theory*, New York: Van Nostrand Reinhold, 1982.

9. R. J. Schwarz and B. Friedland, *Linear Systems*, New York: McGraw-Hill, 1965.

10. E. I. Jury, *Theory and Application of z-Transform Method*, New York: Wiley, 1964.

11. B. Noble, *Applied Linear Algebra*, Englewood Cliffs, N.J.: Prentice-Hall, 1969.

12. D. D. Siljak, Stability criteria for two-variable polynomials, *IEEE Trans. Circuits Syst.*, vol. CAS-22, pp. 185–189, March 1975.

13. PRO-MATLAB, *User's Guide*, South Natick, Mass: The Mathworks, 1990.

14. D. D. Siljak, Algebraic criteria for positive realness relatiave to the unit circle, *J. Franklin Inst.*, vol. 295, pp. 469–476, June 1973.

15. J. N. Chiasson, S. D. Brierley, and E. B. Lee, A simplified derivation of the Zeheb-Walach 2-D stability test with applications to time-delay systems, *IEEE Trans. Automat. Contr.*, vol. AC-30, pp. 411–414, April 1985.

16. R. J. Walker, *Algebraic Curves*, New York: Springer-Verlag, 1978.

17. W.-S. Lu and E. B. Lee, Stability analysis for two-dimensional systems, *IEEE Trans. Circuits Syst.*, vol. CAS-30, pp. 455–461, July 1983.

18. T. Kailath, *Linear Systems*, Englewood Cliffs, N.J.: Prentice-Hall, 1980.

19. W.-S. Lu and E. B. Lee, Stability analysis for two-dimensional systems via a Lyapunov approach, *IEEE Trans. Circuits Syst.*, vol. CAS-32, pp. 61–68, Jan. 1985.

20. N. G. El-Agizi and M. M. Fahmy, Two-dimensional digital filters with no overflow oscillations, *IEEE Trans. Acoust., Speech, Signal Process.*, vol. ASSP-27, pp. 465–469, June 1979.

21. B. D. O. Anderson, P. Agathoklis, E. I. Jury, and M. Mansour, Stability and the matrix Lyapunov equations for discrete two-dimensional systems, *IEEE Trans. Circuits Syst.*, vol. CAS-33, pp. 261–267, March 1986.

22. B. T. O'Connor and T. S. Huang, Stability of general two-dimensional recursive filters, *Topics in Applied Physics,* vol. 42, (edited by T. S. Huang), New York: Springer-Verlag, 1981, pp. 85–154.

PROBLEMS

5.1 A filter is characterized by

$$H(z_1, z_2) = \frac{1}{\overline{D}(z_1^{-1}, z_2^{-1})}$$

where

(a) $\overline{D}(z_1^{-1}, z_2^{-1}) = 1 - 1.2z_1^{-1} - 1.5z_2^{-1} + 0.5z_1^{-2} + 1.8z_1^{-1}z_2^{-1}$
$$+ 0.6z_2^{-2} - 0.72z_1^{-1}z_2^{-2} - 0.75z_1^{-2}z_2^{-1}$$
$$+ 0.29z_1^{-2}z_2^{-2}$$

(b) $\overline{D}(z_1^{-1}, z_2^{-1}) = 1 - 1.2z_1^{-1} - 1.5z_2^{-1} + 0.5z_1^{-2} + 1.8z_1^{-1}z_2^{-1}$
$$+ 0.6z_2^{-2} - 0.72z_2^{-1}z_2^{-2} - 0.75z_1^{-2}z_2^{-1}$$
$$+ 0.25z_1^{-2}z_2^{-2}$$

Check the stability of each filter by using Theorems 5.6 and 5.11.

5.2 Check the stability of the filters described in Problem 5.1 by using Theorem 5.8 along with the resultant method.

5.3 Check the stability of the filters described in Problem 5.1 by using the stability criterion described in Sec. 5.5.

5.4 Show that the first-order transfer function

$$H(z_1, z_2) = \frac{1}{\overline{D}(z_1^{-1}, z_2^{-1})}$$

with

$$\overline{D}(z_1^{-1}, z_2^{-1}) = 1 + az_1^{-1} + bz_2^{-1} + cz_1^{-1}z_2^{-1}$$

is stable if and only if

$$|a + b| - 1 < c < 1 - |a - b|$$

5.5 Check the stability of the transfer function given by Eq. (4.5) where

$$A_1 = \begin{bmatrix} 0.5 & 0 \\ 0 & 0.5 \end{bmatrix}, \qquad A_2 = \begin{bmatrix} 1 \\ 5 \end{bmatrix}, \qquad A_3 = [-9.95\,2], \qquad A_4 = 0.7$$

by applying Theorem 5.12.

5.6 Use Theorem 5.14 to check the stability of the filter described in Problem 5.5.

5.7 Use Theorem 5.15 to check the stability of the filter described in Problem 5.5.

5.8 Test the stability of the filters represented by

$$H(z_1, z_2) = \frac{1}{\overline{D}(z_1^{-1}, z_2^{-1})}$$

where

$$\overline{D}(z_1^{-1}, z_2^{-1}) = [1 \; z_1^{-1} \; z_1^{-2}] \mathbf{B} [1 \; z_2^{-1} \; z_2^{-2}]^T$$

and

(a) $\mathbf{B} = \begin{bmatrix} 1 & 0.5 & 0.25 \\ 0.5 & 0.25 & 0 \\ 0 & 0 & 0 \end{bmatrix}$

(b) $\mathbf{B} = \begin{bmatrix} 1 & -0.75 & 0.9 \\ 1.5 & -1.2 & 1.3 \\ 1.2 & 0.9 & 0.5 \end{bmatrix}$

5.9 Repeat Problem 5.8 for the filters represented by

(a) $\mathbf{B} = \begin{bmatrix} 1 & -1.2 & 0.5 \\ -1.5 & 1.8 & -0.75 \\ 0.6 & -0.72 & 0.2718 \end{bmatrix}$

(b) $\mathbf{B} = \begin{bmatrix} 1 & -1.2 & 0.5 \\ -1.5 & 1.8 & -0.75 \\ 0.6 & -0.72 & 0.28 \end{bmatrix}$

5.10 Show that the filter characterized by the transfer function

$$H(z_1, z_2) = \frac{1}{\sum_{i=0}^{N_1} \sum_{j=0}^{N_2} b_{ij} z_1^{-i} z_2^{-j}} = \frac{1}{\sum_{j=0}^{N_2} b_j(z_1^{-1}) z_2^{-j}}$$

is stable if

(i) $b_0(z_1^{-1}) \neq 0$ for $|z_1^{-1}| \leq 1$

and

(ii) $|b_0(z_1^{-1})| > \sum_{j=1}^{N_2} |b_j(z_1^{-1})|$ for $|z_1^{-1}| = 1$

5.11 Use the sufficient condition for stability of the transfer function $H(z_1, z_2)$ given in Problem 5.10 to show that

$$H(z_1, z_2) = \frac{1}{2 + z_1^{-1} + z_2^{-1} + 0.5 z_1^{-1} z_2^{-1} + 0.5 z_1^{-2} + 0.4 z_2^{-2}}$$

is a stable transfer function.

5.12 A 2-D digital filter can be represented by the state-space model in Eq. (2.9) from which the matrix

$$A(z) = \begin{bmatrix} A_1 & A_2 \\ 0 & 0 \end{bmatrix} + z \begin{bmatrix} 0 & 0 \\ A_3 & A_4 \end{bmatrix}$$

can be defined. Show that the transfer function given by Eq. (4.5) is stable if and only if

$$\max_{z \in T} \rho[A(z)] < 1$$

where $\rho[A(z)]$ denotes the spectral radius of $A(z)$ and $T = \{z: |z| = 1\}$.

6
Approximations for Nonrecursive Filters

6.1 INTRODUCTION

The design of 2-D digital filters, like the design of 1-D filters, encompasses four different steps in general, as follows:

- Approximation
- Realization
- Implementation
- Study of quantization effects

The filtering application at hand will impose certain constraints that must be observed and certain requirements that must be satisfied, which may relate to the amplitude or phase response and possibly to the space-domain response of the filter. Once certain specifications are prescribed, a decision must be made as to whether a nonrecursive or recursive filter is more appropriate for the required specifications. When the choice is made, the approximation problem must be solved. The approximation step is the process of finding a stable transfer function with real coefficients such that the required specifications imposed on the amplitude, phase, and/ or space-domain response are satisfied. This is one of the more time-consuming parts of the design process to the extent that the solution of the approximation problem for the filter is often said to be the design of the filter.

With a suitable transfer function available, the realization problem can be solved. This is the process of converting the transfer function or some other characterization of the filter into a digital-filter network or structure. The choice here might be a cascade or parallel realization and possibly a state-space realization (see Sec. 1.8 and Chap. 2).

The implementation of the filter depends on the application and may be in terms of software or hardware. In a software implementation, the equations that characterize the digital-filter network are used to write a computer program, usually in some high-level language, which can be used to calculate the output array produced by the filter for any specified input array. In a hardware implementation, on the other hand, the digital-filter network is converted directly into specialized hardware. Evidently, a software implementation is more likely to be appropriate in nonreal-time applications where the data to be processed are available in recorded form and where high speed of operation is not crucial.

The last step in the design of digital filters is to study the effects brought about by the use of finite-precision arithmetic in the implementation. The use of finite-length registers for coefficients introduces errors in the amplitude response, while the use of finite-length registers for signals introduces noise at the output of the filter known as roundoff noise. In software implementations where general-purpose computers or work stations are used, the register length is sufficiently large to render quantization effects negligible. However, in hardware implementations where the word length must be kept as low as possible to minimize the cost of hardware and to maximize the speed of operation, the study of quantization effects is mandatory.

This chapter is concerned with the solution of the approximation problem in the case of nonrecursive filters. As will be demonstrated, this problem can be solved by using 2-D Fourier series in conjunction with 2-D window functions [1] or by applying transformations to 1-D discrete transfer functions [2–5].

Methods for the solution of the approximation problem in the case of recursive filters will be considered in Chap. 7; the application of the optimization techniques for the solution of the approximation problem in recursive and nonrecursive filters will be considered in Chaps. 8 and 9, respectively. The realization, the study of quantization effects, and the implementation of 2-D digital filters will be considered in Chaps. 10, 11, and 12, respectively.

6.2 PROPERTIES OF 2-D NONRECURSIVE FILTERS

Nonrecursive filters have three important advantages. First, they are always stable and hence the application of the complicated stability tests of Chap.

5 is unnecessary. Second, a linear phase response with respect to the frequency variables ω_1 and ω_2 can easily be achieved. Third, owing to their finite impulse response, these filters can be implemented in terms of fast Fourier transforms. A linear phase response is particularly important in the processing of images, as was demonstrated by Huang et al. [6].

This section is concerned with some of the general properties of nonrecursive filters.

6.2.1 Linear-Phase Filters

Consider a 2-D causal nonrecursive filter characterized by the transfer function

$$H(z_1, z_2) = \sum_{n_1=0}^{N_1-1} \sum_{n_2=0}^{N_2-1} h(n_1T_1, n_2T_2)z_1^{-n_1}z_2^{-n_2} \tag{6.1}$$

The frequency response of the filter is given by

$$H(e^{j\omega_1T_1}, e^{j\omega_2T_2}) = \sum_{n_1=0}^{N_1-1} \sum_{n_2=0}^{N_2-1} h(n_1T_1, n_2T_2)e^{j(\omega_1 n_1 T_1 + \omega_2 n_2 T_2)}$$

$$= M(\omega_1, \omega_2)e^{j\theta(\omega_1,\omega_2)} \tag{6.2}$$

where

$$M(\omega_1, \omega_2) = |H(e^{j\omega_1T_1}, e^{j\omega_2T_2})| \quad \text{and} \quad \theta(\omega_1, \omega_2) = \arg H(e^{j\omega_1T_1}, e^{j\omega_2T_2})$$

are the amplitude and phase responses of the filter, respectively. If the phase response is linear with respect to frequencies ω_1 and ω_2, then

$$\theta(\omega_1, \omega_2) = -\tau_1\omega_1 - \tau_2\omega_2 + \theta_0 \tag{6.3}$$

where

$$\tau_1 = -\frac{\partial\theta(\omega_1, \omega_2)}{\partial\omega_1} \quad \text{and} \quad \tau_2 = -\frac{\partial\theta(\omega_1, \omega_2)}{\partial\omega_2}$$

are the group delays of the filter and θ_0 is a constant.

If $\theta_0 = 0$, Eqs. (6.2) and (6.3) give

$$\theta(\omega_1, \omega_2) = -\tau_1\omega_1 - \tau_2\omega_2$$

$$= \tan^{-1}\left(-\frac{\sum_{n_1=0}^{N_1-1} \sum_{n_2=0}^{N_2-1} h(n_1T_1, n_2T_2)\sin(\omega_1 n_1 T_1 + \omega_2 n_2 T_2)}{\sum_{n_1=0}^{N_1-1} \sum_{n_2=0}^{N_2-1} h(n_1T_1, n_2T_2)\cos(\omega_1 n_1 T_1 + \omega_2 n_2 T_2)}\right)$$

and consequently

$$\sum_{n_1=0}^{N_1-1} \sum_{n_2=0}^{N_2-1} h(n_1 T_1, n_2 T_2) \sin[(\tau_1 - n_1 T_1)\omega_1 + (\tau_2 - n_2 T_2)\omega_2] = 0 \quad (6.4)$$

The solution of this equation can be shown to be

$$\tau_1 = (N_1 - 1)T_1/2, \qquad \tau_2 = (N_2 - 1)T_2/2 \qquad (6.5)$$

with

$$h(n_1 T_1, n_2 T_2) = h[(N_1 - 1 - n_1)T_1, (N_2 - 1 - n_2)T_2] \qquad (6.6)$$

for $0 \le n_1 \le (N_1 - 1)$, $0 \le n_2 \le (N_2 - 1)$. Evidently, for odd values of N_1 and N_2, a linear phase response can be achieved with respect to the entire baseband by ensuring that the impulse response is symmetrical about point $[(N_1 - 1)T_1/2, (N_2 - 1)T_2/2]$.

If $\theta_0 = \pm \pi/2$, Eq. (6.4) is replaced by

$$\sum_{n_1=0}^{N_1-1} \sum_{n_2=0}^{N_2-1} h(n_1 T_1, n_2 T_2) \cos[(\tau_1 - n_1 T_1)\omega_1 + (\tau_2 - n_2 T_2)\omega_2] = 0 \quad (6.7)$$

The solution of this equation can be shown to be

$$\tau_1 = (N_1 - 1)T_1/2, \qquad \tau_2 = (N_2 - 1)T_2/2 \qquad (6.8)$$

with

$$h(n_1 T_1, n_2 T_2) = -h[(N_1 - 1 - n_1)T_1, (N_2 - 1 - n_2)T_2] \qquad (6.9)$$

for $0 \le n_1 \le (N_1 - 1)$, $0 \le n_2 \le (N_2 - 1)$. In this case, for odd values N_1 and N_2, a linear phase response can be achieved by ensuring that the impulse response is antisymmetrical about point $[(N_1 - 1)T_1/2, (N_2 - 1)T_2/2]$.

6.2.2 Frequency Response

Equations (6.6) and (6.9) lead to some relatively simple expressions for the frequency response of a linear-phase nonrecursive filter, which are often quite useful. For a symmetrical impulse response with N_1 and N_2 odd, Eq. (6.1) can be written as [7]

$$H(z_1, z_2) = z_1^{-(N_1-1)/2} z_2^{-(N_2-1)/2} \left\{ \sum_{n_1=0}^{N_1-1} \sum_{n_2=0}^{(N_2-3)/2} h(n_1 T_1, n_2 T_2) \right.$$

$$\times \left(z_1^{(N_1-1)/2 - n_1} z_2^{(N_2-1)/2 - n_2} + z_1^{-[(N_1-1)/2 - n_1]} z_2^{-[(N_2-1)/2 - n_2]} \right)$$

$$+ \sum_{n_1=0}^{(N_1-3)/2} h[n_1T_1, (N_2 - 1)T_2/2](z_1^{(N_1-1)/2-n_1} + z_1^{-[(N_1-1)/2-n_1]})$$

$$+ \, h[(N_1 - 1)T_1/2, (N_2 - 1)T_2/2 \Big\}$$

and after some manipulation, the frequency response can be obtained as

$$H(e^{j\omega_1 T_1}, e^{j\omega_2 T_2}) = e^{-j[(N_1-1)\omega_1 T_1/2 + (N_2-1)\omega_2 T_2/2]}$$

$$\cdot \Bigg\{ 2 \sum_{n_1=0}^{(N_1-3)/2} \sum_{n_2=0}^{(N_2-3)/2} h(n_1T_1, n_2T_2)$$

$$\times \cos\left[\left(\frac{N_1 - 1}{2} - n_1 \right) T_1\omega_1 \right.$$

$$+ \left(\frac{N_2 - 1}{2} - n_2 \right) T_2\omega_2 \Bigg]$$

$$+ \, h[(N_1 - 1 - n_1)T_1, n_2T_2]$$

$$\times \cos\left[\left(\frac{N_1 - 1}{2} - n_1 \right) T_1\omega_1 \right.$$

$$- \left(\frac{N_2 - 1}{2} - n_2 \right) T_2\omega_2 \Bigg]$$

$$+ \, 2 \sum_{n_1=0}^{(N_1-3)/2} h\left[n_1T_1, \frac{(N_2 - 1)T_2}{2} \right]$$

$$\times \cos\left[\left(\frac{N_1 - 1}{2} - n_1 \right) T_1\omega_1 \right]$$

$$+ \, 2 \sum_{n_2=0}^{(N_2-3)/2} h\left[\frac{(N_1 - 1)T_1}{2}, n_2T_2 \right]$$

$$\times \cos\left[\left(\frac{N_2 - 1}{2} - n_2 \right) T_2\omega_2 \right]$$

$$+ \, h[(N_1 - 1)T_1/2, (N_2 - 1)T_2/2] \Bigg\} \quad (6.10)$$

Similarly, the frequency response for the case of antisymmetrical impulse response with N_1 and N_2 odd is given by

$$H(e^{j\omega_1 T_1}, e^{j\omega_2 T_2}) = e^{-j[(N_1-1)\omega_1 T_1/2 + (N_2-1)\omega_2 T_2/2 - \pi/2]}$$

$$\times \left\{ 2 \sum_{n_1=0}^{(N_1-3)/2} \sum_{n_2=0}^{(N_2-3)/2} h(n_1 T_1, n_2 T_2) \right.$$

$$\times \sin\left[\left(\frac{N_1-1}{2} - n_1 \right) T_1 \omega_1 + \left(\frac{N_2-1}{2} - n_2 \right) T_2 \omega_2 \right]$$

$$- h[(N_1 - 1 - n_1)T_1, n_2 T_2]$$

$$\times \sin\left[\left(\frac{N_1-1}{2} - n_1 \right) T_1 \omega_1 \right.$$

$$\left. - \left(\frac{N_2-1}{2} - n_2 \right) T_2 \omega_2 \right] + 2 \sum_{n_1=0}^{(N_1-3)/2} h\left[n_1 T_1, \frac{(N_2-1)T_2}{2} \right]$$

$$\times \sin\left[\left(\frac{N_1-1}{2} - n_1 \right) T_1 \omega_1 \right]$$

$$+ 2 \sum_{n_2=0}^{(N_2-3)/2} h\left[\frac{(N_1-1)}{2} T_1, n_2 T_2 \right]$$

$$\times \sin\left[\left(\frac{N_2-1}{2} - n_2 \right) T_2 \omega_2 \right] \tag{6.11}$$

6.3 DESIGN BASED ON FOURIER SERIES

One of the successful methods for the design of 1-D nonrecursive filters is based on the application of the Fourier series. In this method, it is observed that the frequency response of a nonrecursive filter is a periodic function of frequency with a period equal to the sampling frequency and, consequently, it can be expressed in terms of the Fourier series. The Fourier series by itself does not lead to satisfactory results but by using the Fourier series in conjunction with a special class of functions known as window functions, good results can be achieved. Approximations obtained by this method are always suboptimal but the amounts of design effort and computation required are relatively insignificant.

By using the 2-D Fourier series along with 2-D window functions, the design of 2-D nonrecursive digital filters can also be accomplished. As will be shown in this section, the 2-D Fourier series and 2-D window functions are straightforward extensions of their 1-D counterparts.

6.3.1 Design of 1-D Filters

Let us assume that a 1-D filter with a frequency response $H_1(e^{j\omega T})$ is required. By applying the Fourier series, we can write

$$H_1(e^{j\omega T}) = \sum_{n=-\infty}^{\infty} h_1(nT)e^{-j\omega nT} \qquad (6.12)$$

where

$$h_1(nT) = \frac{1}{\omega_s} \int_{-\omega_s/2}^{\omega_s/2} H_1(e^{j\omega T})e^{j\omega nT}\, d\omega \qquad (6.13)$$

and $\omega_s = 2\pi/T$, and on replacing $e^{j\omega T}$ by z, we obtain

$$H_1(z) = \sum_{n=-\infty}^{\infty} h_1(nT)z^{-n} \qquad (6.14)$$

Therefore, if an expression is available for the frequency response, a transfer function can be obtained, which happens to be noncausal and of infinite order.

A finite-order transfer function can be obtained by truncating the Fourier series in Eq. (6.12). This can be accomplished by letting

$$h_1(nT) = 0 \qquad \text{for } |n| > (N - 1)/2$$

but, as was shown elsewhere [8], the truncation of the Fourier series tends to introduce *Gibbs' oscillations* in the frequency response of the filter. These oscillations are particularly pronounced if the desired frequency response has abrupt discontinuities, which cause slow convergence in the Fourier series.

A causal transfer function can be obtained by introducing a delay of $(N - 1)T/2$ in the impulse response of the filter. This is accomplished by modifying the transfer function as

$$H_1'(z) = z^{-(N-1)/2}H_1(z)$$

This modification does not change the amplitude response of the filter since the value of the magnitude of $z^{-(N-1)/2}$ is unity on the unit circle $|z| = 1$.

Note that if the required frequency is an even function with respect to ω, that is,

$$H_1(e^{-j\omega T}) = H_1(e^{j\omega T})$$

then $h_1(nT)$ turns out to be an even function with respect to nT, as can be easily demonstrated:

$$h_1(-nT) = h_1(nT)$$

Consequently, the filter represented by $H_1(z)$ has zero phase response and the filter represented by $H_1'(z)$ has linear phase response.

6.3.2 One-Dimensional Window Functions

The amplitudes of Gibbs' oscillations can be reduced by using discrete window functions. A 1-D window function, represented by $w_1(nT)$, has the following time-domain properties:

1. $w_1(n) = 0$ for $|n| > (N - 1)/2$.
2. It is symmetrical with respect to the nT axis, that is, $w_1(-nT) = w_1(nT)$.

Its amplitude spectrum, on the other hand, consists of a main lobe and several side lobes, such that the area of the side lobes is a small proportion of the area of the main lobe, as illustrated in Fig. 6.1. The parameters of a window function are its order N, the width of its main lobe, and its ripple

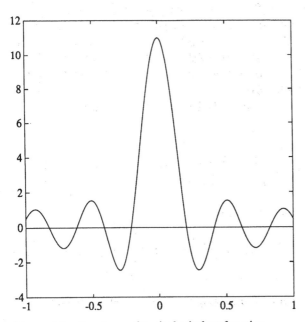

Figure 6.1 Spectrum of typical window function.

ratio, which is the ratio of the maximum side-lobe amplitude to the main-lobe amplitude, usually as a percentage [8].

The application of a window function consists of multiplying the impulse response obtained by applying the Fourier series by the window function to obtain a modified impulse response as

$$h_{W1}(nT) = w_1(nT)h_1(nT) \tag{6.15}$$

Since the window function is of finite duration, a finite-order transfer function given by

$$H_{W1}(z) = \sum_{n=-(N-1)/2}^{(N-1)/2} w_1(nT)h_1(nT)z^{-n} \tag{6.16}$$

is obtained. Further, if $h_1(nT)$ is an even function with respect to nT, then $w_1(nT)h_1(nT)$ is also an even function with respect to nT by virtue of property (2) above and, therefore, the filter obtained has zero phase response.

Through the application of the 1-D complex convolution, the frequency response of the modified filter can be expressed as

$$H_{W1}(e^{j\omega T}) = \frac{T}{2\pi} \int_0^{2\pi/T} H_1(e^{j\Omega T})W_1(e^{j(\omega-\Omega)T}) \, d\Omega \tag{6.17}$$

As demonstrated elsewhere [8], the application of the window function has two effects on the amplitude response of the filter. First, the amplitudes of Gibbs' oscillations in the passbands and stopbands are directly related to the ripple ratio of the window. Second, transition bands are introduced between passbands and stopbands whose widths are directly related to the main-lobe width of the window.

The most frequently used window functions are as follows:

1. *Rectangular window:*

$$w_R(nT) = \begin{cases} 1 & \text{for } |n| \le (N-1)/2 \\ 0 & \text{otherwise} \end{cases}$$

2. *von Hann and Hamming windows:*

$$w_H(nT) = \begin{cases} \alpha + (1-\alpha)\cos\dfrac{2\pi n}{N-1} & \text{for } |n| \le (N-1)/2 \\ 0 & \text{otherwise} \end{cases}$$

where $\alpha = 0.5$ in the von Hann window and $\alpha = 0.54$ in the Hamming window.

3. *Blackman window*:

$$w_B(nT) =$$

$$\begin{cases} 0.42 + 0.5 \cos\dfrac{2\pi n}{N-1} + 0.08 \cos\dfrac{4\pi n}{N-1} & \text{for } |n| \leq (N-1)/2 \\ 0 & \text{otherwise} \end{cases}$$

The parameters of these windows for $N = 21$, 41, and 61 are summarized in Table 6.1. As can be seen, the ripple ratio depends on the window but is relatively independent of N. The rectangular window corresponds to direct truncation of the Fourier series and, as can be seen in Table 6.1, its ripple ratio is quite large.

An alternative window function which has the attractive property that its ripple ratio can be adjusted continuously from the low value of the Blackman window to the high value of the rectangular window by changing a window parameter is the Kaiser window function. This is given by

$$w_K(nT) = \begin{cases} \dfrac{I_0(\beta)}{I_0(\alpha)} & \text{for } |n| \leq (N-1)/2 \\ 0 & \text{otherwise} \end{cases}$$

where α is an independent parameter and

$$\beta = \alpha \left[1 - \left(\frac{2n}{N-1} \right)^2 \right]^{1/2}$$

Function $I_0(x)$ is the zeroth-order modified Bessel function of the first kind and can be evaluated by using the rapidly converging series

$$I_0(x) = 1 + \sum_{k=1}^{\infty} \left[\frac{1}{k!} \left(\frac{x}{2} \right)^k \right]^2$$

Table 6.1 Parameters of Window Functions

Type of window	Main-lobe width	Ripple ratio (%)		
		$N = 21$	$N = 41$	$N = 61$
Rectangular	$2\omega_s/N$	21.89	21.77	21.74
von Hann	$4\omega_s/N$	2.67	2.67	2.67
Hamming	$4\omega_s/N$	0.93	0.78	0.76
Blackman	$6\omega_s/N$	0.12	0.12	0.12

As in other window functions, the main-lobe width of the Kaiser window function can be adjusted by changing the value of N.

Another variable window function is one developed by Dolph for the design of antenna arrays [9]. This window is based on the properties of Chebyshev polynomials and is often referred to as the Dolph–Chebyshev window. For odd values of N, it is given by [10,11]

$$w_{DC}(nT) =$$

$$\begin{cases} \dfrac{1}{N}\left\{ \dfrac{1}{r} + 2\sum_{i=1}^{(N-1)/2} C_{N-1}\left[x_0 \cos\left(\dfrac{i\pi}{N}\right) \right] \cos\left(\dfrac{2i\pi n}{N}\right) \right\} & \text{for } |n| \le (N-1)/2 \\ 0 & \text{otherwise} \end{cases}$$

where

$$r = \frac{\text{amplitude of side lobes}}{\text{amplitude of main lobe}}$$

$$x_0 = \cosh\left[\frac{1}{(N-1)} \cosh^{-1}\left(\frac{1}{r}\right) \right]$$

where $C_k(x)$ is Chebyshev polynomial given by

$$C_k(x) = \begin{cases} \cos(k \cos^{-1} x) & \text{for } |x| \le 1 \\ \cosh(k \cosh^{-1} x) & \text{for } |x| > 1 \end{cases}$$

The most significant properties of the Dolph–Chebyshev window are as follows:

1. The amplitudes of the side lobes are all equal.
2. The main-lobe width is minimum for a given ripple ratio.
3. The ripple ratio can be independently assigned.

6.3.3 Design of 2-D Filters

Since the frequency response $H_2(e^{j\omega_1 T_1}, e^{j\omega_2 T_2})$ of a 2-D digital filter is a periodic function with respect to variables (ω_1, ω_2) with fundamental period $(2\pi/T_1, 2\pi/T_2)$, we can write

$$H_2(e^{j\omega_1 T_1}, e^{j\omega_2 T_2}) = \sum_{n_1=-\infty}^{\infty} \sum_{n_2=-\infty}^{\infty} h_2(n_1 T_1, n_2 T_2) e^{-j(\omega_1 n_1 T_1 + \omega_2 n_2 T_2)} \quad (6.18)$$

where

$$h_2(n_1T_1, n_2T_2) = \frac{1}{\omega_{s1}\omega_{s2}} \int_{-\omega_{s1}/2}^{\omega_{s1}/2} \int_{-\omega_{s2}/2}^{\omega_{s2}/2}$$
$$H_2(e^{j\omega_1 T_1}, e^{j\omega_2 T_2})e^{j(\omega_1 n_1 T_1 + \omega_2 n_2 T_2)} \, d\omega_1 \, d\omega_2 \quad (6.19)$$

provided that the preceding double summation converges. With the substitutions $e^{j\omega_1 T_1} = z_1$ and $e^{j\omega_2 T_2} = z_2$, we obtain

$$H_2(z_1, z_2) = \sum_{n_1=-\infty}^{\infty} \sum_{n_2=-\infty}^{\infty} h_2(n_1T_1, n_2T_2)z_1^{-n_1}z_2^{-n_2} \quad (6.20)$$

Hence, as with 1-D filters, if an expression is available for the frequency response, a transfer function can be obtained, which happens to be noncausal and of infinite order.

A finite-order transfer function can be obtained by letting

$$h_2(n_1T_1, n_2T_2) = 0 \quad \text{for } |n_1| > (N_1 - 1)/2 \text{ or } |n_2| > (N_2 - 1)/2$$

in Eq. (6.18) but as with 1-D filters, the truncation of the Fourier series tends to introduce Gibbs' oscillations.

A causal transfer function can be obtained by introducing delays of $(N_1 - 1)T_1/2$ and $(N_2 - 1)T_2/2$ in the impulse response with respect to the n_1T_1 and n_2T_2 axes. This is accomplished by modifying the transfer function as

$$H_2'(z_1, z_2) = z_1^{-(N_1-1)/2}z_2^{-(N_2-1)/2}H_2(z_1, z_2) \quad (6.21)$$

If the required frequency response is an even function with respect to ω_1 and ω_2, then $h_2(n_1T_1, n_2T_2)$ turns out to be an even function with respect to n_1T_1 and n_2T_2. Consequently, the filter represented by $H_2(z_1, z_2)$ has zero phase response and the filter represented by $H_2'(z_1, z_2)$ has linear phase response.

6.3.4 2-D Window Functions

The principles described in Sec. 6.3.2 can readily be applied for the elimination of Gibbs' oscillations in digital filters designed by using the 2-D Fourier series. This is accomplished by using 2-D window functions.

A 2-D window function, represented by $w_2(n_1T_1, n_2T_2)$, is a 2-D discrete function with the following space-domain properties:

1. It has a finite region of support, that is,

$$w_2(n_1T_1, n_2T_2) = 0 \quad \text{for } |n_1| > (N_1 - 1)/2 \text{ or } |n_2| > (N_2 - 1)/2$$

2. It is symmetrical with respect to the n_1T_1 and n_2T_2 axes, that is,

$$w_2(-n_1T_1, -n_2T_2) = w_2(n_1T_1, n_2T_2)$$

On the other hand, its amplitude spectrum consists of a 3-D main lobe centered at the origin of the (ω_1, ω_2) plane and a number of 3-D side lobes such that the volume of the side lobes is a small proportion of the volume of the main lobe.

The application of 2-D window functions is analogous to that in the case of 1-D filters. The impulse response obtained by applying the 2-D Fourier series is multiplied by the 2-D window function to obtain a modified impulse response as

$$h_{w2}(n_1T_1, n_2T_2) = w_2(n_1T_1, n_2T_2)h_2(n_1T_1, n_2T_2) \qquad (6.22)$$

and since the window function has finite support by virtue of property (1) above, a finite-order transfer function given by

$$H_{W2}(z_1, z_2)$$

$$= \sum_{n_1=-(N_1-1)/2}^{(N_1-1)/2} \sum_{n_2=-(N_2-1)/2}^{(N_2-1)/2} w_2(n_1T_1, n_2T_2)h_2(n_1T_1, n_2T_2)z_1^{-n_1}z_2^{-n_2} \qquad (6.23)$$

is obtained. If $h_2(n_1T_1, n_2T_2)$ is an even function with respect to n_1T_1 and n_2T_2, then $h_{w2}(n_1T_1, n_2T_2)$ is also an even function with respect to n_1T_1 and n_2T_2 by virtue of property (2) and, consequently, the filter represented by $H_{W2}(z_1, z_2)$ has zero phase response.

Through the application of the 2-D complex convolution given in Sec. 3.2.5, the frequency response of the modified filter can be expressed as

$$H_{w2}(e^{j\omega_1 T_1}, e^{j\omega_2 T_2}) = \frac{T_1 T_2}{4\pi^2} \int_0^{2\pi/T_2} \int_0^{2\pi/T_1} H_2(e^{j\Omega_1 T_1}, e^{j\Omega_2 T_2})$$

$$\times \; W_2(e^{j(\omega_1-\Omega_1)T_1}, e^{j(\omega_2-\Omega_2)T_2}) \; d\Omega_1 \, d\Omega_2 \qquad (6.24)$$

If $w_{1A}(n_1T_1)$ and $w_{1B}(n_2T_2)$ are 1-D windows of the type described in Sec. 6.3.2, then the function

$$w_2(n_1T_1, n_2T_2) = w_{1A}(n_1T_1)w_{1B}(n_2T_2) \qquad (6.25)$$

has the required properties and, therefore, it can serve as a 2-D window function.

Another way to construct satisfactory 2-D window functions was proposed by Huang [1]. In this approach, a 1-D window $w_1(nT)$ is transformed into a 2-D window $w_2(n_1T_1, n_2T_2)$ by letting

$$w_2(n_1T_1, n_2T_2) = w_1(nT)\big|_{nT=\sqrt{(n_1T_1)^2+(n_2T_2)^2}} \qquad (6.26)$$

The validity of Huang's approach is demonstrated in the following analysis.

Theorem 6.1 *Let $W_1(z)$ and $W_2(z_1, z_2)$ be the z transforms of $w_1(nT)$ and $w_2(n_1T_1, n_2T_2)$, respectively, and assume that $w_2(n_1T_1, n_2T_2)$ is obtained from $w_1(nT)$ by using Eq. (6.26). If $A_1(z_1)$ and $A_2(z_1, z_2)$ are allpass functions such that*

$$A_1(z_1) = A_2(z_1, z_2) = 1$$

for all z_1 and z_2, then

$$\frac{T_2}{2\pi} W_2(e^{j\omega_1 T_1}, e^{j\omega_2 T_2}) \otimes A_2(e^{j\omega_1 T_1}, e^{j\omega_2 T_2}) = W_1(e^{j\omega_1 T_1}) \otimes A_1(e^{j\omega_1 T_1})$$

where the symbol $\otimes$ represents convolution.

Proof. From Eq. (6.26), we note that

$$w_2(n_1T_1, 0) = w_1(n_1T_1) \tag{6.27}$$

and since

$$w_2(n_1T_1, n_2T_2) = \frac{1}{(2\pi j)^2} \oint_{\Gamma_2} \oint_{\Gamma_1} W_2(z_1, z_2) z_1^{n_1-1} z_2^{n_2-1} \, dz_1 \, dz_2 \tag{6.28}$$

and

$$w_1(n_1T_1) = \frac{1}{2\pi j} \oint_{\Gamma_1} W_1(z_1) z_1^{n_1-1} \, dz_1 \tag{6.29}$$

Eqs. (6.27)–(6.29) give

$$\frac{1}{(2\pi j)^2} \oint_{\Gamma_2} \oint_{\Gamma_1} W_2(z_1, z_2) z_1^{n_1-1} z_2^{-1} \, dz_1 \, dz_2 = \frac{1}{2\pi j} \oint_{\Gamma_1} W_1(z_1) z_1^{n_1-1} \, dz_1$$

or

$$\oint_{\Gamma_1} \left[\frac{1}{2\pi j} \oint_{\Gamma_2} W_2(z_1, z_2) z_2^{-1} \, dz_2 \right] z_1^{n_1-1} \, dz_1 = \oint_{\Gamma_1} W_1(z_1) z_1^{n_1-1} \, dz_1$$

and so

$$W_1(z_1) = \frac{1}{2\pi j} \oint_{\Gamma_2} W_2(z_1, z_2) z_2^{-1} \, dz_2$$

From the 1-D complex convolution, we can write

$$W_1(z_1) \otimes A_1(z_1) = \frac{1}{2\pi j} \oint_{\Gamma_1} W_1(v_1) A_1\left(\frac{z_1}{v_1}\right) \frac{dv_1}{v_1} = \frac{1}{2\pi j} \oint_{\Gamma_1} W_1(v_1) v_1^{-1} \, dv_1$$

and, similarly, from the 2-D complex convolution (see Sec. 3.2.5), we have

$$
W_2(z_1, z_2) \otimes A_2(z_1, z_2) = \frac{1}{(2\pi j)^2} \oint_{\Gamma_2} \oint_{\Gamma_1} W_2(v_1, v_2) A_2\left(\frac{z_1}{v_1}, \frac{z_2}{v_2}\right) \frac{dv_1}{v_1} \frac{dv_2}{v_2}
$$

$$
= \frac{1}{2\pi j} \oint_{\Gamma_1} \left[\frac{1}{2\pi j} \oint_{\Gamma_2} W_2(v_1, v_2) A_2\left(\frac{z_1}{v_1}, \frac{z_2}{v_2}\right) \frac{dv_2}{v_2} \right] \frac{dv_1}{v_1}
$$

$$
= \frac{1}{2\pi j} \oint_{\Gamma_1} \left[\frac{1}{2\pi j} \oint_{\Gamma_2} W_2(v_1, v_2) v_2^{-1} \, dv_2 \right] v_1^{-1} \, dv_1
$$

$$
= \frac{1}{2\pi j} \oint_{\Gamma_1} W_1(v_1) v_1^{-1} \, dv_1
$$

$$
= W_1(z_1) \otimes A_1(z_1)
$$

Hence, if we let $z_1 = e^{j\omega_1 T_1}$, $z_2 = e^{j\omega_2 T_2}$, $v_1 = e^{j\Omega_1 T_1}$, and $v_2 = e^{j\Omega_2 T_2}$, we obtain

$$
\frac{T_2}{2\pi} \int_0^{2\pi/T_2} \int_0^{2\pi/T_1} W_2(e^{j\Omega_1 T_1}, e^{j\Omega_2 T_2}) A_2(e^{j(\omega_1 - \Omega_1)T_1}, e^{j(\omega_2 - \Omega_2)T_2}) \, d\Omega_1 \, d\Omega_2
$$

$$
= \int_0^{2\pi/T_1} W_1(e^{j\Omega_1 T_1}) A_1(e^{j(\omega_1 - \Omega_1)T_1}) \, d\Omega_1
$$

or

$$
\frac{T_2}{2\pi} W_2(e^{j\omega_1 T_1}, e^{j\omega_2 T_2}) \otimes A_2(e^{j\omega_1 T_1}, e^{j\omega_2 T_2}) = W_1(e^{j\omega_1 T_1}) \otimes A_1(e^{j\omega_1 T_1}) \quad \blacksquare
$$

It will now be demonstrated that Eq. (6.26) provides a mechanism for generating good 2-D windows for the broader class of 2-D circularly symmetric filters with arbitrary piecewise-constant amplitude responses. Assume that $T_1 = T_2 = T$ and let

$$
H_1(e^{j\omega_1 T}) = \begin{cases} g & \text{for } \omega_{C1} \le |\omega_1| \le \omega_{C2} \\ 0 & \text{otherwise} \end{cases} \tag{6.30}
$$

be the frequency response of an arbitrary 1-D ideal bandpass digital filter, where g is a constant, $\omega_{C1} \ge 0$, and $\omega_{C2} \le \pi/T$, and let

$$
H_2(e^{j\omega_1 T}, e^{j\omega_2 T}) = \begin{cases} g & \text{for } \omega_{C1} \le \sqrt{\omega_1^2 + \omega_2^2} \le \omega_{C2} \\ 0 & \text{otherwise} \end{cases} \tag{6.31}
$$

be the frequency response of a corresponding 2-D circularly symmetric bandpass digital filter. If $w_1(nT)$ is a 1-D window function with a main-lobe width B such that $B \ll \omega_{C2} - \omega_{C1}$ and $w_2(n_1 T, n_2 T)$ is the 2-D window obtained by using Eq. (6.26), then Eqs. (6.15)–(6.17) and Eqs.

(6.22)–(6.24) yield the modified 1-D and 2-D frequency responses

$$H_{W1}(e^{j\omega_1 T}) = W_1(e^{j\omega_1 T}) \otimes H_1(e^{j\omega_1 T})$$

and

$$H_{W2}(e^{j\omega_1 T}, e^{j\omega_2 T}) = W_2(e^{j\omega_1 T}, e^{j\omega_2 T}) \otimes H_2(e^{j\omega_1 T}, e^{j\omega_2 T})$$

respectively. Under these circumstances, if $\omega_1 = \omega_{1k}$ is a frequency in the passband of the 1-D filter such that $\omega_{C1} + B/2 \leq \omega_{1k} \leq \omega_{C2} - B/2$, one can readily show that

$$H_{W1}(e^{j\omega_1 T}) \approx g W_1(e^{j\omega_1 T}) \otimes A_1(e^{j\omega_1 T})$$

(see Fig. 9.6 of [8]). Similarly, for any pair of frequencies (ω_1, ω_2) such that $\sqrt{\omega_1^2 + \omega_2^2} = \omega_{1k}$ and $\omega_{C1} + B/2 \leq \omega_{1k} \leq \omega_{C2} - B/2$, one can show that

$$H_{W2}(e^{j\omega_1 T}, e^{j\omega_2 T}) \approx g W_2(e^{j\omega_1 T}, g^{j\omega_2 T}) \otimes A_2(e^{j\omega_1 T}, e^{j\omega_2 T})$$

Therefore, from Theorem 6.1 we obtain

$$\frac{T}{2\pi} H_{W2}(e^{j\omega_1 T}, e^{j\omega_2 T}) \approx H_{W1}(e^{\omega_{1k} T})\big|_{\omega_{1k} = \sqrt{\omega_1^2 + \omega_2^2}}$$

In other words, if the 1-D window $w_1(nT)$ yields a good 1-D bandpass filter, then the 2-D window obtained by using Eq.(6.26) yields a good 2-D bandpass filter. Note that Eqs. (6.30)–(6.31) hold for lowpass filters (i.e., $\omega_{C1} = 0$ and $\omega_{C2} = \omega_C$) and, as can be readily demonstrated, they are also applicable to each passband in the case of filters with several passbands. Consequently, the window method can be used to design circularly symmetric filters with arbitrary piecewise-constant amplitude responses. The method is also quite effective for the design of other types of quadrantally symmetric filters, as will be demonstrated in Sec. 6.3.7.

6.3.5 Design of 2-D Circularly Symmetric Filters

In this section, the design of circularly symmetric lowpass, highpass, bandpass and bandstop filters is considered.

Lowpass Filters

Circularly symmetric lowpass filters can be designed by assuming an idealized frequency response of the form

$$H_I(e^{j\omega_1 T}, e^{j\omega_2 T}) = \begin{cases} 1 & \text{for } \sqrt{\omega_1^2 + \omega_2^2} \leq \omega_C \\ 0 & \text{otherwise} \end{cases} \tag{6.32}$$

where ω_C is the cutoff frequency. The impulse response of the filter can be determined as

$$h_l(n_1 T_1, n_2 T) = \frac{1}{(2\pi j)^2} \oint_{|z_2|=1} \oint_{|z_1|=1}$$
$$H_l(z_1, z_2) z_1^{n_1-1} z_2^{n_2-1} \, dz_1 \, dz_2$$

$$= \frac{T_2}{4\pi^2} \int_{-\pi/T}^{\pi/T} \int_{-\pi/T}^{\pi/T}$$
$$H_l(e^{j\omega_1 T}, e^{j\omega_2 T}) e^{j(\omega_1 n_1 T + \omega_2 n_2 T)} \, d\omega_1 \, d\omega_2 \qquad (6.33)$$

Using the variable transformation

$$\omega = \sqrt{\omega_1^2 + \omega_2^2}, \qquad \phi = \tan^{-1}\left(\frac{\omega_2}{\omega_1}\right)$$

and letting

$$\theta = \tan^{-1}\left(\frac{n_2}{n_1}\right)$$

the integral in Eq. (6.33) becomes (see [12], pp. 30–31)

$$h_l(n_1 T, n_2 T) = \frac{1}{4\pi^2} \int_0^{\omega_c T} \int_0^{2\pi} \omega e^{j\omega \sqrt{n_1^2 + n_2^2}\cos(\theta-\phi)} \, d\phi \, d\omega$$

$$= \frac{1}{2\pi} \int_0^{\omega_c T} \omega J_0(\omega \sqrt{n_1^2 + n_2^2}) \, d\omega$$

$$= \frac{\omega_c T}{2\pi} \frac{J_1(\omega_c T \sqrt{n_1^2 + n_2^2})}{\sqrt{n_1^2 + n_2^2}}$$

where

$$J_i(x) = x^i \sum_{k=0}^{\infty} \frac{(-1)^k x^{2k}}{2^{2k+i} k!(k+i)!}$$

is the ith-order Bessel function of the first kind. Since

$$\lim_{x \to 0} \frac{J_1(x)}{x} = 0.5$$

the impulse response of the lowpass filter can be put in the form

$$h_l(n_1 T, n_2 T) = \begin{cases} \dfrac{\omega_C^2 T^2}{4\pi} & (n_1, n_2) = (0, 0) \\[2ex] \dfrac{\omega_c T}{2\pi} \dfrac{J_1(\omega_c T \sqrt{n_1^2 + n_2^2})}{\sqrt{n_1^2 + n_2^2}} & \text{otherwise} \end{cases} \qquad (6.34)$$

Now by generating a 2-D window function by applying Eq. (6.26) to one of the 1-D window functions given in Sec. 6.3.2, the transfer function of the filter can be obtained as

$$H_{LP}(z_1, z_2) = \sum_{n_1 = -(N_1-1)/2}^{(N_1-1)/2} \sum_{n_2 = -(N_2-1)/2}^{(N_2-1)/2} h_{LP}(n_1 T_1, n_2 T_2) z_1^{-n_1} z_2^{-n_2} \quad (6.35)$$

where

$$h_{LP}(n_1 T_1, n_2 T_2) = w_2(n_1 T_1, n_2 T_2) h_I(n_1 T_1, n_2 T_2) \quad (6.36)$$

As can be seen in Eq. (6.34), $h_I(n_1 T, n_2 T)$ is an even function with respect to $n_1 T$ and $n_2 T$ and, therefore, $H_{LP}(z_1, z_2)$ represents a zero-phase filter. A causal transfer function can be formed as in Eq. (6.21).

Highpass Filters

If $H_{LP}(z_1, z_2)$ is the transfer function of a zero-phase circularly symmetric lowpass filter designed by the preceding approach, then

$$H_{LP}(e^{j\omega_1 T}, e^{j\omega_2 T}) \approx \begin{cases} 1 & \text{for } \sqrt{\omega_1^2 + \omega_2^2} \leq \omega_C \\ 0 & \text{otherwise} \end{cases} \quad (6.37)$$

Consequently, the transfer function

$$H_{HP}(z_1, z_2) = 1 - H_{LP}(z_1, z_2) \quad (6.38)$$

represents a zero-phase circularly symmetric highpass filter with frequency response

$$H_{HP}(e^{j\omega_1 T}, e^{j\omega_2 T}) \approx \begin{cases} 0 & \text{for } \sqrt{\omega_1^2 + \omega_2^2} \leq \omega_C \\ 1 & \text{otherwise} \end{cases}$$

Since the inverse z transform of unity is the impulse function, Eq. (6.38) gives the impulse response of the highpass filter as

$$h_{HP}(n_1 T, n_2 T) = \delta(n_1 T, n_2 T) - h_{LP}(n_1 T, n_2 T)$$

where $h_{LP}(n_1 T, n_2 T)$ is given by Eq. (6.36). Hence the transfer function of the filter can be obtained as

$$H_{HP}(z_1, z_2) = \sum_{n_1 = -(N_1-1)/2}^{(N_1-1)/2} \sum_{n_2 = -(N_2-1)/2}^{(N_2-1)/2} h_{HP}(n_1 T_1, n_2 T_2) z_1^{-n_1} z_2^{-n_2} \quad (6.39)$$

where

$$h_{HP}(n_1 T_1, n_2 T_2) = \begin{cases} 1 - h_{LP}(n_1 T_1, n_2 T_2) & \text{for } (n_1, n_2) = (0, 0) \\ -h_{LP}(n_1 T_1, n_2 T_2) & \text{otherwise} \end{cases}$$

(6.40)

Bandpass Filters

If $H_{LP1}(z_1, z_2)$ and $H_{LP2}(z_1, z_2)$ are the transfer functions of two zero-phase circularly symmetric lowpass filters with cutoff frequencies ω_{C1} and ω_{C2} such that $\omega_{C1} < \omega_{C2}$, then the transfer function

$$H_{BP}(z_1, z_2) = H_{LP2}(z_1, z_2) - H_{LP1}(z_1, z_2)$$

(6.41)

represents a zero-phase circularly symmetric bandpass filter such that

$$H_{BP}(e^{j\omega_1 T}, e^{j\omega_2 T}) \approx \begin{cases} 1 & \text{for } \omega_{C1} \leq \sqrt{\omega_1^2 + \omega_2^2} \leq \omega_{C2} \\ 0 & \text{otherwise} \end{cases}$$

Therefore, a bandpass filter can be obtained by designing two lowpass filters.

Bandstop Filters

If $H_{LP1}(z_1, z_2)$ and $H_{LP2}(z_1, z_2)$ represent two lowpass filters as in the case of bandpass filters, then from Eq. (6.38) the transfer function

$$H_{HP2}(z_1, z_2) = 1 - H_{LP2}(z_1, z_2)$$

(6.42)

represents a highpass filter with a cutoff frequency ω_{C2}. If $\omega_{C1} < \omega_{C2}$, then

$$H_{BS}(z_1, z_2) = H_{HP2}(z_1, z_2) + H_{LP1}(z_1, z_2)$$

(6.43)

represents a zero-phase circularly symmetric bandstop filter with a frequency response

$$H_{BP}(e^{j\omega_1 T}, e^{j\omega_2 T}) \approx \begin{cases} 0 & \text{for } \omega_{C1} \leq \sqrt{\omega_1^2 + \omega_2^2} \leq \omega_{C2} \\ 1 & \text{otherwise} \end{cases}$$

Hence, like bandpass filters, bandstop filters can be obtained by designing appropriate lowpass filters.

6.3.6   Design of 2-D Circularly Symmetric Filters Satisfying Prescribed Specifications

In many applications, filters are required that satisfy prescribed specifications. In these applications, the lowest filter order that would yield a satisfactory design should be used in order to minimize the amount of computation or the complexity of the hardware. Such a design can be obtained by means of a trial-and-error approach but if it were possible to predict the minimum filter order and the filter parameters that would yield the most economical design, considerable design effort would be eliminated.

Lowpass Filters

The design of 2-D circularly symmetric, nonrecursive, lowpass filters satisfying prescribed specifications can be carried out by using the Kaiser window in conjunction with prediction formulas for parameter α of the window and for the required filter order, as in the 1-D case [8,13]. Empirical prediction formulas for the 2-D case have been reported by Speake and Mersereau [14].

Let δ_{pl}, δ_{al}, ω_{pl}, and ω_{al} be the maximum passband error, the maximum stopband error, the passband edge, and the stopband edge of a circularly symmetric, nonrecursive, lowpass filter, respectively, as illustrated in Fig. 6.2, and assume that $\omega_{s1} = \omega_{s2} = \omega_s = 2\pi/T$. A filter of this type satisfying the set of specifications

$$S_{LP} = \{\delta_{pl}, \delta_{al}, \omega_{pl}, \omega_{al}, \omega_{s1}, \omega_{s2}\} \tag{6.44}$$

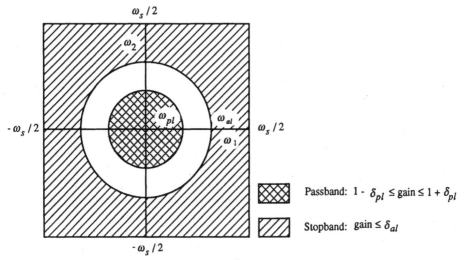

Figure 6.2 Specifications of lowpass filter.

can be obtained by the following step-by-step procedure:

1. Let

$$\omega_c = \tfrac{1}{2}(\omega_{pl} + \omega_{al})$$

and obtain the impulse response of an idealized lowpass filter using Eq. (6.34).

2. Choose window parameter α as follows [14]:

   (i) If the window function in Eq. (6.25) is used, let

   $$\alpha = \begin{cases} 0.42(A - 19.3)^{0.4} + 0.089(A - 19.3) & \text{if } 20 \le A \le 60 \\ 0 & \text{if } A < 20 \end{cases}$$

   where

   $$A = -20 \log\sqrt{\delta_p \delta_a}$$

   (ii) If the window function in Eq. (6.26) is used, let

   $$\alpha = \begin{cases} 0.56(A - 20.2)^{0.4} + 0.083(A - 20.2) & \text{if } 20 \le A \le 60 \\ 0 & \text{if } A < 20 \end{cases}$$

3. (i) If the window in Eq. (6.25) is used, select the lowest odd value of N such that

   $$N \ge \frac{(A - 8.0)\omega_s}{13.1B_t}$$

   where

   $$B_t = (\omega_{al} - \omega_{pl})$$

   (ii) If the window in Eq. (26) is used, select the lowest odd value of N such that

   $$N \ge \frac{(A - 7.0)\omega_s}{13.6B_t}$$

4. Form $w_2(n_1T, n_2T)$ using Eq. (6.25) or (6.26) where $w_1(nT)$, $w_{1A}(n_1T)$, and $w_{1B}(n_1T)$ are 1-D Kaiser window functions (see Sec. 6.3.2).

5. Obtain the transfer function of the lowpass filter using Eq. (6.35).

For circularly symmetric filters, the 2-D window function obtained by using Eq. (6.26) gives somewhat better results, as was demonstrated in [14], and is preferred.

Example 6.1 Design a circularly symmetric lowpass nonrecursive filter satisfying the following specifications:

- Maximum passband error $\leq 8.6 \times 10^{-3}$
- Maximum stopband error $\leq 1.0 \times 10^{-2}$
- Passband edge $= 1.5$ rad/s
- Stopband edge $= 2.5$ rad/s
- Sampling frequencies $= 10$ rad/s

Solution. The required lowpass filter must satisfy the specifications $\delta_{pl} = 8.6 \times 10^{-3}$, $\delta_{al} = 1.0 \times 10^{-2}$, $\omega_{pl} = 1.5$ rad/s, $\omega_{al} = 2.5$ rad/s, and $\omega_{s1} = \omega_{s2} = 10$ rad/s. From step 1 of the design procedure

$$T = \frac{2\pi}{\omega_s} = \frac{\pi}{5} \quad \text{and} \quad \omega_c = 2 \text{ rad/s}$$

and hence

$$h_I(n_1 T_1, n_2 T_2) = \begin{cases} 0.04\pi & \text{for } (n_1, n_2) = (0, 0) \\ \dfrac{0.2 J_1(0.4\pi\sqrt{n_1^2 + n_2^2})}{\sqrt{n_1^2 + n_2^2}} & \text{otherwise} \end{cases}$$

If the window function in Eq. (6.26) is used, step 2(ii) gives

$$A = -20 \log\sqrt{8.6 \times 10^{-3} \times 0.01} = 40.655 \text{ db}$$

and

$$\alpha = 0.56(40.655 - 20.2)^{0.4} + 0.083(40.655 - 20.2) = 3.57$$

From step 3(ii)

$$N \geq \frac{(40.655 - 7.0) \times 10}{13.6} = 24.75$$

and hence $N = 25$. Finally, steps 4 and 5 give

$$h_{LP}(n_1 T_1, n_2 T_2) =$$
$$\begin{cases} 0.04\pi & \text{for } (n_1, n_2) = (0, 0) \\ \dfrac{J_2}{\sqrt{n_1^2 + n_2^2} I_0(3.57)} & \text{for } \sqrt{n_1^2 + n_2^2} \leq 12, (n_1, n_2) \neq (0, 0) \\ 0 & \text{otherwise} \end{cases}$$

where

$$J_2 = 0.2 J_1(0.4\pi\sqrt{n_1^2 + n_2^2}) I_0[3.57\sqrt{1 - (n_1^2 + n_2^2)/144}]$$

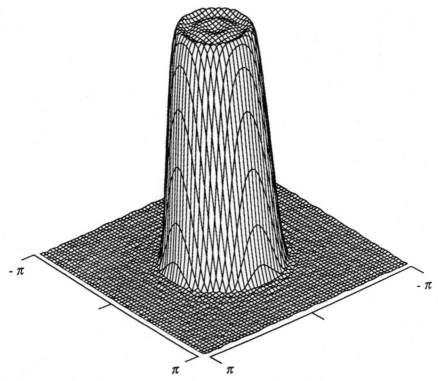

Figure 6.3 Amplitude response of lowpass filter (Example 6.1).

The amplitude response achieved is illustrated in Fig. 6.3. The maximum passband and stopband errors are 8.4×10^{-3} and 9.0×10^{-3}, respectively, and are less than the corresponding specified errors.

Highpass Filters

If $H_{LP}(z_1, z_2)$ is the transfer function of a zero-phase circularly symmetric lowpass filter with specifications S_{LP} as in Eq. (6.44), then from Fig. 6.4 we note that the transfer function of Eq. (6.38) represents a zero-phase circularly symmetric highpass filter with specifications

$$S_{HP} = \{\delta_{ah}, \delta_{ph}, \omega_{ah}, \omega_{ph}, \omega_{s1}, \omega_{s2}\}$$

where

$$\delta_{ph} = \delta_{al}, \qquad \delta_{ah} = \delta_{pl}$$

$$\omega_{ph} = \omega_{al}, \qquad \omega_{ah} = \omega_{pl}$$

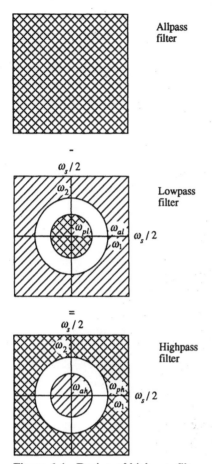

Figure 6.4 Design of highpass filters satisfying prescribed specifications.

Consequently, a highpass filter satisfying specifications S_{HP} can readily be obtained by designing a lowpass filter that would satisfy the specifications

$$S_{LP} = \{\delta_{ah}, \delta_{ph}, \omega_{ah}, \omega_{ph}, \omega_{s1}, \omega_{s2}\}$$

Example 6.2 Design a circularly symmetric highpass nonrecursive filter satisfying the following specifications:

- Maximum passband error $\leq 1.0 \times 10^{-2}$
- Maximum stopband error $\leq 8.6 \times 10^{-3}$
- Passband edge = 2.5 rad/s
- Stopband edge = 1.5 rad/s
- Sampling frequencies = 10 rad/s

Solution. The required lowpass filter must satisfy the specifications $\delta_{pl} =$ 8.6×10^{-3}, $\delta_{al} = 1.0 \times 10^{-2}$, $\omega_{pl} = 1.5$ rad/s, $\omega_{al} = 2.5$ rad/s, and ω_{s1} $= \omega_{s2} = 10$ rad/s and it is given in Example 6.1. The impulse response of the highpass filter can be obtained from Eq. (6.40). The amplitude response achieved is illustrated in Fig. 6.5 The maximum passband and stopband errors are 9.0×10^{-3} and 8.4×10^{-3}, respectively, and are less than the corresponding specified errors.

Bandpass Filters

If $H_{LP1}(z_1, z_2)$ and $H_{LP2}(z_1, z_2)$ are the transfer functions of two zero-phase circularly symmetric lowpass filters with specifications

$$S_{LP1} = \{\delta_{pl1} \, \delta_{al1}, \omega_{pl1}, \omega_{al1}, \omega_{s1}, \omega_{s2}\}$$

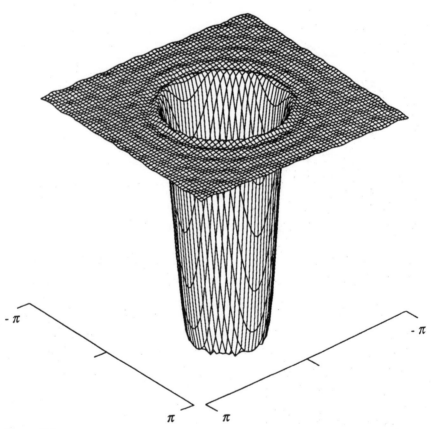

$-\pi$

$-\pi$

π π

Figure 6.5 Amplitude response of highpass filter (Example 6.2).

and

$$S_{LP2} = \{\delta_{p l2} \, \delta_{al2}, \, \omega_{p l2}, \, \omega_{al2}, \, \omega_{s1}, \, \omega_{s2}\}$$

respectively, where $\omega_{p l1} < \omega_{al1} < \omega_{p l2} < \omega_{al2}$, then the transfer function in Eq. (6.41) represents a zero-phase circularly symmetric bandpass filter. If its specifications are assumed to be

$$S_{BP} = \{\delta_{a1}, \, \delta_p, \, \delta_{a2}, \, \omega_{a1}, \, \omega_{p1}, \, \omega_{p2}, \, \omega_{a2}, \, \omega_{s1}, \, \omega_{s2}\}$$

where δ_p is the maximum passband error, δ_{a1} and δ_{a2} are the maximum stopband errors in the lower and upper stopbands, ω_{p1} and ω_{p2} are the lower and upper passband edges, ω_{a1} and ω_{a2} are the lower and upper stopband edges, respectively, then from Fig. 6.6

$$\delta_p \le \delta_{p l2} + \delta_{al1} \tag{6.45a}$$

$$\delta_{a1} \le \delta_{p l1} + \delta_{p l2} \tag{6.45b}$$

$$\delta_{a2} \le \delta_{al1} + \delta_{al2} \tag{6.45c}$$

$$\omega_{a1} = \omega_{p l1}, \qquad \omega_{p1} = \omega_{al1}$$

$$\omega_{p2} = \omega_{p l2}, \qquad \omega_{a2} = \omega_{al2}$$

Now if the maximum passband and stopband errors of the filter are required not to exceed $\hat{\delta}_p$, $\hat{\delta}_{a1}$, and $\hat{\delta}_{a2}$, Eqs. (6.45a)–(6.45c) yield

$$\delta_p \le \delta_{p l2} + \delta_{al1} \le \hat{\delta}_p$$

$$\delta_{a1} \le \delta_{p l1} + \delta_{p l2} \le \hat{\delta}_{a1}$$

$$\delta_{a2} \le \delta_{al1} + \delta_{al2} \le \hat{\delta}_{a2}$$

On assigning

$$\delta_{p l1} = \delta_{p l2} = \delta_{p l} \qquad \text{and} \qquad \delta_{al1} = \delta_{al2} = \delta_{al}$$

satisfactory values for $\delta_{p l}$ and δ_{al} can be determined as

$$\delta_{p l} = 0.5\gamma\hat{\delta}_{a1}, \qquad \delta_{al} = 0.5\gamma\hat{\delta}_{a2}$$

where

$$\gamma = \min\left(1, \frac{2\hat{\delta}_p}{\hat{\delta}_{a1} + \hat{\delta}_{a2}}\right)$$

The design can be completed by designing the two lowpass filters specified by S_{LP1} and S_{LP2}.

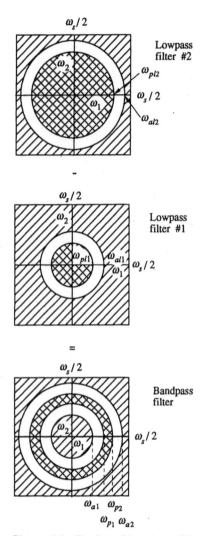

Figure 6.6 Design of bandpass filters satisfying prescribed specifications.

Example 6.3 Design a circularly symmetric bandpass nonrecursive filter satisfying the following specifications:

- Maximum passband error $\leq 1.84 \times 10^{-2}$
- Maximum stopband errors in lower and upper stopbands $\leq 2.0 \times 10^{-2}$
- Passband edges = 1.86, 3.0 rad/s
- Stopband edges = 1.0, 3.86 rad/s
- Sampling frequencies = 10 rad/s

Solution. The required lowpass filters must satisfy the specifications $\delta_{p l1}$ = $\delta_{a l1}$ = 9.2×10^{-3}, $\omega_{p l1}$ = 1.0 rad/s, $\omega_{a l1}$ = 1.86 rad/s, and $\delta_{p l2}$ = $\delta_{a l2}$ = 9.2×10^{-3}, $\omega_{p l2}$ = 3.0 rad/s, $\omega_{a l2}$ = 3.86 rad/s, respectively. From step 2(ii) of the design procedure, we obtain α = 3.58 and from step 3(ii) N = 29 for each filter. Therefore, Eqs. (6.36) and (6.41) yield

$$h_{BP}(n_1 T_1, n_2 T_2) =$$

$$\begin{cases} \dfrac{J_2 - J_3}{\sqrt{n_1^2 + n_2^2} I_0(3.58)} & \text{for } \sqrt{n_1^2 + n_2^2} \leq 14,\ (n_1, n_2) \neq (0, 0) \\ 0.3054 & \text{for } (n_1, n_2) = (0, 0) \\ 0 & \text{otherwise} \end{cases}$$

where

$$J_2 = 0.343\, J_1(0.686\pi\sqrt{n_1^2 + n_2^2}) I_0[3.58\sqrt{1 - (n_1^2 + n_2^2)/196}]$$

$$J_3 = 0.143\, J_1(0.286\pi\sqrt{n_1^2 + n_2^2}) I_0[3.58\sqrt{1 - (n_1^2 + n_2^2)/196}]$$

The amplitude response of the bandpass filter is illustrated in Fig. 6.7. The maximum errors in the lower stopband, passband, and upper stopband are 2.0×10^{-2}, 8.3×10^{-3}, and 1.22×10^{-2}, respectively, and are equal to or less than the corresponding specified errors.

Bandstop Filters

If $H_{LP1}(z_1, z_2)$ and $H_{LP2}(z_1, z_2)$ represent two lowpass filters as in the case of bandpass filters, then the transfer function $H_{HP2}(z_1, z_2)$ in Eq. (6.42) represents a highpass filter with passband and stopband edges $\omega_{a l2}$ and $\omega_{p l2}$, respectively, as depicted in Fig. 6.8. On the other hand, the transfer function in Eq. (6.43) represents a zero-phase circularly symmetric bandstop filter, as was shown earlier. If the specifications of the bandstop filter are assumed to be

$$S_{BS} = \{\delta_{p1}, \delta_a, \delta_{p2}, \omega_{p1}, \omega_{a1}, \omega_{a2}, \omega_{p2}, \omega_{s1}, \omega_{s2}\}$$

then from Fig. 6.8

$$\delta_{p1} \leq \delta_{p l1} + \delta_{p l2} \qquad (6.46a)$$

$$\delta_{p2} \leq \delta_{a l1} + \delta_{a l2} \qquad (6.46b)$$

$$\delta_a \leq \delta_{a l1} + \delta_{p l2} \qquad (6.46c)$$

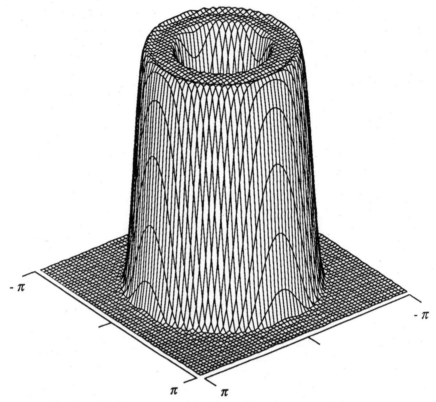

Figure 6.7 Amplitude response of bandpass filter (Example 6.3).

$$\omega_{p1} = \omega_{pl1}, \qquad \omega_{a1} = \omega_{al1}$$

$$\omega_{a2} = \omega_{pl2}, \qquad \omega_{p2} = \omega_{al2}$$

and if the maximum passband and stopband errors are required not to exceed $\hat{\delta}_{p1}$, $\hat{\delta}_{p2}$, and $\hat{\delta}_a$, Eqs. (6.46a)–(6.46c) yield

$$\delta_{p1} \leq \delta_{pl1} + \delta_{pl2} \leq \hat{\delta}_{p1}$$

$$\delta_{p2} \leq \delta_{al1} + \delta_{al2} \leq \hat{\delta}_{p2}$$

$$\delta_a \leq \delta_{al1} + \delta_{pl2} \leq \hat{\delta}_a$$

On assigning

$$\delta_{pl1} = \delta_{pl2} = \delta_{pl} \qquad \text{and} \qquad \delta_{al1} = \delta_{al2} = \delta_{al}$$

satisfactory values for δ_{pl} and δ_{al} can be determined as

$$\delta_{pl} = 0.5\gamma\hat{\delta}_{p1}, \qquad \delta_{al} = 0.5\gamma\hat{\delta}_{p2}$$

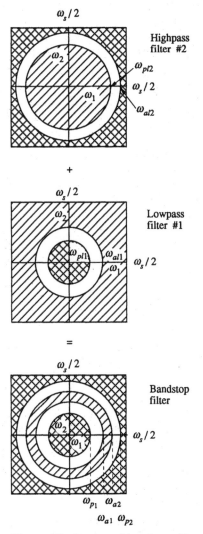

Figure 6.8 Design of bandstop filters satisfying prescribed specifications.

where

$$\gamma = \min\left(1, \frac{2\hat{\delta}_a}{\hat{\delta}_{p1} + \hat{\delta}_{p2}}\right)$$

As in the case of bandpass filters, the design can be completed by designing the two lowpass filters specified by S_{LP1} and S_{LP2}.

6.3.7 Design of Fan Filters

The window method is also very useful for the design of other types of quadrantally symmetric filters such as fan filters [15], as will now be demonstrated.

Consider an ideal fan filter characterized by the amplitude response illustrated in Fig. 6.9. If $v_1 = \omega_1 T_1$ and $v_2 = \omega_2 T_2$, then the impulse response of the filter is given by

$$h_I(n_1 T_1, n_2 T_2) = \frac{1}{(2\pi)^2} \int_{-\pi}^{\pi} \int_{-\pi}^{\pi} H_I(e^{jv_1}, e^{jv_2}) e^{j(n_1 v_1 + n_2 v_2)} \, dv_1 \, dv_2$$

where

$$H_I(e^{jv_1}, e^{jv_2}) = \begin{cases} 1 & \text{for } -kv_1 \le v_2 \le kv_1, \, v_1 \ge 0 \\ & \text{or } kv_1 \le v_2 \le -kv_1, \, v_1 < 0 \\ 0 & \text{otherwise} \end{cases}$$

with $k = \tan \theta$. Thus

$$h_I(n_1 T_1, n_2 T_2) = \frac{1}{(2\pi)^2} \left[\int_{-\pi}^{0} e^{jn_1 v_1} \left(\int_{kv_1}^{-kv_1} e^{jn_2 v_2} \, dv_2 \right) dv_1 \right.$$

$$\left. + \int_{0}^{\pi} e^{jn_1 v_1} \left(\int_{-kv_1}^{kv_1} e^{jn_2 v_2} dv_2 \right) dv_1 \right]$$

$$= \begin{cases} k/2 & \text{for } (n_1, n_2) = (0, 0) \\ J_2 & n_1 \ne 0, \, n_2 = 0 \\ 0 & \text{for } n_2 \ne 0, \, n_1^2 - n_2^2 k^2 = 0 \\ J_3 & \text{otherwise} \end{cases} \quad (6.47)$$

where

$$J_2 = \frac{k[(-1)^{n_1} - 1]}{n_1^2 \pi^2}$$

$$J_3 = \frac{k}{(n_2^2 k^2 - n_1^2) \pi^2} - \frac{1}{2n_2 \pi^2} \left\{ \frac{\cos[(n_2 k - n_1)\pi]}{n_2 k - n_1} + \frac{\cos[(n_2 k + n_1)\pi]}{n_2 k + n_1} \right\}$$

Since the filter being designed is quadrantally symmetric, it appears appropriate to use the rectangular 2-D window of Eq. (6.25). Hence we can write

$$h_F(n_1 T_1, n_2 T_2) = w_{1A}(n_1 T_1) w_{1B}(n_2 T_2) h_I(n_1 T_1, n_2 T_2) \quad (6.48)$$

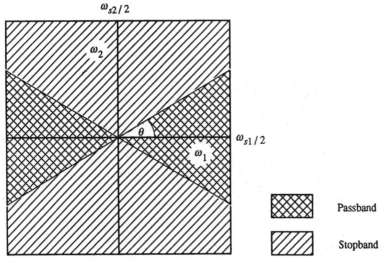

Figure 6.9 Ideal amplitude response of fan filter.

Example 6.4 Design a 29×29 fan filter with $k = 0.35$ using sampling frequencies of 10 rad/s. Use the Kaiser window function with $\alpha = 4.0$.

Solution. The impulse response of the filter is obtained from Eqs. (6.47) and (6.48). The amplitude response achieved is illustrated in Fig. 6.10. The maximum passband and stopband errors are 4.38×10^{-2} and 2.33×10^{-2}, respectively.

6.4 DESIGN BASED ON TRANSFORMATIONS

Another approach to the design of nonrecursive filters consists of designing a 1-D nonrecursive filter and then applying an appropriate transformation to the transfer function obtained. A notable design method of this category that has been investigated by several researchers [4,5,16] is based on the McClellan transformation [3]. The method is simple to apply and leads to linear-phase quadrantally symmetric nonrecursive filters that can be realized very efficiently as was shown by McClellan and Chan [17] (see Sec. 10.2).

6.4.1 The McClellan Transformation

Consider a 1-D zero-phase nonrecursive filter characterized by the transfer function

$$H_1(z) = \sum_{n=-(N-1)/2}^{(N-1)/2} h_1(nT)z^{-n}$$

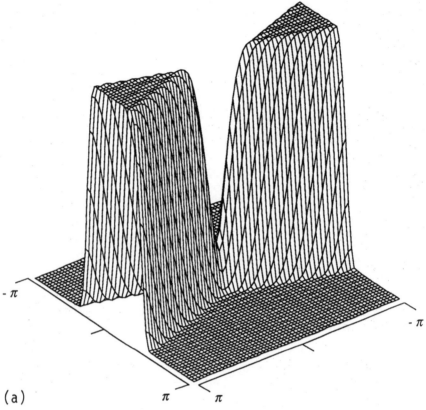

-π

-π

(a)   π   π

Figure 6.10 Amplitude response of fan filter (Example 6.4): (a) 3-D pilot; (b) contour plot.

where

$$h_1(nT) = h_1(-nT), \qquad n = 0, 1, \ldots, (N-1)/2$$

The frequency response of the filter can be expressed as

$$H_1(e^{j\omega T}) = \sum_{n=0}^{(N-1)/2} a_0(nT)\cos \omega nT \qquad (6.49)$$

where

$$a_0(nT) = \begin{cases} h_1(0) & \text{for } n = 0 \\ 2h_1(nT) & \text{otherwise} \end{cases}$$

For any integer n, $\cos \omega nT$ can be expressed in terms of an nth-order polynomial of $x = \cos \omega T$ known as a Chebyshev polynomial and usually

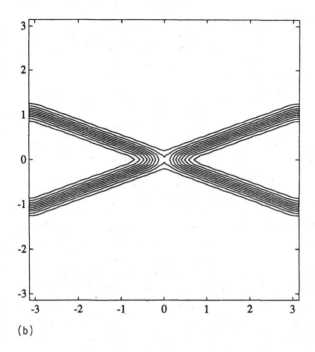

(b)

denoted by $C_n(x)$. Consequently, Eq. (6.49) can be written as

$$H_1(e^{j\omega T}) = \sum_{n=0}^{(N-1)/2} a_0(nT)C_n(\cos \omega T)$$

$$= \sum_{n=0}^{(N-1)/2} a_1(nT)(\cos \omega T)^n \quad (6.50)$$

where constants $a_1(nT)$ for $n = 0, 1, \ldots, (N + 1)/2$ can be obtained from constants $a_0(nT)$. The McClellan transformation is given by

$$\cos \omega T = A \cos \omega_1 T_1 + B \cos \omega_2 T_2$$
$$+ C \cos \omega_1 T_1 \cos \omega_2 T_2 + D \quad (6.51)$$

$$= F(\omega_1 T_1, \omega_2 T_2)$$

where A, B, C, and D are constants. Applying the McClellan transformation to the frequency response of a 1-D nonrecursive filter given in Eq. (6.50) yields

$$H_2(e^{j\omega_1 T_1}, e^{j\omega_2 T_2}) = \sum_{n_1=0}^{(N-1)/2} \sum_{n_2=0}^{(N-1)/2} a_2(nT)(\cos \omega_1 T_1)^{n_1}(\cos \omega_2 T_2)^{n_2} \quad (6.52)$$

and since $(\cos \omega_i T^i)^{n_i}$, for $i = 1, 2$, can be expressed as linear combinations of $\cos \omega_i T_i$, $\cos 2\omega_i T_i$, ..., $\cos n_i \omega_i T_i$, Eq. (6.52) can be written as

$$H_2(e^{j\omega_1 T_1}, e^{j\omega_2 T_2}) = \sum_{n_1=0}^{(N-1)/2} \sum_{n_2=0}^{(N-1)/2} a_3(nT) \cos n_1 \omega_1 T_1 \cos n_2 \omega_2 T_2 \quad (6.53)$$

This expression may be considered as the frequency response of a 2-D zero-phase nonrecursive digital filter. A corresponding 2-D transfer function $H_2(z_1, z_2)$ can be obtained by substituting $(z_i^{n_i} + z_i^{-n_i})/2$ for $\cos n_i \omega_i T_i$.

From Eq. (6.51), ω_2 can be expressed as a function of ω and ω_1 as

$$\omega_2 = \frac{1}{T_2} \cos^{-1}\left(\frac{\cos \omega T - A \cos \omega_1 T_1 - D}{B + C \cos \omega_1 T_1}\right) \quad (6.54)$$

For a fixed ω, this equation describes a curve in the (ω_1, ω_2) plane. If the McClellan transformation is used to derive a 2-D digital filter from a corresponding 1-D digital filter, then the frequency response of the 2-D filter at every point on the curve defined by Eq. (6.54) will be constant and equal to the frequency response of the 1-D filter at frequency ω. As ω is varied in the range 0 to $\omega_s/2$, a family of contours is produced that describes the frequency response of the 2-D digital filter. By choosing suitable values for constants A, B, C, and D, the transformation can be forced to map the frequency interval $0 \leq \omega \leq \omega_s/2$ onto the region $0 \leq \omega_1 \leq \omega_{s1}/2$, $0 \leq \omega_2 \leq \omega_{s2}/2$ of the (ω_1, ω_2) plane. For example, if $A = B = C = -D = 0.5$, the McClellan transformation maps the frequency inteval $0 \leq \omega \leq \omega_s/2$ onto a set of contours that appear to be circular at low frequencies, as depicted in Fig. 6.11.

6.4.2 Choice of Transformation Parameters

As was demonstrated in the previous section, the type of mapping produced by the McClellan transformation and, in turn, the frequency response of the 2-D filter obtained tend to depend heavily on the choice of transformation parameters. In this section, a technique for the choice of the transformation parameters due to Mersereau, Mecklenbraüker, and Quatieri [4] is described.

Let us assume that a 1-D nonrecursive lowpass filter with passband edge ω_p is available and a corresponding 2-D nonrecursive lowpass filter is to be designed such that the contours of the amplitude response are concentric ellipses about the origin of the (ω_1, ω_2) plane. The frequency point $\omega = \omega_p$ in the 1-D filter will map onto a contour in the (ω_1, ω_2) plane if

$$\cos \omega_p T = A \cos \omega_1 T_1 + B \cos \omega_2 T_2$$
$$+ C \cos \omega_1 T_1 \cos \omega_2 T_2 + D \quad (6.55)$$

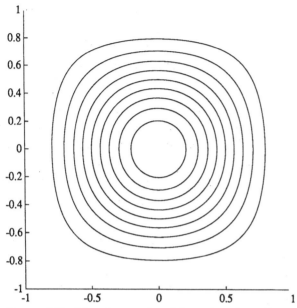

Figure 6.11 Mapping produced by using $A = B = C = -D = 0.5$ in the McClellan transformation.

If the passband boundary in the 2-D filter is assumed to be the ellipse described by

$$\frac{\omega_1^2}{\omega_{p1}^2} + \frac{\omega_2^2}{\omega_{p2}^2} = 1 \qquad (6.56)$$

then

$$\omega_2 = \omega_{p2}\sqrt{1 - (\omega_1/\omega_{p1})^2}$$

and if the frequency point $\omega = \omega_p$ in the 1-D filter is required to map onto the ellipse in Eq. (6.56), then

$$\cos \omega_p T = A \cos \omega_1 T_1 + B \cos[T_2\omega_{p2}\sqrt{1 - (\omega_1/\omega_{p1})^2}]$$

$$+ C \cos \omega_1 T_1 \cos[T_2\omega_{p2}\sqrt{1 - (\omega_1/\omega_{p1})^2}] + D \qquad (6.57)$$

The McClellan transformation should, in addition, map the point $\omega = 0$ onto the origin of the (ω_1, ω_2) plane, and hence Eq. (6.51) gives

$$1 = A + B + C + D \qquad (6.58)$$

and on eliminating C in Eq. (6.57), the error function

$$E(\omega_1) = \cos \omega_p T - A \cos \omega_1 T_1 - B \cos[T_2\omega_{p2}\sqrt{1 - (\omega_1/\omega_{p1})^2}]$$

$$+ (1 - A - B - D) \cos \omega_1 T_1 \cos[T_2\omega_{p2}\sqrt{1 - (\omega_1/\omega_{p1})^2}] - D \tag{6.59}$$

can be constructed. Choosing parameters A, B, ... so as to reduce the preceding error function with respect to some range of ω_1 results in a transformation that has the desired properties in the chosen frequency range. If a lowpass filter with elliptical amplitude-response contours within the boundary defined by Eq. (6.56) is required, the parameters A, B, ... can be determined by minimizing

$$e_\infty = \lim_{p \to \infty} \left[\int_0^{\omega_{p1}} |E(\omega_1)|^p \, d\omega_1 \right]^{1/p}$$

$$= \max_{0 \le \omega_1 \le \omega_{p1}} |E(\omega_1)| \tag{6.60}$$

or

$$e_2 = \int_0^{\omega_{p1}} E^2(\omega_1) \, d\omega_1 \tag{6.61}$$

The quantity e_∞ is a nonlinear function of the parameters A, B, ... and can be minimized by using one of the standard iterative minimax optimization techniques [18,19] (see Sec. 8.3). Quantity e_2, on the other hand, turns out to be a quadratic function of the transformation parameters, which can be readily minimized in closed form. If we let

$$\alpha = \cos \omega_p T$$

$$\beta = \cos \omega_1 T_1$$

$$\gamma = \cos[T_2\omega_{p2}\sqrt{1 - (\omega_1/\omega_{p1})^2}]$$

$$\delta = \cos \omega_1 T_1 \cos[T_2\omega_{p2}\sqrt{1 - (\omega_1/\omega_{p1})^2}]$$

Eq. (6.59) gives

$$\begin{aligned}
E^2(\omega_1) = & (\beta - \delta)^2 A^2 + (\gamma - \delta)^2 B^2 + (1 - \delta)^2 D^2 \\
& + 2(\beta\gamma + \delta^2 - \beta\delta - \gamma\delta)AB + 2(\beta + \delta^2 - \beta\delta - \delta)AD \\
& + 2(\gamma + \delta^2 - \gamma\delta - \delta)BD + 2(\beta\delta - \alpha\beta - \delta^2 + \alpha\delta)A \\
& + 2(\gamma\delta - \alpha\gamma - \delta^2 + \alpha\delta)B + 2(\delta - \alpha - \delta^2 + \alpha\delta)D \\
& + (\alpha^2 - \alpha\delta + \delta^2)
\end{aligned}$$

Therefore, from Eq. (6.61)

$$e_2 = \tfrac{1}{2}\mathbf{x}^T\mathbf{Q}\mathbf{x} - \mathbf{b}^T\mathbf{x} + k \tag{6.62}$$

where $\mathbf{x} = [A \quad B \quad D]^T$, $\mathbf{Q} = \{q_{ij}\}$ is a 3×3 positive-definite symmetric matrix, $\mathbf{b} = -[b_1 \quad b_2 \quad b_3]^T$, and

$$k = \int_0^{\omega_{p1}} (\alpha^2 - \alpha\delta + \delta^2)\, d\omega_1$$

with

$$q_{11} = 2\int_0^{\omega_{p1}} (\beta - \delta)^2 d\omega_1, \qquad q_{12} = 2\int_0^{\omega_{p1}} (\beta\gamma + \delta^2 - \beta\delta - \gamma\delta)\, d\omega_1$$

$$q_{13} = 2\int_0^{\omega_{p1}} (\beta + \delta^2 - \beta\delta - \delta)\, d\omega_1, \qquad q_{22} = 2\int_0^{\omega_{p1}} (\gamma - \delta)^2\, d\omega_1$$

$$q_{23} = 2\int_0^{\omega_{p1}} (\gamma + \delta^2 - \gamma\delta - \delta)\, d\omega_1, \qquad q_{33} = 2\int_0^{\omega_{p1}} (1 - \delta)^2\, d\omega_1$$

$$b_1 = 2\int_0^{\omega_{p1}} (\beta\delta - \alpha\beta - \delta^2 + \alpha\delta)\, d\omega_1,$$

$$b_2 = 2\int_0^{\omega_{p1}} (\gamma\delta - \alpha\gamma - \delta^2 + \alpha\delta)\, d\omega_1$$

$$b_3 = 2\int_0^{\omega_{p1}} (\delta - \alpha - \delta^2 + \alpha\delta)\, d\omega_1$$

The preceding integrals can be evaluated by using one of the standard numerical methods. On applying Newton's method, the solution of the minimization problem is obtained as

$$\mathbf{x} = \mathbf{Q}^{-1}\mathbf{b} \tag{6.63}$$

This equation along with Eq. (6.58) give the required transformation parameters.

While Eq. (6.51) maps points in the frequency range $0 \le \omega \le \omega_s/2$ onto contours of the (ω_1, ω_2) plane, sometimes there are regions of the (ω_1, ω_2) plane that do not correspond to real frequencies in the range $0 \le \omega \le \omega_s/2$. This problem can arise if the right-hand side of the McClellan transformation does not satisfy the constraint

$$-1 \le F(\omega_1 T_1, \omega_2 T_2) \le 1 \tag{6.64}$$

The problem can be overcome, however, without considerable difficulty by modifying function $F(\omega_1 T_1, \omega_2 T_2)$ somewhat. If we let

$$F'(\omega_1 T_1, \omega_2 T_2) = c_1 F(\omega_1 T_1, \omega_2 T_2) - c_2 \qquad (6.65)$$

where

$$c_1 = \frac{2}{F_{max} - F_{min}}, \qquad c_2 = c_1 F_{max} - 1$$

and F_{max} and F_{min} are the maximum and minimum values of $F(\omega_1 T_1, \omega_2 T_2)$, respectively, for $|\omega_1| \le \omega_{s1}/2$ and $|\omega_2| \le \omega_{s2}/2$, then we have

$$- 1 \le F'(\omega_1 T_1, \omega_2 T_2) \le 1$$

as required. Now applying the transformation

$$\cos \omega' T = F'(\omega_1 T_1, \omega_2 T_2) \qquad (6.66)$$

to $H_1(e^{j\omega' T})$ yields a 2-D digital filter, as before. Furthermore, like the original transformation, Eq. (6.66) will map points in the frequency range $0 \le \omega \le \omega_s/2$ onto concentric contours. Since c_1 and c_2 are constants, the new contours will be scaled versions of the contours produced by the original transformation, as can be verified. Consequently, the passband edge of the 1-D filter that was used to calculate the transformation parameters, namely ω_p, no longer corresponds to the passband boundary of the 2-D filter given by Eq. (6.56). This problem can be overcome by determining ω_p', the required scaled passband edge, in terms of ω_p. From Eqs. (6.66), (6.65), and (6.51)

$$\cos \omega' T = c_1 F(\omega_1 T_1, \omega_2 T_2) - c_2$$

$$= c_1 \cos \omega T - c_2$$

and, therefore,

$$\omega_p' = \frac{1}{T} \cos^{-1}(c_1 \cos \omega_p T - c_2) \qquad (6.67)$$

As may have been noted, the 1-D prototype filter is not needed in the determination of the transformation parameters. The designer simply guesses the required value of ω_p and proceeds with the design. Once parameters $A, B, \ldots$ are determined, the maximum and minimum values of $F(\omega_1 T_1, \omega_2 T_2)$ are calculated. If the constraint in Eq. (6.64) is satisfied, then the assumed value of ω_p is correct. On the other hand, if the constraint in Eq. (6.64) is violated, then the transformation is modified and the new passband edge ω_p', is determined.

6.4.3 Design

The design of a 2-D lowpass filter with concentric amplitude-response contours and a passband boundary defined by Eq. (6.56) can be completed by using the following steps:

1. Assume a passband edge ω_p for the 1-D filter and determine parameters $A, B, \ldots$ using the technique of the previous section.
2. Calculate the maximum and minimum values of $F(\omega_1 T_1, \omega_2 T_2)$. If the constraint in Eq. (6.64) is not satisfied, recalculate the passband edge of the 1-D filter using Eq. (6.67).
3. Design a 1-D nonrecursive zero-phase filter of order $N - 1$ with passband edge ω_p or ω'_p, as appropriate.
4. Calculate parameters $a_0(nT)$ in Eq. (6.49).
5. Calculate parameters $a_1(nT)$ in Eq. (6.50).
6. Apply the McClellan transformation to Eq. (6.50) and calculate parameters $a_2(nT)$ in Eq. (6.52).
7. Calculate parameters $a_3(nT)$ in Eq. (6.53).
8. Replace $\cos n_i \omega_i T_i$ by $(z_i^{n_i} + z_i^{-n_i})/2$ in Eq. (6.53) to obtain the transfer function $H_2(z_1, z_2)$.

In step 5, it is necessary to determine the coefficients of the Chebyshev polynomial $C_n(x)$. This can be done by noting that

$$C_0 = 1, \qquad C_1 = x$$

and then using the recursion formula

$$C_i(x) = 2xC_{i-1}(x) - C_{i-2}(x) \qquad \text{for } i = 2, 3, \ldots, n$$

(see Sec. 6.3.2). The design is illustrated by the following example.

Example 6.5 Design a 2-D nonrecursive circularly symmetric lowpass filter with a passband boundary defined by Eq. (6.56) with $\omega_{p1} = \omega_{p2} = 2$ rad/s. Assume that $\omega_p = 2.5$ rad/s and $\omega_s = \omega_{s1} = \omega_{s2} = 10$ rad/s.

Solution. Numerical integration yields

$$Q = \begin{bmatrix} 0.8445 & 0.1096 & 1.0441 \\ 0.1096 & 0.2651 & 0.4360 \\ 1.0441 & 0.4360 & 1.6617 \end{bmatrix} \qquad \text{and} \qquad b = \begin{bmatrix} 0.5514 \\ 0.2525 \\ 0.9135 \end{bmatrix}$$

The column vector x that minimizes e_2 given by Eq. (6.62) is obtained from Eq. (6.63) as

$$x = \begin{bmatrix} A \\ B \\ D \end{bmatrix} = \begin{bmatrix} 0.9386 \\ 0.9386 \\ -1.3858 \end{bmatrix}$$

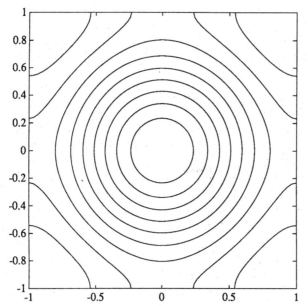

Figure 6.12 Contour plot for McClellan transformation of Example 6.5.

and from Eq. (6.58)

$$C = 0.5086$$

Hence Eq. (6.51) becomes

$$\cos 0.2\pi\omega = 0.9386 \cos 0.2\pi\omega_1 + 0.9386 \cos 0.2\pi\omega_2 \qquad (6.68)$$
$$+ 0.5086 \cos 0.2\pi\omega_1 \cos 0.2\pi\omega_2 - 1.3858$$

The maximum and minimum values of $F(\omega_1 T_1, \omega_2 T_2)$ can be obtained as

$$F_{max} = 1.0 \qquad \text{and} \qquad F_{min} = -2.7544$$

and hence a modified transformation must be used with

$$c_1 = 0.5327 \qquad \text{and} \qquad c_2 = -0.4673$$

On using Eqs. (6.65), (6.66), and (6.68) and then replacing ω' by ω, we have

$$\cos 0.2\pi\omega = 0.5 \cos 0.2\pi\omega_1 + 0.5 \cos 0.2\pi\omega_2$$
$$+ 0.2709 \cos 0.2\pi\omega_1 \cos 0.2\pi\omega_2 - 0.2709$$

The contour plot achieved is illustrated in Fig. 6.12.

The new passband edge of the 1-D filter ω_p' is obtained from Eq. (6.67) as 1.7261 rad/s. On designing a 1-D nonrecursive zero-phase filter with this passband edge using any one of the standard methods, the design of the 2-D filter can be accomplished.

REFERENCES

1. T. S. Huang, Two-dimensional windows, *IEEE Trans. Audio Electroacoust.*, vol. AU-20, pp. 88–89, March 1972.
2. J. H. McClellan and T. W. Parks, Equiripple approximation of fan filters, *Geophysics*, vol. 7, pp. 573–583, 1972.
3. J. H. McClellan, The design of two-dimensional filters by transformations, *Proc. 7th Annual Princeton Conf. Information Sciences and Systems*, pp. 247–251, 1973.
4. R. M. Mersereau, W. F. G. Mecklenbräuker, and T. F. Quatieri, Jr., McClellan Transformations for two-dimensional digital filtering: I—Design, *IEEE Trans. Circuits Syst.*, vol. CAS-23, pp. 405–413, July 1976.
5. W. F. G. Mecklenbräuker and R. M. Mersereau, McClellan Transformations for two-dimensional digital filtering: II—Implementation, *IEEE Trans. Circuits Syst.*, vol. CAS-23, pp. 414–422, July 1976.
6. T. S. Huang, J. W. Burnett, and A. G. Deczky, The importance of phase in image processing filters, *IEEE Trans. Acoust., Speech, Signal Process.*, vol. ASSP-23, pp. 529–542, Dec. 1975.
7. C. Charalambous, The performance of an algorithm of minimax design of two-dimensional linear phase FIR digital filters, *IEEE Trans. Circuit Syst.*, vol. CAS-32, pp. 1016–1028, Oct. 1985.
8. A. Antoniou, *Digital Filters: Analysis and Design*, New York: McGraw-Hill, 1979 (2nd ed. in press).
9. C. L. Dolph, A current distribution of broadside arrays which optimizes the relationship between beamwidth and side lobe levels, *Proc. IRE*, vol. 34, pp. 335–348, June 1946, and vol. 35, pp. 489–492, May 1947.
10. D. Barbiere, A method for calculating the current distribution of Tschebyscheff arrays, *Proc. IRE*, vol. 40, pp. 78–82, Jan. 1952.
11. R. J. Stegen, Excitation coefficients and beamwidth of Tschebyscheff arrays, *Proc. IRE*, vol. 41, pp. 1671–1674, Nov. 1953.
12. D. E. Dudgeon and R. M. Mersereau, *Multidimensional Digital Signal Processing*, Englewood Cliffs, N.J.: Prentice-Hall, 1984.
13. J. F. Kaiser, Nonrecursive digital filter design using the I_0-sinh window function, *Proc. 1974 IEEE Int. Symp. Circuit Theory*, pp. 20–23.
14. T. C. Speake and R. M. Mersereau, A note on the use of windows for two-dimensional FIR filter design, *IEEE Trans. Acoust., Speech, Signal Process.*, vol. ASSP-29, pp. 125–127, Feb. 1981.
15. A. Antoniou and W.-S. Lu, Design of 2-D nonrecursive filters using window method, *IEE Proc.*, vol. 137, pt. G, pp. 247–250, Aug. 1990.

16. D. T. Nguyen and M. N. S. Swamy, Approximation design of 2-D digital filters with elliptical magnitude response of arbitrary orientation, *IEEE Trans. Circuits Syst.*, vol. CAS-33, pp. 597–603, June 1986.

17. J. H. McClellan and D. K. S. Chan, A 2-D FIR filter structure derived from the Chebyshev recursion, *IEEE Trans. Circuits Syst.*, vol. CAS-24, pp. 372–378, July 1977.

18. C. Charalambous, A unified review of optimization, *IEEE Trans. Microwave Theory Tech.*, vol. MTT-22, pp. 289–300, March 1974.

19. C. Charalambous, Acceleration of the least pth algorithm for minimax optimization with engineering applications, *Mathematical Programming*, vol. 17, pp. 270–297, 1979.

PROBLEMS

6.1 Verify Eqs. (6.10) and (6.11).

6.2 Design a circularly symmetric lowpass nonrecursive filter satisfying the following specifications:

Maximum passband error $\leq 5.0 \times 10^{-2}$
Maximum stopband error $\leq 5.0 \times 10^{-2}$
Passband edge = 1.5 rad/s
Stopband edge = 2.5 rad/s
Sampling frequencies = 10 rad/s

Compare the results with those obtained in Example 6.1.

6.3 Design a circularly symmetric lowpass nonrecursive filter satisfying the following specifications:

Maximum passband error $\leq 5.0 \times 10^{-3}$
Maximum stopband error $\leq 5.0 \times 10^{-3}$
Passband edge = 1.5 rad/s
Stopband edge = 2.5 rad/s
Sampling frequencies = 10 rad/s

Compare the results with those obtained in Example 6.1.

6.4 Design a circularly symmetric highpass nonrecursive filter satisfying the following specifications:

Maximum passband error $\leq 5.0 \times 10^{-2}$
Maximum stopband error $\leq 5.0 \times 10^{-2}$
Passband edge = 2.5 rad/s
Stopband edge = 1.5 rad/s
Sampling frequencies = 10 rad/s

Compare the results with those obtained in Example 6.2.

6.5 Design a circularly symmetric bandpass nonrecursive filter satisfying the following specifications:

Maximum passband error $\leq 5.0 \times 10^{-2}$
Maximum stopband error $\leq 5.0 \times 10^{-2}$
Passband edge $= 2.0, 3.0$ rad/s
Stopband edge $= 1.2, 3.8$ rad/s
Sampling frequencies $= 10$ rad/s

6.6 Design a 21×21 fan filter with $k = 0.35$ using sampling frequencies of 10 rad/s. Use the Kaiser window function with $\alpha = 4.0$.

6.7 Design a circularly symmetric lowpass filter with a passband boundary defined by Eq. (6.56) with $\omega_{p1} = \omega_{p2} = 3.5$ rad/s. Assume that $\omega_p = 3.0$ rad/s and $\omega_s = \omega_{s1} = \omega_{s2} = 10$ rad/s.

6.8 Modify, if necessary, the design procedure described in Sec. 6.4.3 to make it applicable for the design of circularly symmetric highpass filters.

6.9 Modify the design procedure described in Sec. 6.4.3 to make it applicable for the design of circularly symmetric bandpass filters.

7

Approximations for Recursive Filters

7.1 INTRODUCTION

Nonrecursive filters are always stable and can readily be designed to have linear phase response, as was demonstrated in Chap. 6. Unfortunately, the selectivity that can be achieved with these filters is limited; consequently, in applications where rapid transitions are required between passbands and stopbands, large filter orders would be required to meet the specifications. As a result, the amount of computation involved in the application of these filters can be prohibitively high. This problem can be overcome by using recursive instead of nonrecursive filters, which are capable of providing much higher selectivity. The increased selectivity in these filters is achieved by utilizing the degrees of freedom provided by the coefficients of the denominator polynomial of the transfer function.

The approximation problem for recursive filters can be solved by applying transformations to 1-D continuous or discrete transfer functions [1–9], by using optimization techniques [10,11] or by applying transformations in conjunction with optimization techniques [12]. Methods based on transformations are easy to apply and involve an insignificant amount of computational effort relative to other methods. Therefore, if they are applicable to the filter design problem at hand, these methods are preferred.

In the first half of this chapter, methods for the design of recursive filters whose passbands or stopbands are rectangular domains or combinations

176

of rectangular domains are considered. Then methods for designing filters whose passbands or stopbands are circular domains are considered. These filters are used in applications where the processing is required to be independent of direction with respect to the origin of the (ω_1, ω_2) plane. The chapter concludes by describing a collection of 1-D transformations due to Constantinides [13] that are also applicable in the design of 2-D digital filters.

7.2 TRANSFORMATIONS

Methods based on transformations can readily be used for the design of filters having piecewise-constant amplitude responses with quadrantal or half-plane symmetry. Methods of this class are often extensions of 1-D methods and can lead to a variety of designs [2,7]. The most frequently used transformation is the bilinear transformation. However, other types of transformations are also used, ranging from linear transformations that map frequency axes onto frequency axes and transformations that can rotate a given coordinate plane. In this section, some of the available transformations are examined.

7.2.1 Bilinear Transformation

Given a 1-D continuous transfer function $H_A(s)$, a corresponding 1-D discrete transfer function $H_D(z)$ can be generated by applying a transformation of the form

$$s = f(z) \tag{7.1}$$

to $H_A(s)$, that is,

$$H_D(z) = H_A(s)|_{s=f(z)}$$

The preceding transformation will yield a useful 1-D discrete transfer function if the following requirements are satisfied:

1. A stable 1-D continuous transfer function (i.e., one that represents a stable analog filter) yields a stable 1-D discrete transfer function.
2. A realizable 1-D continuous transfer function (i.e., one with real coefficients) yields a realizable 1-D digital filter.
3. Some basic features of the 1-D continuous transfer function are preserved in the 1-D discrete transfer function (e.g., a passband at low frequencies or a certain minimum attenuation in a given frequency range).

If these requirements are satisfied, then an inverse function of the form $z = f'(s)$ exists, and the transformation in Eq. (7.1) will map one or more regions of the s plane onto one or more regions of the z plane and vice versa. An important transformation of this class is the bilinear transformation given by

$$s = \frac{2}{T}\left(\frac{z-1}{z+1}\right) \tag{7.2}$$

As can be easily shown [14], the bilinear transformation maps

1. The left half of the s plane onto the interior of the unit circle of the z plane
2. The imaginary axis of the s plane onto the unit circle of the z plane
3. The right half of the s plane onto the exterior of the unit circle of the z plane

as depicted in Fig. 7.1. Under these circumstances, if

$$M_{1i} \leq |H_A(j\omega)| \leq M_{2i} \quad \text{for } \omega_{1i} \leq \omega \leq \omega_{2i}$$

then

$$M_{1i} \leq |H_D(e^{j\Omega T})| \leq M_{2i} \quad \text{for } \Omega_{1i} \leq \Omega \leq \Omega_{2i}$$

for $i = 1, 2, \ldots , n_B$, where n_B is the number of frequency bands of interest and

$$\Omega_{1i} = \frac{2}{T_1} \tan^{-1} \frac{\omega_{1i}T_1}{2} \quad \text{and} \quad \Omega_{2i} = \frac{2}{T_2} \tan^{-1} \frac{\omega_{2i}T_2}{2}$$

(see Fig. 7.8 of Antoniou [14]). In effect

1) Passbands in the analog filter translate into passbands in the digital filter.
2) Stopbands in the analog filter translate into stopbands in the digital filter.
3) The magnitudes of passband ripples in the analog filter are preserved in the digital filter.
4) The minimum stopband attenuations in the analog filter are preserved in the digital filter.

For these reasons, the bilinear transformation method is one of the most effective methods for the design of 1-D recursive digital filters.

7.2.2 Linear Transformations

In the design of 2-D recursive filters, linear transformations are often required that can be used to map frequency axes onto frequency axes in

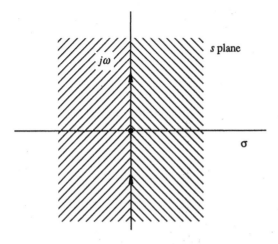

(a)

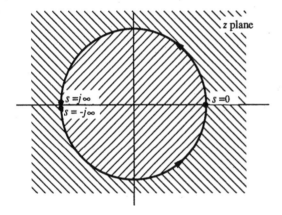

(b)

Figure 7.1 Bilinear transformation: (a) s plane; (b) z plane.

the (ω_1, ω_2) plane [15,16]. Transformations of this class are of the form

$$\begin{bmatrix} \bar{\omega}_1 \\ \bar{\omega}_2 \end{bmatrix} := \begin{bmatrix} \sigma_1 \bar{\omega}_1 \\ \sigma_2 \bar{\omega}_2 \end{bmatrix} \qquad (7.3a)$$

and

$$\begin{bmatrix} \bar{\omega}_1 \\ \bar{\omega}_2 \end{bmatrix} := \begin{bmatrix} \sigma_2 \bar{\omega}_2 \\ \sigma_1 \bar{\omega}_1 \end{bmatrix} \qquad (7.3b)$$

where the symbol $:=$ stands for *is replaced by*, σ_1 and σ_2 are integers that can assume values of -1 or $+1$, and $\overline{\omega}_1$ and $\overline{\omega}_2$ are normalized frequencies given by

$$\overline{\omega}_1 = \frac{\omega_1}{\omega_{s1}} \quad \text{and} \quad \overline{\omega}_2 = \frac{\omega_2}{\omega_{s2}}$$

(ω_{s1} and ω_{s2} are the sampling frequencies). Evidently, eight distinct transformations are possible and each can be expressed as

$$\begin{bmatrix} \overline{\omega}_1 \\ \overline{\omega}_2 \end{bmatrix} := \mathbf{D}(T) \begin{bmatrix} \overline{\omega}_1 \\ \overline{\omega}_2 \end{bmatrix} \tag{7.4}$$

where $\mathbf{D}(T)$ is a 2×2 unitary matrix representing transformation T. The eight transformations can be named after their effect on the $(\overline{\omega}_1, \overline{\omega}_2)$ plane and each can be represented by a distinct symbol, as follows:

1. Identity (I)
2. Reflection about the $\overline{\omega}_1$ axis ($\rho_{\overline{\omega}1}$)
3. Reflection about the $\overline{\omega}_2$ axis ($\rho_{\overline{\omega}2}$)
4. Reflection about the ψ_1 axis ($\rho_{\psi1}$)
5. Reflection about the ψ_2 axis ($\rho_{\psi2}$)
6. Counterclockwise rotation by $90°$ (R_4)
7. Counterclockwise rotation by $180°$ (R_4^2)
8. Counterclockwise rotation by $270°$ (R_4^3)

In the preceding symbolic representation, ψ_1 and ψ_2 represent axes that are rotated by $45°$ in the counterclockwise sense relative to the $\overline{\omega}_1$ and $\overline{\omega}_2$ axes, respectively, and R_k denotes rotation by $360°/k$ in the counterclockwise sense. Successive applications of a transformation can be represented by powers of operators (e.g., $\rho_{\overline{\omega}1}^2$, R_4^2). The matrices for the various transformations are as follows:

$$\mathbf{D}(I) = \begin{bmatrix} 1 & 0 \\ 0 & 1 \end{bmatrix}, \quad \mathbf{D}(\rho_{\overline{\omega}1}) = \begin{bmatrix} 1 & 0 \\ 0 & -1 \end{bmatrix},$$

$$\mathbf{D}(\rho_{\overline{\omega}2}) = \begin{bmatrix} -1 & 0 \\ 0 & 1 \end{bmatrix}, \quad \mathbf{D}(\rho_{\psi1}) = \begin{bmatrix} 0 & 1 \\ 1 & 0 \end{bmatrix} \tag{7.5a}$$

$$\mathbf{D}(\rho_{\psi2}) = \begin{bmatrix} 0 & -1 \\ -1 & 0 \end{bmatrix}, \quad \mathbf{D}(R_4) = \begin{bmatrix} 0 & -1 \\ 1 & 0 \end{bmatrix},$$

$$\mathbf{D}(R_4^2) = \begin{bmatrix} -1 & 0 \\ 0 & -1 \end{bmatrix}, \quad \mathbf{D}(R_4^3) = \begin{bmatrix} 0 & 1 \\ -1 & 0 \end{bmatrix} \tag{7.5b}$$

If transformation γ is applied to a function $F(\bar{\omega}_1, \bar{\omega}_2)$, a new function $\hat{F}(\bar{\omega}_1, \bar{\omega}_2)$ can be obtained as

$$\hat{F}(\bar{\omega}_1, \bar{\omega}_2) = T_\gamma[F(\bar{\omega}_1, \bar{\omega}_2)]$$

where T_γ is an operator. For example, if $\gamma = \rho_{\bar{\omega}1}$, then

$$\hat{F}(\bar{\omega}_1, \bar{\omega}_2) = T_{\rho_{\bar{\omega}1}}[F(\bar{\omega}_1, \bar{\omega}_2)] = F(\bar{\omega}_1, -\bar{\omega}_2)$$

If two transformations, say γ and δ, are applied to $F(\bar{\omega}_1, \bar{\omega}_2)$ in sequence, we obtain

$$\hat{F}(\bar{\omega}_1, \bar{\omega}_2) = T_\gamma[T_\delta[F(\bar{\omega}_1, \bar{\omega}_2)]] \tag{7.6}$$

It can readily be shown that for any pair in the preceding set of transformations

$$\hat{F}(\bar{\omega}_1, \bar{\omega}_2) = T_\gamma T_\delta[F(\bar{\omega}_1, \bar{\omega}_2)] \tag{7.7}$$

where $T_\gamma T_\delta$ represents the single transformation characterized by $D(\gamma) \cdot D(\delta)$ and, therefore, Eqs. (7.5) and (7.6) give

$$T_\gamma T_\delta[F(\bar{\omega}_1, \bar{\omega}_2)] = T_\gamma[T_\delta[F(\bar{\omega}_1, \bar{\omega}_2)]]$$

This equation defines an associative operation that can be referred to as "multiplication." By using Eq. (7.7), a multiplication table can be constructed for the preceding set of transformations, as illustrated in Table 7.1.

The following observations can be made at this point:

1. A set of eight elements has been identified that includes an identity element by definition.
2. An associative operation, referred to as multiplication, has been defined for the set.

Table 7.1 Multiplication Table of Group

	I	$\rho_{\bar{\omega}1}$	$\rho_{\bar{\omega}2}$	$\rho_{\psi1}$	$\rho_{\psi2}$	R_4	R_4^2	R_4^3
I	I	$\rho_{\bar{\omega}1}$	$\rho_{\bar{\omega}2}$	$\rho_{\psi1}$	$\rho_{\psi2}$	R_4	R_4^2	R_4^3
$\rho_{\bar{\omega}1}$	$\rho_{\bar{\omega}1}$	I	R_4^2	R_4^3	R_4	$\rho_{\psi2}$	$\rho_{\bar{\omega}2}$	$\rho_{\psi1}$
$\rho_{\bar{\omega}2}$	$\rho_{\bar{\omega}2}$	R_4^2	I	R_4	R_4^3	$\rho_{\psi1}$	$\rho_{\bar{\omega}1}$	$\rho_{\psi2}$
$\rho_{\psi1}$	$\rho_{\psi1}$	R_4	R_4^3	I	R_4^2	$\rho_{\bar{\omega}1}$	$\rho_{\psi2}$	$\rho_{\bar{\omega}2}$
$\rho_{\psi2}$	$\rho_{\psi2}$	R_4^3	R_4	R_4^2	I	$\rho_{\bar{\omega}2}$	$\rho_{\psi1}$	$\rho_{\bar{\omega}1}$
R_4	R_4	$\rho_{\psi1}$	$\rho_{\psi2}$	$\rho_{\bar{\omega}2}$	$\rho_{\bar{\omega}1}$	R_4^2	R_4^3	I
R_4^2	R_4^2	$\rho_{\bar{\omega}2}$	$\rho_{\bar{\omega}1}$	$\rho_{\psi2}$	$\rho_{\psi1}$	R_4^3	I	R_4
R_4^3	R_4^3	$\rho_{\psi2}$	$\rho_{\psi1}$	$\rho_{\bar{\omega}1}$	$\rho_{\bar{\omega}2}$	I	R_4	R_4^2

3. A multiplication table (Table 7.1) has been constructed for the set in which each row and each column include the identity element. Hence, an inverse element exists for each element. For example, if

$$T_\gamma T_\delta[F(\bar{\omega}_1, \bar{\omega}_2)] = T_l[F(\bar{\omega}_1, \bar{\omega}_2)] \tag{7.8}$$

we can write

$$\mathbf{D}(\gamma) \cdot \mathbf{D}(\delta) = \mathbf{D}(I)$$

By premultiplying by $\mathbf{D}^{-1}(\gamma)$ or postmultiplying by $\mathbf{D}^{-1}(\delta)$, we have

$$\mathbf{D}(\delta) = \mathbf{D}^{-1}(\gamma) \quad \text{or} \quad \mathbf{D}(\gamma) = \mathbf{D}^{-1}(\delta)$$

and so

$$T_\delta[F(\bar{\omega}_1, \bar{\omega}_2)] = T_\gamma^{-1}[F(\bar{\omega}_1, \bar{\omega}_2)]$$

or

$$T_\gamma[F(\bar{\omega}_1, \bar{\omega}_2)] = T_\delta^{-1}[F(\bar{\omega}_1, \bar{\omega}_2)]$$

that is, if Eq. (7.8) is satisfied, T_δ is the inverse of T_γ, and vice versa.
4. Each and every product is an element of the set, as can be seen in Table 7.1, and hence the set is closed.

Therefore, the set of transformations, together with the multiplication operation, constitutes a group in the algebraic sense since all the required postulates (associativity, closure, existence of identity element, existence of inverse elements) are satisfied. The group defined is isomorphic to the dihedral group of order 8 under multiplication [17,18].

An important property of the group is that each transformation distributes over a product of functions of $\bar{\omega}_1$ and $\bar{\omega}_2$, that is,

$$T_\gamma\left[\prod_{i=1}^{K}F_i(\bar{\omega}_1, \bar{\omega}_2)\right] = \prod_{i=1}^{K}T_\gamma[F_i(\bar{\omega}_1, \bar{\omega}_2)] \tag{7.9}$$

The validity of this property follows from the definition of the transformations.

The preceding frequency transformations translate readily into equivalent space-domain transformations. Let $G(z_1, z_2)$ be the z transform of $g(n_1, n_2)$. If $z_1 = e^{j\omega_1 T_1}$, $z_2 = e^{j\omega_2 T_2}$, and T_γ is a transformation of the type given by Eq. (7.3a), then

$$T_\gamma G(e^{j\omega_1 T_1}, e^{j\omega_2 T_2}) = \sum_{n_1=-\infty}^{\infty}\sum_{n_2=-\infty}^{\infty} g(n_1, n_2)T_\gamma[e^{-j\omega_1 n_1 T_1}e^{-j\omega_2 n_2 T_2}]$$

$$= \sum_{n_1=-\infty}^{\infty}\sum_{n_2=-\infty}^{\infty} g(n_1, n_2)T_\gamma(e^{-j2\pi\bar{\omega}_1 n_1}e^{-j2\pi\bar{\omega}_2 n_2}]$$

$$= \sum_{n_1=-\infty}^{\infty} \sum_{n_2=-\infty}^{\infty} g(n_1, n_2) e^{-j2\pi\sigma_1\bar{\omega}_1 n_1} e^{-j2\pi\sigma_2\bar{\omega}_2 n_2}$$

$$= \sum_{n_1=-\infty}^{\infty} \sum_{n_2=-\infty}^{\infty} g(\sigma_1 n_1, \sigma_2 n_2) e^{-j2\pi\bar{\omega}_1 n_1} e^{-j2\pi\bar{\omega}_2 n_2}$$

$$= \sum_{n_1=-\infty}^{\infty} \sum_{n_2=-\infty}^{\infty} T'_\gamma g(n_1, n_2) e^{-j\omega_1 n_1 T_1} e^{-j\omega_2 n_2 T_2}$$

$$= \mathcal{L} T'_\gamma [g(n_1, n_2)]|_{z_1 = e^{-j\omega_1 T_1}, \ z_2 = e^{-j\omega_2 T_2}}$$

where T'_γ is a transformation of the form

$$\begin{bmatrix} n_1 \\ n_2 \end{bmatrix} := \begin{bmatrix} \sigma_1 n_1 \\ \sigma_2 n_2 \end{bmatrix} \tag{7.10a}$$

Similarly, if T_δ is a transformation of the type given by Eq. (7.3b), it can be shown that

$$T_\delta G(e^{j\omega_1 T_1}, e^{j\omega_2 T_2}) = \mathcal{L} T'_\delta [g(n_1, n_2)]|_{z_1 = e^{-j\omega_1 T_1}, \ z_2 = e^{-j\omega_2 T_2}}$$

where T'_δ is a transformation of the form

$$\begin{bmatrix} n_1 \\ n_2 \end{bmatrix} := \begin{bmatrix} \sigma_2 n_2 \\ \sigma_1 n_1 \end{bmatrix} \tag{7.10b}$$

If $H(z_1, z_2)$ represents a causal filter with impulse response $h(n_1, n_2)$, then the filter represented by $H(z_1^{-1}, z_2^{-1})$ has an impulse response

$$\mathcal{L}^{-1} H(z_1^{-1}, z_2^{-1}) = \mathcal{L}^{-1}[T_{R_4^2} H(z_1, z_2)]$$

$$= T_{R_4^2} \mathcal{L}^{-1} H(z_1, z_2)$$

$$= h(-n_1, -n_2)$$

and is, therefore, noncausal since $h(-n_1, -n_2) \neq 0$ for $n_1 < 0$ and $n_2 < 0$. Such a filter can be implemented in terms of the causal transfer function $H(z_1, z_2)$, as will now be demonstrated. The noncausal filter of Fig. 7.2a can be represented by the equation

$$Y(z_1, z_2) = H(z_1^{-1}, z_2^{-1}) X(z_1, z_2) \tag{7.11}$$

From Table 7.1, we note that

$$T_{R_4^2} T_{R_4^2} = T_I$$

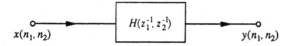

$x(n_1, n_2)$ $y(n_1, n_2)$

(a)

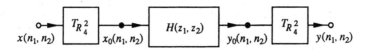

$x(n_1, n_2)$ $x_0(n_1, n_2)$ $y_0(n_1, n_2)$ $y(n_1, n_2)$

(b)

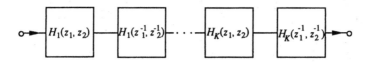

(c)

Figure 7.2 Design of zero-phase filters: (a) noncausal filter; (b) equivalent causal implementation; (c) cascade zero-phase filter.

Hence, Eqs. (7.9) and (7.11) give

$$Y(z_1, z_2) = T_{R_4^2}T_{R_4^2}[H(z_1^{-1}, z_2^{-1})X(z_1, z_2)]$$
$$= T_{R_4^2}[H(z_1, z_2)T_{R_4^2}X(z_1, z_2)]$$
$$= T_{R_4^2}[H(z_1, z_2)X_0(z_1, z_2)]$$
$$= T_{R_4^2}Y_0(z_1, z_2)$$

where

$$X_0(z_1, z_2) = T_{R_4^2}X(z_1, z_2)$$

and

$$Y_0(z_1, z_2) = H(z_1, z_2)X_0(z_1, z_2)$$

From Eq. (7.10a), we note that $x_0(n_1, n_2) = x(-n_1, -n_2)$ and $y(n_1, n_2)$ $= y_0(-n_1, -n_2)$ and, therefore, the noncausal filter can be implemented by rotating the n_1 and n_2 axes of the input signal by 180°, processing the rotated signal by the causal filter, and then rotating the axes of the output signal by 180°, as depicted in Fig. 7.2b. If the data are given in matrix form, then the n_1 and n_2 axes of the input signal can be rotated 180° by reversing, first, the order of rows and, second, the order of columns, or vice versa. Evidently, this type of processing can be carried out only if all the data to be processed are available at the start of the processing.

Noncausal filters can be used for the realization of zero-phase filters [1]. Consider a cascade arrangement of K pairs of filters whose transfer functions are $H_i(z_1, z_2)$ and $H_i(z_1^{-1}, z_2^{-1})$ for $i = 1, 2, \ldots, K$, as depicted in Fig 7.2c. The overall transfer function of the cascade arrangement can be expressed as

$$H(z_1, z_2) = \prod_{i=1}^{K} H_i(z_1, z_2)H_i(z_1^{-1}, z_2^{-1})$$

and if $z_1 = e^{j\omega_1 T_1}$ and $z_2 = e^{j\omega_2 T_2}$, then

$$H(e^{j\omega_1 T_1}, e^{j\omega_2 T_2}) = \prod_{i=1}^{K} |H_i(e^{j\omega_1 T_1}, e^{j\omega_2 T_2})|^2$$

since

$$H_i(e^{-j\omega_1 T_1}, e^{-j\omega_2 T_2}) = H_i^*(e^{j\omega_1 T_1}, e^{j\omega_2 T_2})$$

Evidently, the phase shift of the filter is zero and, consequently, the cascade arrangement of Fig. 7.2c, like a linear-phase filter, can be used in applications where phase distortion is objectionable. As was stated earlier, all the data must be available at the start of the processing and, furthermore, the output arrays of the various filter sections must be large enough to ensure that the effect of truncating the outputs of the sections is negligible. The sizes of the output arrays can be kept small by using sections whose impulse responses reduce to zero fairly quickly.

7.2.3 Analog-Filter Transformations

The design of analog filters is usually accomplished by applying analog-filter transformations to normalized continuous lowpass transfer functions like those obtained by using the Bessel, Butterworth, Chebyshev, and elliptic approximations. These transformations are of the form

$$s = f(\bar{s})$$

and can be used to design lowpass, highpass, bandpass, or bandstop filters satisfying arbitrary piecewise-constant amplitude-response specifications. Through the application of the bilinear transformation, corresponding 1-D digital filters can be designed (see pp. 196–214 of [14]), and since 2-D digital filters can be designed in terms of 1-D digital filters, these transformations are of considerable importance in the design of 2-D digital filters as well.

Another group of transformations of interest in the design of 2-D digital filters consists of transformations that can be used to derive 2-D continuous transfer functions from corresponding 1-D continuous transfer functions. These are of the form

$$s = g(s_1, s_2)$$

Two transformations of this type whose application for the design of 2-D filters was first suggested by Shanks, Treitel, and Justice [1] are

$$s = g_1(s_1, s_2) = -s_1 \sin \beta + s_2 \cos \beta \tag{7.12a}$$

$$s = g_2(s_1, s_2) = s_1 \cos \beta + s_2 \sin \beta \tag{7.12b}$$

If $H_{A1}(s)$ is a 1-D continuous transfer function, then a corresponding 2-D continuous transfer function can be generated as

$$H_{A2}(s_1, s_2) = H_{A1}(s)|_{s = g_1(s_1, s_2)} \tag{7.13}$$

Now if $s = j\omega$, $s_1 = j\omega_1$, and $s_2 = j\omega_2$, we have

$$|H_{A2}(j\omega_1, j\omega_2)| = |H_{A1}(j\omega)|$$

provided that

$$\omega_2 = \omega_1 \tan \beta + \frac{\omega}{\cos \beta}$$

Evidently, if $\omega = \hat{\omega}_i$ for $i = 1, 2, \ldots$, are specific frequencies of interest in the amplitude response of the 1-D analog filter, as depicted in Fig. 7.3a, then each $\hat{\omega}_i$ will give rise to a contour in the amplitude response of the 2-D analog filter that is a straight line rotated by an angle β with respect to the ω_1 axis. The contours obtained for $\beta = 0°$ and $45°$ are illustrated in Fig. 7.3b and c, respectively. For $\beta = 0$, Eqs. (7.12a) and (7.13) give

$$H_{A2}(s_1, s_2) = H_{A1}(s_2)$$

and, therefore, for $\beta \neq 0$, $H_{A2}(s_1, s_2)$ may be regarded as a rotated version of the 2-D transfer function $H_{A1}(s_2)$ (which happens to be independent of s_1).

Similarly, if the transformation in Eq. (7.12b) is used, frequencies $\hat{\omega}_i$ give rise to contours in the amplitude response of the 2-D analog filter that

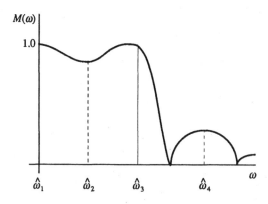

(a)

Figure 7.3 1-D to 2-D analog-filter transformation: (a) amplitude response of 1-D analog filter; (b) contour plot of 2-D digital filter, $\beta = 0°$; (c) contour plot of 2-D digital filter, $\beta = 45°$.

are straight lines rotated by an angle β with respect to the ω_2 axis, as can easily be demonstrated. In this case, $H_{A2}(s_1, s_2)$ may be regarded as a rotated version of $H_{A1}(s_1)$. Filters obtained by using the transformations in Eq. (7.12) are sometimes said to be rotated filters.

As can be seen in Fig. 7.3c, a 1-D analog lowpass filter will yield a 2-D analog filter that has a lowpass amplitude response with respect to both the ω_1 and ω_2 axes. Consequently, if the bilinear transformation of Eq. (7.2) is applied to $H_{A2}(s_1, s_2)$, we obtain

$$H_{D2}(z_1, z_2) = H_{A2}(s_1, s_2)\big|_{s_1 = \frac{2}{T_1}\left(\frac{z_1-1}{z_1+1}\right),\ s_2 = \frac{2}{T_2}\left(\frac{z_2-1}{z_2+1}\right)} \qquad (7.14)$$

Since the bilinear transformation will map the $j\omega_1$ and $j\omega_2$ axes of the s_1 and s_2 planes onto the unit circles of the z_1 and z_2 planes, respectively, the 2-D digital filter obtained will have a lowpass amplitude response on the Ω_1 and Ω_2 axes of the (Ω_1, Ω_2) plane, where

$$\Omega_1 = \frac{2}{T_1} \tan^{-1} \frac{\omega_1 T_1}{2} \qquad \text{and} \qquad \Omega_2 = \frac{2}{T_2} \tan^{-1} \frac{\omega_2 T_2}{2}$$

The mapping of the amplitude response of the 1-D analog filter is illustrated in Fig. 7.4. As can be seen, the contours in the amplitude response of the digital filter are not linear, owing to the warping effect [14] and, furthermore, they converge onto the Nyquist point $(\omega_{s1}/2, \omega_{s2}/2)$ since the infinite ω_1 and ω_2 axes map onto the finite frequency ranges $-\omega_{s1}/2$ to $\omega_{s1}/2$ and $-\omega_{s2}/2$ to $\omega_{s2}/2$, respectively.

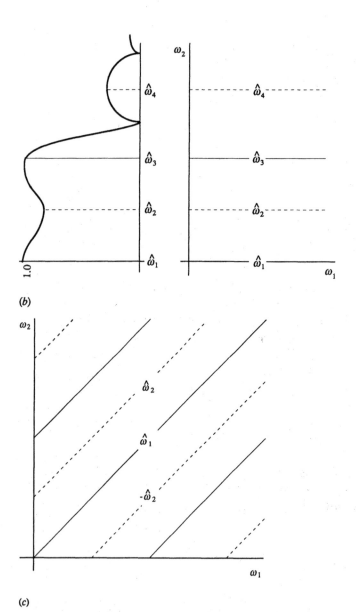

(b)

(c)

Figure 7.3 Continued

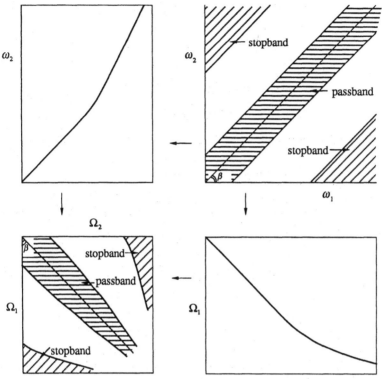

Figure 7.4 Application of the bilinear transformation to a 2-D continuous transfer function.

Another transformation that can be used for the derivation of 2-D continuous transfer functions is one suggested by Ahmadi, Constantinides, and King [4]. This is given by

$$s = g_3(s_1, s_2) = \frac{a_1 s_1 + a_2 s_2}{1 + c s_1 s_2} \qquad (7.15)$$

where a_1, a_2, and c are positive constants.

If a transformation transforms a 1-D transfer function into a corresponding 2-D transfer function of the same type (e.g., a 1-D lowpass into a 2-D lowpass transfer function) with respect to the entire first quadrant of the (ω_1, ω_2) plane, it is said to be a globally type-preserving transformation [5]. On the other hand, if the transformation preserves the type of the filter over a subset of the (ω_1, ω_2) plane, it is said to be a locally type-preserving transformation. If we let

$$\Gamma(\omega_1, \omega_2) = \frac{g(j\omega_1, j\omega_2)}{j}$$

and

$$\Omega = \{(\omega_1, \omega_2) : \omega_1 \geq 0, \omega_2 \geq 0\} \qquad (7.16)$$

then it can be shown that $g(s_1, s_2)$ is globally type preserving if and only if

1. $\Gamma(\omega_1, \omega_2)$ is continuous on Ω.
2. $\Gamma(0, 0) = 0$.
3. $\Gamma(\omega_1, \omega_2)$ is a positive monotonic increasing function in each variable in Ω, such that

$$\lim_{\omega_1 \to \infty} \Gamma(\omega_1, \omega_2) = \infty \qquad \text{for any } \omega_2$$

and

$$\lim_{\omega_2 \to \infty} \Gamma(\omega_1, \omega_2) = \infty \qquad \text{for any } \omega_1$$

On the other hand, if the 1-D continuous transfer function represents a filter of a given type over the frequency range $[0, \omega_0]$ where $\omega_0 > 0$, then the transformation is locally type preserving if and only if

1. $\Gamma(\omega_1, \omega_2)$ is continuous on Ω_I.
2. $\Gamma(0, 0) = 0$ and $[0, \omega_0] \subseteq \Gamma(\Omega_I)$.
3. $\Gamma(\omega_1, \omega_2)$ is a positive monotonic increasing function in each variable in Ω_I.

It can readily be shown that the transformations in Eqs. (7.12a) and (7.12b) are globally type preserving whereas that in Eq. (7.15) is locally type preserving with Ω_I given by

$$\Omega_I = \{(\omega_1, \omega_2) : \omega_1 \geq 0, \omega_2 \geq 0, \omega_1\omega_2 \geq 1/c\}$$

7.3   METHOD OF HIRANO AND AGGARWAL

In applications where the passband or stopband of the required filter is a rectangular domain or a combination of rectangular domains, the design can be accomplished by using a method proposed by Hirano and Aggarwal [7]. The method can be used to design filters with quadrantal or half-plane symmetry.

7.3.1 Filters with Quadrantal Symmetry

Consider two 1-D bandpass digital filters F_1 and F_2 characterized by the transfer functions $H_1(z)$ and $H_2(z)$, respectively, and let $z = z_1$ in the first one and $z = z_2$ in the second. If $H_1(z_1)$ and $H_2(z_2)$ have passbands in the frequency ranges $\omega_{12} \leq \omega_1 \leq \omega_{13}$ and $\omega_{22} \leq \omega_2 \leq \omega_{23}$, respectively, and stopbands in the frequency ranges $0 \leq \omega_1 \leq \omega_{11}$, $\omega_{14} \leq \omega_1 \leq \infty$ and $0 \leq \omega_2 \leq \omega_{21}$, $\omega_{24} \leq \omega_2 \leq \infty$, respectively, then

$$|H_1(e^{j\omega_1 T_1})| \approx \begin{cases} 0 & 0 \leq \omega_1 \leq \omega_{11} \\ 1 & \omega_{12} \leq \omega_1 \leq \omega_{13} \\ 0 & \omega_{14} \leq \omega_1 \leq \infty \end{cases} \qquad (7.17)$$

and

$$|H_2(e^{j\omega_2 T_2})| \approx \begin{cases} 0 & 0 \leq \omega_2 \leq \omega_{21} \\ 1 & \omega_{22} \leq \omega_2 \leq \omega_{23} \\ 0 & \omega_{24} \leq \omega_2 \leq \infty \end{cases} \qquad (7.18)$$

Since $H_1(z_1)$ is independent of z_2, it can be regarded as a 2-D transfer function whose passband is a vertical strip in the (ω_1, ω_2) plane of width $\omega_{13} - \omega_{12}$, as depicted in Fig. 7.5a. Likewise, since $H_2(z_2)$ is independent of z_1, it can be regarded as a 2-D transfer function whose passband is a horizontal strip of width $\omega_{23} - \omega_{22}$, as depicted in Fig. 7.5b. If these two filters are connected in cascade, a 2-D filter is obtained whose transfer function is given by

$$H(z_1, z_2) = H_1(z_1)H_2(z_2)$$

and since

$$|H(e^{j\omega_1 T_1}, e^{j\omega_2 T_2})| = |H_1(e^{j\omega_1 T_1})| \, |H_2(e^{j\omega_2 T_2})|$$

Eqs. (7.17) and (7.18) give

$$|H(e^{j\omega_1 T_1}, e^{j\omega_2 T_2})| \approx \begin{cases} 1 & \omega_{12} \leq \omega_1 \leq \omega_{13} \quad \text{and} \quad \omega_{22} \leq \omega_2 \leq \omega_{23} \\ 0 & \text{otherwise} \end{cases}$$

Evidently, the 2-D filter obtained will pass frequency components which are in the passband of $H_1(z_1)$ *and* the passband of $H_2(z_2)$; that is, the passband $H(z_1, z_2)$ will be a rectangle with sides $\omega_{13} - \omega_{12}$ and $\omega_{23} - \omega_{22}$, as depicted in Fig. 7.5c. On the other hand, frequency components that are in the stopband of the first filter *or* the stopband of the second filter will be rejected. Hence, the stopband of $H(z_1, z_2)$ consists of the domain obtained by combining the stopbands of the two filters, as depicted in Fig. 7.5c. If each of the two filters is allowed to be a lowpass, bandpass, or highpass 1-D filter, then nine different rectangular passbands can be achieved,

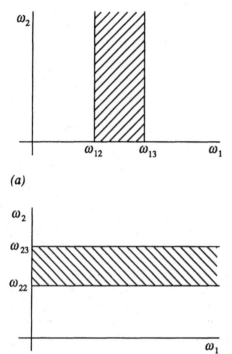

(a)

(b)

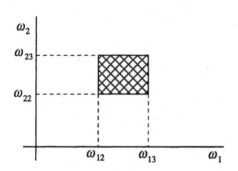

(c)

Figure 7.5 Idealized amplitude responses: (a) filter F_1; (b) filter F_2; (c) 2-D filter obtained by cascading filters F_1 and F_2.

as depicted in Fig. 7.6. The cascade arrangement of any two of these filters may be referred to as a generalized bandpass filter.

Now let us consider the configuration of Fig. 7.7 where each of the two products $H_1(z_1)H_2(z_2)$ is a generalized bandpass filter of the type already described, $H_{A1}(z_1)$ and $H_{A2}(z_2)$ are allpass 1-D filters whose poles are the poles of $H_1(z_1)$ and $H_2(z_2)$, respectively, and k is an integer equal to -1, 0, or 1. From Fig. 7.7

$$H(z_1, z_2) = H_{A1}(z_1)H_{A2}(z_2) - e^{-jk\pi}[H_1(z_1)H_2(z_2)]^2 \quad (7.19)$$

If

$$H_1(z_1) = \frac{N_1(z_1)}{D_1(z_1)} = \frac{\sum_{i=0}^{N} a_{1i}z_1^i}{\sum_{i=0}^{N} b_{1i}z_1^i}$$

and

$$H_2(z_2) = \frac{N_2(z_2)}{D_2(z_2)} = \frac{\sum_{i=0}^{M} a_{2i}z_2^i}{\sum_{i=0}^{M} b_{2i}z_2^i}$$

then the 1-D allpass transfer functions can be constructed as

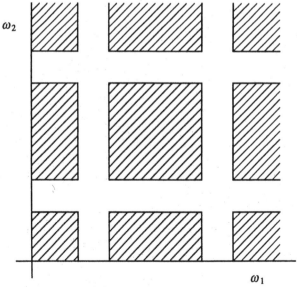

Figure 7.6 Idealized amplitude response of generalized bandpass filter.

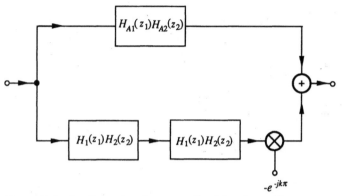

Figure 7.7 Configuration for generalized bandstop filter.

$$H_{A1}(z_1) = \frac{z_1^N D_1(z_1^{-1})}{D_1(z_1)} = \frac{\sum_{i=0}^{N} b_{1(N-i)} z_1^i}{\sum_{i=0}^{N} b_{1i} z_1^i}$$

and

$$H_{A2}(z_2) = \frac{z_2^M D_2(z_2^{-1})}{D_2(z_2)} = \frac{\sum_{i=0}^{M} b_{2(M-i)} z_2^i}{\sum_{i=0}^{M} b_{2i} z_2^i}$$

If we let

$$N_1(e^{j\omega_1 T_1}) = M_{N1}(\omega_1) e^{j\theta_{N1}(\omega_1)}$$
$$N_2(e^{j\omega_2 T_2}) = M_{N2}(\omega_2) e^{j\theta_{N2}(\omega_2)}$$
$$D_1(e^{j\omega_1 T_1}) = M_{D1}(\omega_1) e^{j\theta_{D1}(\omega_1)}$$

and

$$D_2(e^{j\omega_2 T_2}) = M_{D2}(\omega_2) e^{j\theta_{D2}(\omega_2)}$$

then Eq. (7.19) yields

$$H(e^{j\omega_1 T_1}, e^{j\omega_2 T_2}) = e^{j\psi_1(\omega_1, \omega_2)} - \frac{M_{N1}(\omega_1) M_{N2}(\omega_2)}{M_{D1}(\omega_1) M_{D2}(\omega_2)} e^{j\psi_2(\omega_1, \ \omega_2)} \quad (7.20)$$

where

$$\psi_1(\omega_1, \omega_2) = N\omega_1 T_1 + M\omega_2 T_2 - 2[\theta_{D1}(\omega_1) + \theta_{D2}(\omega_2)]$$

and

$$\psi_2(\omega_1, \omega_2) = -k\pi + 2[\theta_{N1}(\omega_1) + \theta_{N2}(\omega_2)] - 2[\theta_{D1}(\omega_1) + \theta_{D2}(\omega_2)]$$

Now if

$$\theta_{N1}(\omega_1) + \theta_{N2}(\omega_2) = \tfrac{1}{2}(k\pi + N\omega_1 T_1 + M\omega_2 T_2) \quad (7.21)$$

then

$$\psi_1(\omega_1, \omega_2) = \psi_2(\omega_1, \omega_2)$$

and so Eq. (7.20) assumes the form

$$H(e^{j\omega_1 T_1}, e^{j\omega_2 T_2}) = \{1 - [|H_1(e^{j\omega_1 T_1})| \, |H_2(e^{j\omega_2 T_2})|]^2\} e^{j\psi_1(\omega_1, \omega_2)} \quad (7.22)$$

Let us now assume that the 1-D filters characterized by $H_1(z_1)$ and $H_2(z_2)$ are bandpass filters. From Eqs. (7.17), (7.18), and (7.22) we obtain

$$|H(e^{j\omega_1 T_1}, e^{j\omega_2 T_2})| \approx \begin{cases} 0 & \omega_{12} \le \omega_1 \le \omega_{13} \quad \text{and} \quad \omega_{22} \le \omega_2 \le \omega_{23} \\ 1 & \text{otherwise} \end{cases}$$

and, therefore, the configuration of Fig. 7.7 realizes a 2-D filter whose amplitude response is complementary to that of the generalized bandpass filter described earlier in that all frequency components within the passbands of filters F_1 and F_2 will be rejected and all other components will be passed. If each of the two cascade filters is allowed to be a lowpass, highpass, or bandpass filter, nine rectangular stopbands can be achieved, as depicted in Fig. 7.8. Under these circumstances, the configuration of Fig. 7.7 can be referred to as a generalized bandstop filter.

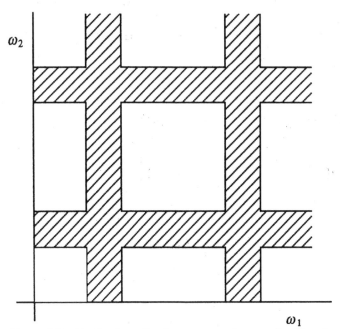

Figure 7.8 Idealized amplitude response of generalized bandstop filter.

The configuration of Fig. 7.7 will complement the amplitude response of the two cascaded filters F_1 and F_2 only if the phase angles of the numerator polynomials $N_1(z_1)$ and $N_2(z_2)$ are linear with respect to ω_1 and ω_2, respectively, as in Eq. (7.21). Following the approach in pp. 218–219 of [14], it can be shown that Eq. (7.21) will be satisfied under the following circumstances:

1. The coefficients of $N_1(z_1)$ and $N_2(z_2)$ have mirror-image symmetry, that is, $a_{1i} = a_{1(N-i)}$ for $i = 0, 1, 2, \ldots, N$, and $k = 0$ in Fig. 7.7.
2. The coefficients of $N_1(z_1)$ and $N_2(z_2)$ have mirror-image antisymmetry, that is, $a_{1i} = -a_{1(N-i)}$ for $i = 0, 1, 2, \ldots, N$, and $k = 0$ in Fig. 7.7.
3. The coefficients of $N_1(z_1)$ [or $N_2(z_2)$] have mirror-image symmetry and those of $N_2(z_2)$ [or $N_1(z_1)$] have mirror-image antisymmetry, and $k = -1$ or $+1$ in Fig 7.7.

For 1-D discrete transfer functions obtained by applying the bilinear transformation to classical 1-D continuous transfer functions with zeros on the $j\omega$ axis or at the origin of the s plane (e.g., Bessel, Butterworth, Chebyshev, and elliptic lowpass, highpass, and bandpass transfer functions) one of the preceding conditions is always satisfied, as can be easily demonstrated and, consequently, any one of the nine stopbands in Fig. 7.8 can readily be achieved.

Extending these principles, if a 2-D filter is constructed by cascading K 2-D filters with passbands P_i, stopbands S_i, and transfer functions $H_i(z_1, z_2)$, the overall transfer function is obtained as

$$H(z_1, z_2) = \prod_{i=1}^{K} H_i(z_1, z_2)$$

and if

$$H_i(e^{j\omega_1 T_1}, e^{j\omega_2 T_2}) = M_i(\omega_1, \omega_2)e^{j\theta_i(\omega_1, \omega_2)}$$

then

$$H(e^{j\omega_1 T_1}, e^{j\omega_2 T_2}) = \left[\prod_{i=1}^{K} M_i(\omega_1, \omega_2) \right] e^{j\theta(\omega_1, \omega_2)}$$

where

$$\theta(\omega_1, \omega_2) = \sum_{i=1}^{K} \theta_i(\omega_1, \omega_2)$$

Consequently, the passband P and stopband S of the cascade arrangement are obtained as

$$P = \bigcap_{i=1}^{K} P_i \quad \text{and} \quad S = \bigcup_{i=1}^{K} S_i$$

that is, the only frequency components not to be rejected will be those components that will be passed by each and every filter in the cascade arrangement.

If a 2-D filter is constructed by connecting K 2-D filters in parallel, then

$$H(z_1, z_2) = \sum_{i=1}^{K} H_i(z_1, z_2)$$

and hence

$$H(e^{j\omega_1 T_1}, e^{j\omega_2 T_2}) = \sum_{i=1}^{K} M_i(\omega_1, \omega_2)e^{j\theta_i(\omega_1, \omega_2)}$$

Assuming that all the parallel filters have the same phase shift, that is, $\theta_1(\omega_1, \omega_2) = \theta_2(\omega_1, \omega_2) = \cdots$, then

$$H(e^{j\omega_1 T_1}, e^{j\omega_2 T_2}) = \left[\sum_{i=1}^{K} M_i(\omega_1, \omega_2) \right] e^{j\theta_1(\omega_1, \omega_2)}$$

If, in addition, the passbands P_i of the various filters are not overlapping, that is,

$$\bigcap_{i=1}^{K} P_i = 0$$

then the passband and stopband of the parallel arrangement is given by

$$P = \bigcup_{i=1}^{K} P_i \quad \text{and} \quad S = \bigcap_{i=1}^{K} S_i$$

Parallel recursive filters are more difficult to design than cascade ones, because of the requirement that the phase shifts of the various parallel filters be equal. However, if all the data to be processed are available at the start of the processing, zero-phase filters can be used (see Sec. 7.2.2).

By combining parallel and cascade subfilters, 2-D filters whose pass-bands or stopbands are combinations of rectangular domains can be designed, as illustrated by the following example.

Example 7.1 Design a 2-D digital filter whose passband is the area between two overlapping rectangles, as depicted in Fig. 7.9a.

Solution. Using 1-D lowpass filters, two 2-D lowpass filters whose pass-bands are rectangular, as shown in Fig. 7.9b and c can readily be obtained. By using the second 2-D lowpass filter in the configuration of Fig. 7.7, a 2-D highpass filter can be obtained whose stopband is the unshaded area in Fig. 7.9d. Now if the first 2-D lowpass filter is cascaded with the 2-D

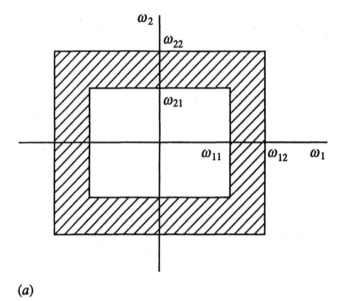

(a)

(b)

Figure 7.9 Amplitude responses for the filter of Example 7.1: (a) specified response; (b) and (c) responses of 2-D lowpass filters; (d) response of highpass filter.

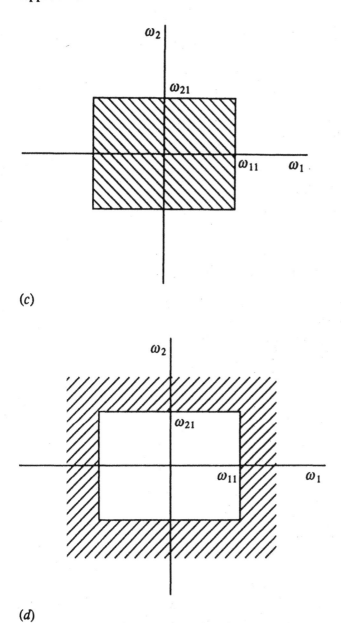

(c)

(d)

Figure 7.9 Continued.

highpass filter, the required filter is obtained since the passband of the resulting cascade arrangement is the intersection of the passbands in Fig. 7.9b and d.

7.3.2 Filters with Half-Plane Symmetry

In certain applications, filter characteristics are required that have half-plane symmetry. In these filters

$$|H(z_1, z_2)| = |H(z_1^*, z_2^*)|$$

as in filters with quadrantal symmetry. On the other hand,

$$|H(z_1, z_2) \neq |H(z_1^*, z_2)|$$
$$\neq |H(z_1, z_2^*)|$$

that is, half-plane symmetry does not imply quadrantal symmetry. The design of these filters can be accomplished by using two quadrant pass filters.

Consider an ideal 2-D lowpass filter with a square passband and passband edges $\omega_{1p} = \omega_{2p} = \pi/2 - \varepsilon$, as depicted in Fig. 7.10a (ε is a small positive constant). A separable realization of this type of characteristic can be achieved by cascading two 1-D lowpass filters, as described in the previous section. Let the transfer function of the filter obtained be

$$H(z_1, z_2) = H_1(z_1)H_2(z_2)$$

On applying the transformations

$$z_1 := z_1 e^{-j\theta_1}$$
$$z_2 := z_2 e^{-j\theta_2}$$

to $H_1(z_1)$ and $H_2(z_2)$, we obtain

$$\tilde{H}_1(z_1, \theta_1) = H_1(z_1)|_{z_1 := z_1 e^{-j\theta_1}}$$
$$\tilde{H}_2(z_2, \theta_2) = H_2(z_2)|_{z_2 := z_2 e^{-j\theta_2}}$$

Consequently

$$\tilde{H}(z_1, \theta_1, z_2, \theta_2) = \tilde{H}_1(z_1, \theta_1)\tilde{H}_2(z_2, \theta_2)$$

where the coefficients of $\tilde{H}_1(z_1, \theta_1)$, $\tilde{H}_2(z_2, \theta_2)$, and $\tilde{H}(z_1, \theta_1, z_2, \theta_2)$ are complex, in general. The effect of the preceding transformations is to shift the amplitude response of the original 2-D filter by θ_1 and θ_2 in the directions of the positive ω_1 axis and the positive ω_2 axis, respectively. If $\theta_1 = \theta_2 = \pi/2$, the passband of the original filter is shifted into the first quadrant of the (ω_1, ω_2) plane, as shown in Fig. 7.10b. The filter obtained is said

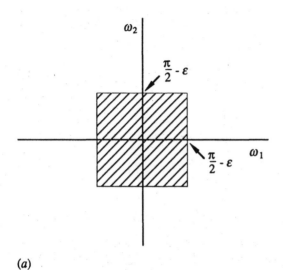

(a)

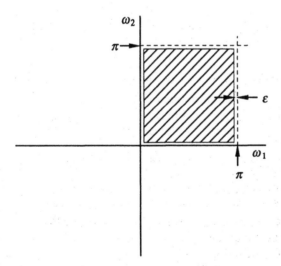

(b)

Figure 7.10 (a) Amplitude response of ideal lowpass filter; (b) effect of transformation on the amplitude response.

to be a first-quadrant pass filter. By shifting the passband of the original
2-D filter into second, third, etc., quadrant, second-, third-, etc., quadrant
pass filters can be obtained. The transfer functions of these filters are given
by

$$\hat{H}_1(z_1, z_2) = \bar{H}\left(z_1, \frac{\pi}{2}, z_2, \frac{\pi}{2}\right) \qquad (7.23a)$$

$$\hat{H}_2(z_1, z_2) = \bar{H}\left(z_1, -\frac{\pi}{2}, z_2, \frac{\pi}{2}\right) \qquad (7.23b)$$

$$\hat{H}_3(z_1, z_2) = \bar{H}\left(z_1, -\frac{\pi}{2}, z_2, -\frac{\pi}{2}\right) \qquad (7.23c)$$

$$\hat{H}_4(z_1, z_2) = \bar{H}\left(z_1, \frac{\pi}{2}, z_2, -\frac{\pi}{2}\right) \qquad (7.23d)$$

where the subscript of $\hat{H}$ identifies the quadrant to which the passband has
been moved. These filters can now be used to synthesize two quadrant
zero-phase pass filters. On using Eqs. (7.23a) and (7.23c) or Eqs. (7.23b)
and (7.23d), we have

$$\hat{H}_{13}(z_1, z_2) = \hat{H}_1(z_1, z_2)\hat{H}_1^*(z_1^{-1}, z_2^{-1}) + \hat{H}_3(z_1, z_2)\hat{H}_3^*(z_1^{-1}, z_2^{-1}) \qquad (7.24a)$$

or

$$\hat{H}_{24}(z_1, z_2) = \hat{H}_2(z_1, z_2)\hat{H}_2^*(z_1^{-1}, z_2^{-1}) + \hat{H}_4(z_1, z_2)\hat{H}_4^*(z_1^{-1}, z_2^{-1}) \qquad (7.24b)$$

where $\hat{H}_1^*(z_1, z_2)$ is the transfer function obtained by replacing the coeffi-
cients in $\hat{H}_1(z_1, z_2)$ by their complex conjugates. Note that $\hat{H}_1^*(z_1^{-1}, z_2^{-1})$ is the complex conjugate of $\hat{H}_1(z_1, z_2)$ and so

$$\hat{H}_1(e^{j\omega_1 T_1}, e^{j\omega_2 T_2})\hat{H}_1^*(e^{-j\omega_1 T_1}, e^{-j\omega_2 T_2}) = |\hat{H}_1(e^{j\omega_1 T_1}, e^{j\omega_2 T_2})|^2.$$

Hence each of the two terms in Eqs. (7.24a) or (7.24b) is real when $z_1 = e^{j\omega_1 T_1}$ and $z_2 = e^{j\omega_2 T_2}$ and, furthermore, each will yield a passband in one
quadrant. Thus the filters characterized by Eqs. (7.24a) and (7.24b) have
passbands in the first and third quadrants and the second and fourth quadrants, respectively, as illustrated in Fig. 7.11

By using the method described in Sec. 7.3.1, arbitrary quadrantally
symmetric filters whose passbands or stopbands are combinations of rectangular domains can readily be designed. By connecting two quadrant
pass filters in cascade with these designs, filters whose characteristics have
half-plane symmetry can also be designed. For example, by connecting the
filter designed in Example 7.1 in cascade with the filter characterized by
Eq. (7.24a), the chracteristic of Fig. 7.12 is obtained.

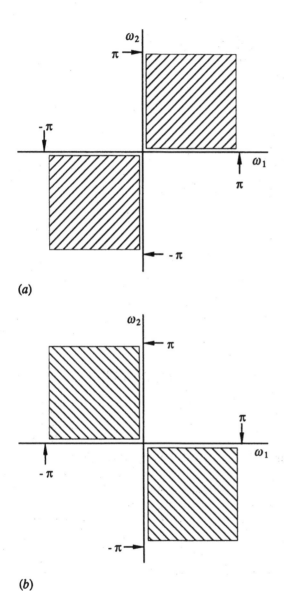

(a)

(b)

Figure 7.11 Two quadrant pass filters: (a) Eq. (7.24a); (b) Eq. (7.24b).

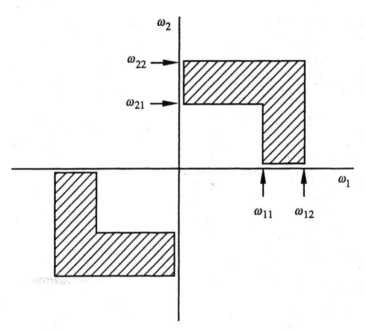

Figure 7.12   Filter with half-plane symmetry.

7.4   DESIGN OF CIRCULARLY SYMMETRIC FILTERS

Transformations can also be used for the design of 2-D nonseparable filters having piecewise-constant amplitude responses with circular symmetry. Noteworthy methods of this class are those described by Costa and Ve-netsanopoulos [2] and Goodman [6]. In these methods, transformations that can produce rotation in the amplitude response of either an analog or a digital filter are used to achieve circular symmetry in the amplitude response. The two methods lead to filters that are, in theory, unstable but by using an alternative transformation proposed by Mendonça et al. [9], this problem can be eliminated.

7.4.1   Method of Costa and Venetsanopoulos

In the method of Costa and Venetsanopoulos, a set of 2-D continuous transfer functions is obtained by applying the transformation of Eq. (7.12a) for several different values of the rotation angle β to a 1-D continuous lowpass transfer function. A set of 2-D discrete lowpass transfer functions is then deduced through the application of the bilinear transformation. The design is completed by connecting the 2-D digital filters obtained in cascade. The steps involved are as follows:

STEP 1: Obtain a stable 1-D continuous lowpass transfer function

$$H_{A1}(s) = \frac{N(s)}{D(s)} = K_0 \frac{\Pi_{i=1}^M(s - z_{ai})}{\Pi_{i=1}^N(s - p_{ai})} \qquad (7.25)$$

where z_{ai} and p_{ai} for $i = 1, 2, \ldots$, are the zeros and poles of $H_{A1}(s)$, respectively, and K_0 is a multiplier constant.

STEP 2: Let β_k for $k = 1, 2, \ldots, K$ be a set of rotation angles given by

$$\beta_k = \begin{cases} [(2k - 1)/2K + 1]\pi & \text{for even } K \\ [(k - 1)/K + 1]\pi & \text{for odd } K \end{cases} \qquad (7.26)$$

STEP 3: Apply the transformation of Eq. (7.12a) to $H_{A1}(s)$ to obtain a 2-D continuous transfer function as

$$H_{A2k}(s_1, s_2) = H_{A1}(s)|_{s = -s_1 \sin \beta_k + s_2 \cos \beta_k} \qquad (7.27)$$

for each rotation angle β_k identified in Step 2.

STEP 4: Apply the double bilinear transformation of Eq. (7.14) to $H_{A2k}(s_1, s_2)$ to obtain

$$H_{D2k}(z_1, z_2) = H_{A2k}(s_1, s_2)|_{s_1 = \frac{2}{T_1}\left(\frac{z_1-1}{z_1+1}\right), \quad s_2 = \frac{2}{T_2}\left(\frac{z_2-1}{z_2+1}\right)} \qquad (7.28)$$

Assuming that $T_1 = T_2 = T$, Eqs. (7.25) and (7.27) yield

$$H_{D2k}(z_1, z_2) = K_1 \frac{\Pi_{i=1}^M(a_{11i} + a_{21i}z_1 + a_{12i}z_2 + a_{22i}z_1z_2)}{\Pi_{i=1}^N(b_{11i} + b_{21i}z_1 + b_{12i}z_2 + b_{22i}z_1z_2)} \qquad (7.29)$$

where

$$K_1 = K_0\left(\frac{T}{2}\right)^{N-M}$$

$$a_{11i} = -\cos\beta_k + \sin\beta_k - Tz_{ai}/2, \quad a_{21i} = -\cos\beta_k - \sin\beta_k - Tz_{ai}/2 \qquad (7.30a)$$

$$a_{12i} = \cos\beta_k + \sin\beta_k - Tz_{ai}/2, \quad a_{22i} = \cos\beta_k - \sin\beta_k - Tz_{ai}/2 \qquad (7.30b)$$

$$b_{11i} = -\cos\beta_k + \sin\beta_k - Tp_{ai}/2, \quad b_{21i} = -\cos\beta_k - \sin\beta_k - Tp_{ai}/2 \qquad (7.30c)$$

$$b_{12i} = \cos\beta_k + \sin\beta_k - Tp_{ai}/2, \quad b_{22i} = \cos\beta_k - \sin\beta_k - Tp_{ai}/2 \qquad (7.30d)$$

STEP 5: Cascade the filters obtained in Step 4 to yield an overall transfer function

$$H(z_1, z_2) = \prod_{k=1}^{K} H_{D2k}(z_1, z_2)$$

In order to illustrate the design procedure, consider the design of a 2-D filter using the second-order Butterworth transfer function

$$H_{A1}(s) = \frac{1}{s^2 + \sqrt{2}s + 1}$$

and assume that $\omega_{s1} = \omega_{s2} = 10$ rad/s and $K = 6$. From Step 2, the rotation angles are 195°, 225°, 255°, 285°, 315°, and 345°. The amplitude response of the 2-D digital subfilter obtained in Step 4 for $k = 4$, that is, for a rotation angle of 285°, is illustrated in Fig. 7.13a. The amplitude response obtained by cascading the subfilters with rotation angles 195°, 225°, and 255° consists of a set of concentric ellipses as shown in Fig. 7.13b and that obtained by cascading the subfilters with rotation angles 285°, 315°, and 345° consists of a set of concentric ellipses as shown in Fig. 7.13c. Now, connecting all these subfilters in cascade yields an overall amplitude response whose contours are concentric circles as depicted in Fig. 7.13d.

The 2-D digital filter designed by the preceding method will be useful in practice only if the transfer function in Eq. (7.29) is stable for each rotation angle β_k. From Eq. (7.29), we can write

$$H_{D2k}(z_1, z_2) = K_1 \prod_{i=1}^{N} H_i(z_1, z_2) \qquad (7.31)$$

where

$$H_i(z_1, z_2) = \frac{a_{11i} + a_{21i}z_1 + a_{12i}z_2 + a_{22i}z_1z_2}{b_{11i} + b_{21i}z_1 + b_{12i}z_2 + b_{22i}z_1z_2} \qquad (7.32)$$

The filter designed will be stable if and only if each $H_i(z_1, z_2)$ represents a stable first-order filter.

At point $(z_1, z_2) = (-1, -1)$, both the numerator and denominator polynomials of $H_i(z_1, z_2)$ assume the value of zero, as can be readily seen from Eqs. (7.30) and (7.32), and hence each $H_i(z_1, z_2)$ has a nonessential singularity of the second kind on the unit bicircle

$$U^2 = \{(z_1, z_2) : |z_1| = 1, |z_2| = 1\}$$

Consequently, the stability methods of Chap. 5 cannot be used to check the stability of the filter designed. Nevertheless, Goodman [6] has shown that in theory $H_i(z_1, z_2)$ represents an unstable subfilter in general by demonstrating that the impulse response of the subfilter is not absolutely summable. In practice, however, the numerator polynomials of $H_i(z_1, z_2)$

are unlikely to be precisely equal to zero, owing to coefficient quantization, and the subfilter may or may not be stable depending on chance.

The nonessential singularity of each $H_i(z_1, z_2)$ can be eliminated and, furthermore, each subfilter can be stabilized by letting

$$b'_{12i} = b_{12i} + \varepsilon b_{11i} \tag{7.33a}$$

$$b'_{22i} = b_{22i} + \varepsilon b_{21i} \tag{7.33b}$$

where ε is a small positive constant. With this modification, the denominator polynomial of each $H_i(z_1, z_2)$ is no longer zero and, furthermore, from Sec. 5.5, the stability of the subfilter can be guaranteed if

1. $|b'_{12i}| < |b'_{22i}|$

2. $\max\left\{\left|\dfrac{b_{21i} + b_{11i}}{b'_{22i} + b'_{12i}}\right|, \left|\dfrac{b_{21i} - b_{11i}}{b'_{22i} - b'_{12i}}\right|\right\} < 1$

From Eqs. (7.30) and (7.33), condition 1 is satisfied if

$$\left|\frac{(1 - \varepsilon)}{(1 + \varepsilon)}\cos\beta_k + \sin\beta_k - \rho_i\right| < \left|\frac{(1 - \varepsilon)}{(1 + \varepsilon)}\cos\beta_k - \sin\beta_k - \rho_i\right| \tag{7.34}$$

where ρ_i is the real part of $Tp_{ai}/2$. Now

$$\left|\frac{b_{21i} + b_{11i}}{b'_{22i} + b'_{12i}}\right| = \left|\frac{b_{21i} + b_{11i}}{b_{22i} + b_{12i} + \varepsilon(b_{21i} + b_{11i})}\right| \le \frac{1}{\left|\dfrac{\cos\beta_k - Tp_{ai}/2}{\cos\beta_k + Tp_{ai}/2}\right| - \varepsilon}$$

since $\varepsilon > 0$, and if we assume that

$$\left|\frac{\cos\beta_k - \rho_i}{\cos\beta_k + \rho_i}\right| > 1 \tag{7.35}$$

we have

$$\left|\frac{\cos\beta_k - Tp_{ai}/2}{\cos\beta_k + Tp_{ai}/2}\right| > 1$$

and on letting

$$\varepsilon = \frac{1}{2}\left(\left|\frac{\cos\beta_k - Tp_{ai}/2}{\cos\beta_k + Tp_{ai}/2}\right| - 1\right)$$

we obtain

$$\left|\frac{b_{21i} + b_{11i}}{b'_{22i} + b'_{12i}}\right| \le \frac{1}{\dfrac{1}{2}\left|\dfrac{\cos\beta_k - Tp_{ai}/2}{\cos\beta_k + Tp_{ai}/2}\right| + \dfrac{1}{2}} < 1$$

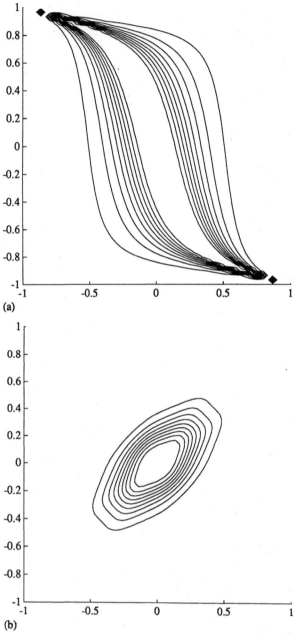

(a)

(b)

Figure 7.13 Contour plots for rotated filters: (a) subfilter for rotation angle of 285°; (b) subfilters for rotation angles of 195°, 225°, and 255° in cascade; (c) subfilters for rotation angles of 285°, 315°, and 345° in cascade; (d) all subfilters in cascade. (In these plots the Nyquist frequencies are normalized to unity.)

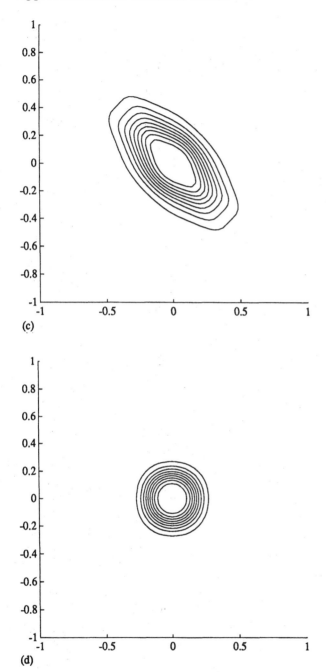

(c)

(d)

On the other hand,

$$\left| \frac{b_{21i} - b_{11i}}{b'_{22i} - b'_{12i}} \right| = \left| \frac{b_{21i} - b_{11i}}{b_{22i} - b_{12i} + \varepsilon(b_{21i} - b_{11i})} \right| = \left| \frac{1}{1 + \varepsilon} \right| < 1$$

and, therefore, if the condition in Eq. (7.35) is satisfied, stability condition 2 above is satisfied.

The set of values β_k and ρ_i that satisfy the conditions in Eqs. (7.34) and (7.35) are

$$S_1 = \left\{ (\beta_k, \rho_i) : \left(\sin \beta_k > 0 \cap \rho_i > \frac{(1 - \varepsilon)}{(1 + \varepsilon)} \cos \beta_k \right) \right.$$
$$\left. \cup \left(\sin \beta_k < 0 \cap \rho_i < \frac{(1 - \varepsilon)}{(1 + \varepsilon)} \cos \beta_k \right) \right\} \qquad (7.36)$$

and

$$S_2 = \{ (\beta_k, \rho_i) : (\rho_i < 0 \cap \cos \beta_k > 0) \cup (\rho_i > 0 \cap \cos \beta_k < 0) \} \qquad (7.37)$$

respectively. Assuming that the 1-D analog filter is stable (i.e., $\rho_i < 0$), then Eqs. (7.36) and (7.37) give the set of permissible rotation angles as

$$S_3 = \{ \beta_k : (\cos \beta_k > 0) \cap (\sin \beta_k < 0) \}$$

that is,

$$270° < \beta_k < 360°$$

For the special cases $\beta_k = 270°$ and $\beta_k = 360°$, $H_{A2k}(s_1, s_2)$ and, in turn, $H_{D2k}(z_1, z_2)$ reduce to 1-D filters that are stable if the original 1-D analog filter is stable. Therefore, in order to assure the stability of the subfilter represented by $H_i(z_1, z_2)$, the rotation angle must be chosen in the range

$$270° \leq \beta_k \leq 360° \qquad (7.38)$$

and, furthermore, its coefficients must be adjusted slightly, as in Eq. (7.33).

As can be seen in Eq. (7.26), half of the rotation angles are in the range $180° < \beta_k < 270°$ and according to the preceding analysis they yield unstable subfilters. However, the problem can easily be overcome by using rotation angles in the range given by Eq. (7.38) and then rotating the transfer function of the subfilter by $-90°$ using the transformations described in Sec. 7.2.2. For example, if an effective rotation angle $\beta_k = 225°$ is required, we can write

$$Y(z_1, z_2) = H_{225°}(z_1, z_2)X(z_1, z_2) \qquad (7.39)$$

From Table 7.1, we note that

$$T_{R_4^3} T_{R4} = T_I$$

Hence, Eqs. (7.9) and (7.39) give

$$
\begin{aligned}
Y(z_1, z_2) &= T_{R_4^3} T_{R4}[H_{225°}(z_1, z_2)X(z_1, z_2) \\
&= T_{R_4^3}[H_{315°}(z_1, z_2)T_{R4}X(z_1, z_2)] \\
&= T_{R_4^3}[H_{315°}(z_1, z_2)X_0(z_1, z_2)] \\
&= T_{R_4^3}Y_0(z_1, z_2)
\end{aligned}
$$

where

$$X_0(z_1, z_2) = T_{R4}X(z_1, z_2)$$

and

$$Y_0(z_1, z_2) = H_{315°}(z_1, z_2)X_0(z_1, z_2)$$

From Eq. (7.10b), we note that $x_0(n_1, n_2) = x(-n_2, n_1)$ and $y(n_1, n_2) = y_0(n_2, -n_1)$ and, therefore, the required rotation can be achieved as depicted in Fig. 7.14, where $H_{315°}(z_1, z_2)$ represents a stable subfilter. In effect, a rotation angle 225° can be achieved by rotating the input data by 90°, filtering using a subfilter rotated by 315°, and then rotating the output data by −90°.

The method of Costa and Venetsanopoulos can readily be extended to the design of zero-phase filters by including cascade subfilters for rotation

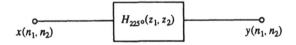

(a)

(b)

Figure 7.14 Realization of subfilter for rotation angle in the range $180° < \beta_k < 270°$.

angles $\pi + \beta_k$ for $k = 1, 2, \ldots, N$. The resulting transfer function is given by

$$H(z_1, z_2) = \prod_{k=1}^{K} H_{D2k}(z_1, z_2)H_{D2k}(z_1^{-1}, z_2^{-1})$$

where noncausal sections can be realized as illustrated in Fig. 7.2.

7.4.2 Method of Goodman

In the method of Goodman, a 1-D discrete transfer function is first obtained by applying the bilinear transformation to a 1-D continuous transfer function. Then, through the application of an allpass transformation that rotates the contours of the magnitude of the 1-D discrete transfer function, a corresponding 2-D discrete transfer function is obtained. The steps involved are as follows:

STEP 1:   Obtain a stable 1-D continuous lowpass transfer function $H_{A1}(s)$ of the form given by Eq. (7.25).

STEP 2:   Apply the bilinear transformation of Eq. (7.2) to $H_{A1}(s)$ to obtain

$$H_{D1}(z) = H_{A1}(s)\big|_{s=\frac{2}{T}\left(\frac{z-1}{z+1}\right)} \tag{7.40}$$

STEP 3:   Let β_k for $k = 1, 2, \ldots, K$ be a set of rotation angles given by Eq. (7.26).

STEP 4:   Apply the allpass transformation given by

$$z = f_k(z_1, z_2) = \frac{1 + c_k z_1 + d_k z_2 + e_k z_1 z_2}{e_k + d_k z_1 + c_k z_2 + z_1 z_2} \tag{7.41}$$

where

$$c_k = \frac{1 + \sin \beta_k + \cos \beta_k}{1 - \sin \beta_k + \cos \beta_k}$$

$$d_k = \frac{1 - \sin \beta_k - \cos \beta_k}{1 - \sin \beta_k + \cos \beta_k}$$

$$e_k = \frac{1 + \sin \beta_k - \cos \beta_k}{1 - \sin \beta_k + \cos \beta_k}$$

to obtain the 2-D discrete transfer functions

$$H_{D2k}(z_1, z_2) = H_{D1}(z)\big|_{z=f_k(z_1, z_2)} \tag{7.42}$$

for $k = 1, 2, \ldots, K$. The procedure yields the 2-D transfer function of Eq. (7.29), as can be easily demonstrated, and by cascading the rotated sections obtained as in Step 5 of the method of Costa and Venetsanopoulos, the design can be completed.

As presented, the method of Goodman is equivalent to that of Costa and Venetsanopoulos and consequently, the resulting filter is subject to the same stability problem due to the nonessential singularity of the second kind at point $(z_1, z_2) = (-1, -1)$. To achieve a stable design, Goodman [6] suggested that the transfer functions $H_{D2k}(z_1, z_2)$ for $k = 1, 2, \ldots,$ K be obtained directly by minimizing an appropriate objective function subject to the constraints

$$c_k + d_k - e_k \leq 1 - \varepsilon$$
$$c_k - d_k + e_k \leq 1 - \varepsilon$$
$$-c_k + d_k + e_k \leq 1 - \varepsilon$$
$$-c_k - d_k - e_k \leq 1 - \varepsilon$$

through the application of nonlinear programming. If ε is a small positive constant, the preceding constraints constitute necessary and sufficient conditions for stability and, therefore, such an approach will yield a stable filter.

7.4.3 Elimination of Nonessential Singularities

Nonessential singularities of the second kind at point $(z_1, z_2) = (-1, -1)$ can be eliminated and stability can be assured in the preceding two methods by using a transformation suggested by Mendonça et al. [9].

Function $g_3(s_1, s_2)$ in the transformation of Ahmadi et al. given in Eq. (7.15) is a reactance function and, as was demonstrated by Karivaratharajan, Reddy, Swamy, and Ramachandran [19], its application to a 1-D continuous transfer function followed by the application of the double bilinear transformation yields a stable 2-D discrete transfer function that is free of nonessential singularities of the second kind. However, as was stated earlier, the transformation is not globally type preserving, and its direct application was found to give unsatisfactory results [20]. In order to overcome the problem of nonglobal-type preservation, 1-D guard filters are often necessary, which increase the complexity of the 2-D digital filter.

Consider the transformation

$$s = g_4(s_1, z_2) = \frac{\cos(\beta_k)s_1 + \sin(\beta_k)s_2}{1 + cs_1s_2} \tag{7.43}$$

which can be obtained by combining the transformations in Eqs. (7.12b) and (7.15). If we ensure that

$$\cos \beta_k > 0, \quad \sin \beta_k > 0 \quad \text{and } c > 0$$

then the application of this transformation followed by the application of the double bilinear transformation yields stable 2-D digital filters that are free of nonessential singularities of the second kind. If, in addition

$$c = \frac{1}{\omega_{max}^2}$$

then local-type preservation can be achieved on set Ω_2 given by

$$\Omega_2 = \{(\omega_1, \omega_2) : \omega_1 \geq 0, \omega_2 \geq 0, \omega_1\omega_2 < \omega_{max}^2\}$$

and if $\omega_{max} \to \infty$, then $\Omega_2 \to \Omega$ where Ω is the set given by Eq. (7.16); that is, through the transformation of Eq. (7.43), global-type preservation can be approached as closely as desired. The transformation does not represent true rotation except in the limit $\omega_{max} = \infty$ and it has, as a result, been said to represent pseudorotation. By using the transformation of Eq. (7.43) instead of that in Eq. (7.12a) in the method of Costa and Venetsanopoulos, the transfer function of Eq. (7.29) becomes

$$H_{D2k}(z_1, z_2) = K_1 P_{D2}(z_2, z_2)$$

$$\times \frac{\Pi_{i=1}^{M}(a_{11i} + a_{21i}z_1 + a_{12i}z_2 + a_{22i}z_1z_2)}{\Pi_{i=1}^{N}(b_{11i} + b_{21i}z_1 + b_{12i}z_2 + b_{22i}z_1z_2)} \quad (7.44)$$

where

$$K_1 = K_0\left(\frac{T}{2}\right)^{N-M}$$

$$P_{D2}(z_1, z_2) = \left[1 + \frac{4c}{T^2} + \left(1 - \frac{4c}{T^2}\right)z_1\right.$$

$$\left. + \left(1 - \frac{4c}{T^2}\right)z_2 + \left(1 + \frac{4c}{T^2}\right)z_1z_2\right]^{N-M}$$

and

$$a_{11i} = -\cos\beta_k - \sin\beta_k - \left(\frac{T}{2} + \frac{2c}{T}\right)z_{ai}, \quad a_{21i} = \cos\beta_k - \sin\beta_k - \left(\frac{T}{2} - \frac{2c}{T}\right)z_{ai}$$

$$a_{12i} = -\cos\beta_k + \sin\beta_k - \left(\frac{T}{2} - \frac{2c}{T}\right)z_{ai}, \quad a_{22i} = \cos\beta_k + \sin\beta_k - \left(\frac{T}{2} + \frac{2c}{T}\right)z_{ai}$$

$$b_{11i} = -\cos\beta_k - \sin\beta_k - \left(\frac{T}{2} + \frac{2c}{T}\right)p_{ai}, \quad b_{21i} = \cos\beta_k - \sin\beta_k - \left(\frac{T}{2} - \frac{2c}{T}\right)p_{ai}$$

$$b_{12i} = -\cos\beta_k + \sin\beta_k - \left(\frac{T}{2} - \frac{2c}{T}\right)p_{ai}, \quad b_{22i} = \cos\beta_k + \sin\beta_k - \left(\frac{T}{2} + \frac{2c}{T}\right)p_{ai}$$

The 2-D discrete transfer function obtained can readily be shown to be free of nonessential singularities of the second kind by examining its behavior at $z_1 = \pm 1$, $z_2 = \pm 1$.

An equivalent design can be obtained by applying the allpass transformation of Eq. (7.41) with

$$c_k = \frac{1 + \cos \beta_k - \sin \beta_k - (4c/T^2)}{1 - \cos \beta_k - \sin \beta_k + (4c/T^2)}$$

$$d_k = \frac{1 - \cos \beta_k + \sin \beta_k - 4c/T^2}{1 - \cos \beta_k - \sin \beta_k + 4c/T^2}$$

$$e_k = \frac{1 + \cos \beta_k + \sin \beta_k + 4c/T^2}{1 - \cos \beta_k - \sin \beta_k + 4c/T^2}$$

in Goodman's method.

7.4.4 Realization of Highpass, Bandpass, and Bandstop Filters

Although the design of circularly symmetric lowpass filters can be achieved by using any one of the preceding methods, the design of circularly symmetric highpass filters is not as straightforward, as will now be demonstrated.

Consider two zero-phase filter sections that have been obtained by rotating a 1-D continuous highpass transfer function by angles $-\beta_1$ and β_1, where $0° < \beta_1 < 90°$. Idealized contour plots for the two sections are shown in Fig. 7.15a and b. If these two sections are cascaded, the amplitude response of the combination is obtained by multiplying the amplitude responses of the two sections at corresponding points of the (ω_1, ω_2) plane. The idealized contour plot of the cascade filter is obtained as shown in Fig. 7.15c. As can be seen, the resulting contour plot does not represent the amplitude response of a 2-D circularly symmetric highpass filter and, therefore, the design of these filters cannot be achieved by cascading rotated sections as with circularly symmetric lowpass filters. Their design can, however, be accomplished through the use of a highpass configuration suggested by Mendonça et al. [9], which comprises a combination of zerophase cascade and parallel sections.

If the preceding rotated sections are connected in parallel, a filter is obtained whose contour plot is shown in Fig. 7.15d. By subtracting the output of the cascade filter from the output of the parallel filter, a filter is obtained whose contour plot is shown in Fig. 7.15e. Evidently, this plot resembles the idealized contour plot of a 2-D circularly symmetric highpass

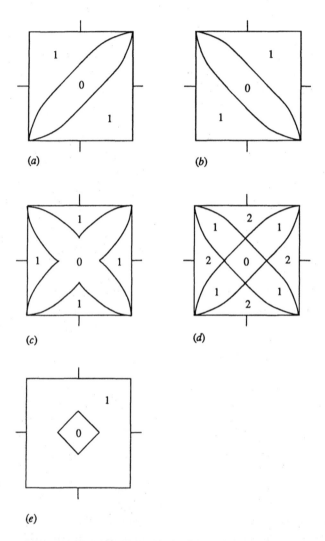

(a)

(b)

(c)

(d)

(e)

Figure 7.15 Derivation of highpass configuration: (a) contour plot of section rotated by angle β_1; (b) contour plot of section rotated by angle $-\beta_1$; (c) contour plot of sections in (a) and (b) connected in cascade; (d) contour plot of sections in (a) and (b) connected in parallel; (e) contour plot obtained by subtracting the amplitude response of the cascade filter in (c) from that of the parallel filter in (d).

filter. The configuration obtained is illustrated in Fig. 7.16a and is characterized by

$$\hat{H}_1 = \tilde{H}_1 = H^{\beta_1} + H^{-\beta_1} - H^{\beta_1}H^{-\beta_1}$$

where

$$H^{\beta_1} = H_1(z_1, z_2)H_1(z_1^{-1}, z_2^{-1})$$

and

$$H^{-\beta_1} = H_1(z_1, z_2^{-1})H_1(z_1^{-1}, z_2)$$

represent zero-phase sections rotated by angles β_1 and $-\beta_1$, respectively.

As in lowpass filters, the degree of circularity can be improved by using several rotation angles. For two different rotation angles β_1 and β_2, the transfer function of the 2-D highpass filter is obtained as

$$\hat{H}_2 = \hat{H}_1 + \tilde{H}_2 - \hat{H}_1\tilde{H}_2$$

where

$$\tilde{H}_2 = H^{\beta_2} + H^{-\beta_2} - H^{\beta_2}H^{-\beta_2}$$

with

$$H^{\beta_2} = H_2(z_1, z_2)H_2(z_1^{-1}, z_2^{-1})$$

and

$$H^{-\beta_2} = H_2(z_1, z_2^{-1})H_2(z_1^{-1}, z_2)$$

Similarly, for N rotation angles, $\hat{H}_N$ is given by the recursive relation

$$\hat{H}_N = \hat{H}_{N-1} + \tilde{H}_N - \hat{H}_{N-1}\tilde{H}_N \tag{7.45}$$

where

$$\tilde{H}_N = H^{\beta_N} + H^{-\beta_N} - H^{\beta_N}H^{-\beta_N}$$

and $\hat{H}_{N-1}$ can be obtained from $\tilde{H}_{N-1}$ and $\hat{H}_{N-2}$, and so on. The configuration obtained is illustrated in Fig. 7.16b, where the realization of $\hat{H}_{N-1}$ is of the same form as that of $\hat{H}_N$.

The rotated sections should be designed using the transformation in Eq. (7.43) in order to avoid the problem of nonessential singularities of the second kind. However, the use of this transformation leads to another problem: the 2-D transfer function obtained has spurious zeros at the Nyquist points. These zeros are due to the fact that the transformation in Eq. (7.43) does not have type preservation in the neighborhoods of the Nyquist points but their presence does not appear to be of serious concern. It should also be mentioned that the complexity of the configuration in

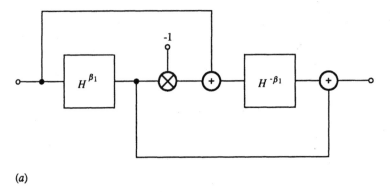

(a)

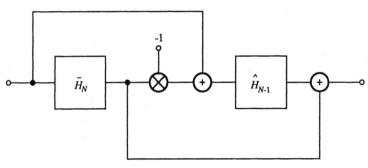

(b)

Figure 7.16 2-D highpass configuration: (a) one rotation angle; (b) N rotation angles.

Fig. 7.16 tends to increase rapidly with the number of rotations and, therefore, their number should be kept to a minimum.

With the availability of circularly symmetric lowpass and highpass filters, bandpass and bandstop filters with circularly symmetric amplitude responses, like those in Fig. 4.6c and d, can readily be obtained. A bandpass filter can be obtained by connecting a lowpass filter and a highpass filter with overlapping passbands in cascade, whereas a bandstop filter can be obtained by connecting a lowpass filter and a highpass filter with overlapping stopbands in parallel.

7.5 CONSTANTINIDES TRANSFORMATIONS

A 1-D discrete transfer function can be transformed into another 1-D discrete function by using a set of transformations due to Constantinides [13]. These transformations are also useful in the design of 2-D digital filters and, as was demonstrated by Pendergrass, Mitra, and Jury [3], their application leads to a variety of amplitude responses. They are of the form

$$z = f(\bar{z}) = e^{jl\pi} \prod_{t=1}^{m} \frac{\bar{z} - a_t^*}{1 - a_t \bar{z}} \qquad (7.46)$$

where l and m are integers and a_t^* is the complex conjugate of a_t. As can readily be demonstrated [14], these transformations map the unit circle $|z| = 1$ of the z plane, its exterior, and its interior onto the unit circle $|\bar{z}| = 1$ of the $\bar{z}$ plane, its interior, and its exterior, respectively. Therefore, passbands and stopbands in the original transfer function give rise to passbands and stopbands, respectively, in the derived transfer function. Furthermore, a stable transfer function results in a stable transfer function. By choosing the parameters l, m, and a_t in Eq. (7.46) properly, a set of four specific transformations can be obtained that can be used to transform a lowpass transfer function into a corresponding lowpass, highpass, bandpass, or a bandstop transfer function. These transformations are summarized in Table 7.2, where subscript i is included to facilitate the application of the transformations to 2-D discrete transfer functions.

Let Ω_i and ω_i for $i = 1, 2$ be the frequency variables in the original and transformed transfer function, respectively. If $H_N(z_1, z_2)$ is a lowpass transfer function with respect z_i with a passband edge Ω_{pi}, then the application of the lowpass-to-lowpass or lowpass-to-highpass transformation of Table 7.2 will yield a lowpass or highpass transfer function with respect to z_i with a passband edge of ω_{pi}. On the other hand, if the lowpass-to-bandpass or lowpass-to-bandstop transformation of Table 7.2 is applied, a bandpass or bandstop filter is obtained with passband edges ω_{p1i} and ω_{p2i}. On applying any two of these transformations to $H_N(z_1, z_2)$, the 2-D transfer function

$$H(\bar{z}_1, \bar{z}_2) = H_N(z_1, z_2) \mid_{z_1 = f_1(\bar{z}_1),\ z_2 = f_2(\bar{z}_2)}$$

is obtained. Some of the possible amplitude responses are illustrated in Fig. 7.17a–f.

7.6 DESIGN OF FILTERS SATISFYING PRESCRIBED SPECIFICATIONS

Practical 2-D digital filters differ from their ideal counterparts in that their passband gain is only approximately equal to unity, their stopband gain is only approximately equal to zero, and transitions between passbands and stopbands are gradual. Practical filters can be made to approach their ideal counterparts as closely as desired by increasing the order of the transfer function, but since this would increase the complexity of the filter, and

Table 7.2 Constantinides Transformations

Type	Transformation	Parameters
LP to LP	$z_i = \dfrac{\bar{z}_i - \alpha_i}{1 - \alpha_i \bar{z}_i}$	$\alpha_i = \dfrac{\sin[(\Omega_{pi} - \omega_{pi})T_i/2]}{\sin[(\Omega_{pi} + \omega_{pi})T_i/2]}$
LP to HP	$z_i = -\dfrac{\bar{z}_i - \alpha_i}{1 - \alpha_i \bar{z}_i}$	$\alpha_i = \dfrac{\cos[(\Omega_{pi} - \omega_{pi})T_i/2]}{\cos[(\Omega_{pi} + \omega_{pi})T_i/2]}$
LP to BP	$z_i = -\dfrac{\bar{z}_i^2 - \dfrac{2\alpha_i k_i}{k_i + 1}\,\bar{z}_i + \dfrac{k_i - 1}{k_i + 1}}{1 - \dfrac{2\alpha_i k_i}{k_i + 1}\,\bar{z}_i + \dfrac{k_i - 1}{k_i + 1}\bar{z}_i^2}$	$\alpha_i = \dfrac{\cos[(\omega_{p2i} + \omega_{p1i})T_i/2]}{\cos[(\omega_{p2i} - \omega_{p1i})T_i/2]}$ $k_i = \tan\dfrac{\Omega_{pi}T_i}{2}\cot\dfrac{(\omega_{p2i} - \omega_{p1i})T_i}{2}$
LP to BS	$z_i = \dfrac{\bar{z}_i^2 - \dfrac{2\alpha_i}{1 + k_i}\,\bar{z}_i + \dfrac{1 - k_i}{1 + k_i}}{1 - \dfrac{2\alpha_i}{1 + k_i}\,\bar{z}_i + \dfrac{1 - k_i}{1 + k_i}\bar{z}_i^2}$	$\alpha_i = \dfrac{\cos[(\omega_{p2i} + \omega_{p1i})T_i/2]}{\cos[(\omega_{p2i} - \omega_{p1i})T_i/2]}$ $k_i = \tan\dfrac{\Omega_{pi}T_i}{2}\tan\dfrac{(\omega_{p2i} - \omega_{p1i})T_i}{2}$

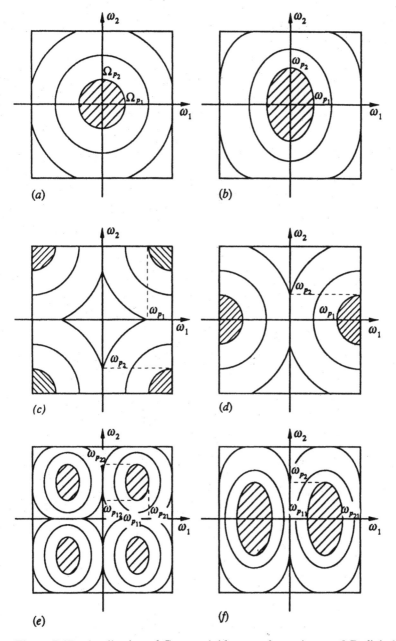

Figure 7.17 Application of Constantinides transformations to 2-D digital filters: (a) circularly symmetric lowpass filter; (b) LP to LP for z_1 and z_2; (c) LP to HP for z_1 and z_2; (d) LP to HP for z_1 and LP to LP for z_2; (e) LP to BP for z_1 and z_2; (f) LP to BP for z_1 and LP to LP for z_2.

hence the amount of computation needed for its implementation, the lowest-order transfer function that satisfies the required specifications with a small safety margin should be used.

The problem of designing filters that satisfy prescribed specifications has to a large extent been solved for the case of 1-D filters, and by extending the available methods [14], 2-D digital filters that satisfy prescribed specifications can also be designed.

7.6.1 Lowpass Filters with Rectangular Passbands

Let us assume that a 2-D lowpass digital filter characterized by $H(z_1, z_2)$ is obtained by cascading two 1-D lowpass digital filters characterized by $H_1(z_1)$ and $H_2(z_2)$. The amplitude response of the cascade filter is given by

$$M(\omega_1, \omega_2) = M_1(\omega_1)M_2(\omega_2) \tag{7.47}$$

where

$$M_i(\omega_i) = |H_i(e^{j\omega_i T_i})|, \quad i = 1, 2$$

If we assume that

$$(1 - \delta_{pi}) \le M_i(\omega_i) \le 1 \quad \text{for } |\omega_i| \le \omega_{pi} \tag{7.48a}$$

$$0 \le M_i(\omega_i) \le \delta_{ai} \quad \text{for } |\omega_i| \ge \omega_{ai} \tag{7.48b}$$

then Eqs. (7.47) and (7.48) yield

$$(1 - \delta_{p1})(1 - \delta_{p2}) \le M(\omega_1, \omega_2) \le 1 \quad \text{for } |\omega_1| \le \omega_{p1} \text{ and } |\omega_2| \le \omega_{p2} \tag{7.49a}$$

$$0 \le M(\omega_1, \omega_2) \le \delta_{a1} \quad \text{for } |\omega_1| \ge \omega_{a1} \tag{7.49b}$$

$$0 \le M(\omega_1, \omega_2) \le \delta_{a2} \quad \text{for } |\omega_2| \ge \omega_{a2} \tag{7.49c}$$

Now if the minimum passband gain and the maximum stopband gain of the cascade arrangement are assumed to be $(1 - \Delta_p)$ and Δ_a, respectively, and that $\delta_{p1} = \delta_{p2} = \delta_p$ and $\delta_{a1} = \delta_{a2} = \delta_a$, Eq. (7.49a) gives

$$(1 - \delta_p)^2 = 1 - \Delta_p$$

and since $\delta_p \ll 1$, we have

$$\delta_p = \tfrac{1}{2}\Delta_p$$

On the other hand, Eqs. (7.49b) and (7.49c) give

$$\delta_a = \Delta_a$$

Consequently, if the maximum allowable passband and stopband errors Δ_p and Δ_a are specified, the maximum passband ripple A_p and the minimum stopband attenuation A_a in dB for the two 1-D filters can be obtained as

$$A_p = 20 \log \frac{1}{1 - \delta_p} = 20 \log \frac{2}{2 - \Delta_p}$$

and

$$A_a = 20 \log \frac{1}{\delta_a} = 20 \log \frac{1}{\Delta_a}$$

Finally, if the passband and stopband edges ω_{pi} and ω_{ai} are also specified, the minimum order and the transfer function of each of the two 1-D filters can readily be obtained using the design procedure in Sec. 8.4 of [14].

7.6.2 Circularly Symmetric Lowpass Filters

A similar approach to that described in the preceding section can be used for the design of circularly symmetric lowpass filters.

Consider a 1-D analog lowpass filter characterized by $H_A(s)$ and assume that its amplitude response satisfies the requirements

$$(1 - \delta_p) \leq M_A(\omega) \leq 1 \quad \text{for } |\omega| \leq \omega_p \quad (7.50a)$$
$$0 \leq M_A(\omega) \leq \delta_a \quad \text{for } |\omega| \geq \omega_a \quad (7.50b)$$

If two 2-D discrete transfer functions $H_1(z_1, z_2)$ and $H_2(z_1, z_2)$ are obtained by using rotation angles β_1 and β_2 in the method of Costa and Venetsanopoulos where $180° \leq \beta_1 \leq 270°$ and $270° < \beta_1 \leq 360°$, the contour plots of the amplitude responses of the two filters will be of the form illustrated in Fig. 7.18a and b, where solid, dashed, and dotted curves are contours of gain equal to unity, $1 - \delta_p$, and δ_a, respectively. If the two filters are connected in cascade, the amplitude response of the cascade arrangement is given by

$$M(\omega_1, \omega_2) = M_1(\omega_1, \omega_2)M_2(\omega_1, \omega_2) \quad (7.51)$$

and its contour plot can be constructed as depicted in Fig. 7.18c. At the intersections of solid curves, the gain is unity; at the intersections of dashed curves, shown as circles, the gain is $(1 - \delta_p)^2$; at the intersections of dotted curves, shown as triangles, the gain is δ_a^2; and at the intersections of solid curves and dotted curves, shown as crosses, the gain is δ_a. Consequently

$$(1 - \delta_p)^2 \leq M(\omega_1, \omega_2) \leq 1 \quad \text{for } \sqrt{(\omega_1^2 + \omega_2^2)} \leq \Omega_p \quad (7.52a)$$
$$0 \leq M(\omega_1, \omega_2) \leq \delta_a \quad \text{for } \sqrt{(\omega_1^2 + \omega_2^2)} \geq \Omega_a \quad (7.52b)$$

where

$$\Omega_p = \frac{2}{T} \tan^{-1} \frac{\omega_p T}{2} \quad \text{and} \quad \Omega_a = \frac{2}{T} \tan^{-1} \frac{\omega_a T}{2}$$

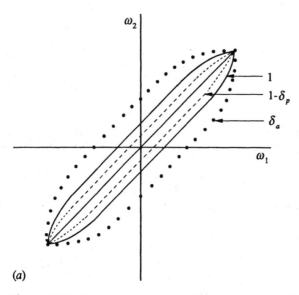

(a)

Figure 7.18 Contour plots of amplitude response: (a) filter section rotated by angle β_1; (b) filter section rotated by angle β_2; (c) both filter sections in cascade.

The parameters Ω_p and Ω_a are the radii of the circular passband and stopband boundaries that would be achieved if a large number of rotation angles were to be used in the method of Costa and Venetsanopoulos. As can be easily demonstrated, they are numerically equal to the passband and stopband edges of the 1-D digital filter that would be obtained by applying the bilinear transformation to $H_A(s)$.

If K rotated filter sections are connected in cascade where half of the rotation angles are in the range 180° to 270° and the other half are in the range 270° to 360°, we obtain

$$(1 - \delta_p)^K \le M(\omega_1, \omega_2) \le 1 \qquad \text{for } \sqrt{(\omega_1^2 + \omega_2^2)} \le \Omega_p \quad (7.53a)$$

$$0 \le M(\omega_1, \omega_2) \le \delta_a^{K/2} \qquad \text{for } \sqrt{(\omega_1^2 + \omega_2^2)} \ge \Omega_a \quad (7.53b)$$

If the minimum passband gain and the maximum stopband gain of the 2-D rotated filter are assumed to be $(1 - \Delta_b)$ and Δ_a, respectively, Eq. (7.53) gives

$$\delta_p = \frac{1}{K} \Delta_p \qquad (7.54a)$$

and

$$\delta_a = \Delta_a^{2/K} \qquad (7.54b)$$

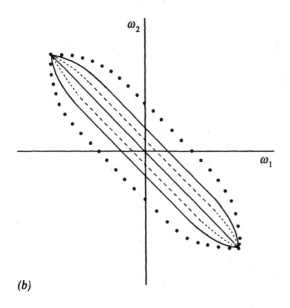

(b)

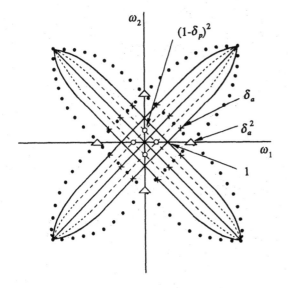

(c)

The lower (or upper bound) of the passband gain given by Eq. (7.53a) would be achieved if all the rotated sections were to have minimum (or maximum) passband gains at the same frequency point. Although it is possible for all the rotated sections to have minimum (or maximum) gains at the origin of the (ω_1, ω_2) plane, the gains are unlikely to be maximum (or minimum) together at some other frequency point and, in effect, the preceding estimate for δ_p is low. A more realistic value for δ_p is

$$\delta_p = \frac{2}{K} \Delta_p \qquad (7.54c)$$

If Δ_p and Δ_a are prescribed, then the passband ripple and minimum stopband attenuation of the analog filter can be obtained from Eqs. (7.54c) and (7.54b) as

$$A_p = 20 \log \frac{K}{K - 2\Delta_p}$$

and

$$A_a = \frac{40}{K} \log \frac{1}{\Delta_a}$$

If Ω_p and Ω_a are also prescribed, the minimum order and the transfer function of the analog filter can be obtained using the method in Sec. 8.4 of [14].

REFERENCES

1. J. L. Shanks, S. Treitel, and J. H. Justice, Stability and synthesis of two-dimensional recursive filters, *IEEE Trans. Audio Electroacoust.*, vol. AU-20, pp. 115–128, June 1972.
2. J. M. Costa and A. N. Venetsanopoulos, Design of circularly symmetric two-dimensional recursive filters, *IEEE Trans. Acoust., Speech, Signal Process.*, vol. ASSP-22, pp. 432–443, Dec. 1974.
3. N. A. Pendergrass, S. K. Mitra, and E. I. Jury, Spectral transformations for two-dimensional digital filters, *IEEE Trans. Circuits Syst.*, vol. CAS-23, pp. 26–35, Jan. 1976.
4. M. Ahmadi, A. G. Constantinides, and R. A. King, Design technique for a class of stable two-dimensional recursive digital filters, *Proc. 1976 IEEE Int. Conf. Acoust., Speech, Signal Process.*, pp. 145–147.
5. S. Chakrabarti and S. K. Mitra, Design of two-dimensional digital filters via spectral transformations, *Proc. IEEE*, vol. 65, pp. 905–914, June 1977.
6. D. M. Goodman, A design technique for circularly symmetric low-pass filters, *IEEE Trans. Acoust., Speech, Signal Process.*, vol. ASSP-26, pp. 290–304, Aug. 1978.

7. K. Hirano and J. K. Aggarwal, Design of two-dimensional recursive digital filters, *IEEE Trans. Circuits Syst.*, vol. CAS-25, pp. 1066–1076, Dec. 1978.
8. A. Antoniou, M. Ahmadi, and C. Charalambous, Design of factorable low-pass 2-dimensional digital filters satisfying prescribed specifications, *IEE Proc.*, vol. 128, pt. G, pp. 53–60, April 1981.
9. G. V. Mendonça, A. Antoniou, and A. N. Venetsanopoulos, Design of two-dimensional pseudorotated digital filters satisfying prescribed specifications, *IEEE Trans. Circuits Syst.*, vol. CAS-34, pp. 1–10, Jan. 1987.
10. G. A. Maria and M. M. Fahmy, An L_p design technique for two-dimensional digital recursive filters, *IEEE Trans. Acoust., Speech, Signal Process.*, vol. ASSP-22, pp. 15–21, Feb. 1974.
11. C. Charalambous, Design of 2-dimensional circularly symmetric digital filters, *IEE Proc.*, vol. 129, pt. G, pp. 47–54, April 1982.
12. P. A. Ramamoorthy and L. T. Bruton, Design of stable two-dimensional analog and digital filters with applications in image processing, *Int. J. Circuit Theory Appl.*, vol. 7, pp. 229–246, April 1979.
13. A. G. Constantinides, Spectral transformations for digital filters, *IEE Proc.*, vol. 117, pp. 1585–1590, Aug. 1970.
14. A. Antoniou, *Digital Filters: Analysis and Design*, New York: McGraw-Hill, 1979 (2nd ed. in press).
15. J. M. Costa and A. N.Venetsanopoulos, A group of linear spectral transformations for two-dimensional digital filters, *IEEE Trans. Acoust., Speech, Signal Process.*, vol. ASSP-24, pp. 424–425, Oct. 1976.
16. K. P. Prasad, A. Antoniou, and B. B. Bhattacharyya, On the properties of linear spectral transformations for 2-dimensional digital filters, *Circuits Systems Signal Process.*, vol. 2, pp. 203–211, 1983.
17. M. Hammermesh, *Group Theory and Its Applications to Physical Problems*, 2nd ed., Reading, Mass.: Addison-Wesley, 1964, pp. 41–43.
18. N. V. V. J. Swamy and M. A. Samuel, *Group Theory Made Easy for Scientists and Engineers*, New York: John Wiley, 1979, pp. 108–114.
19. P. Karivaratharajan, H. C. Reddy, M. N. S. Swamy, and V. Ramachandran, Generalization of two-dimensional digital functions without nonessential singularities of second kind, *IEEE Trans. Acoust., Speech, Signal Process.*, vol. ASSP-28, pp. 216–223, April 1980.
20. A. M. Ali, A. G. Constantinides, and R. A. King, On 2-variable reactance functions for 2-dimensional recursive filters, *Electron. Lett.*, vol. 14, pp. 12–13, Jan. 1978.

PROBLEMS

7.1 Verify the entries in Table 7.1.

7.2 The third-order Butterworth lowpass analog transfer function is given by

$$H_{A1}(s) = \frac{1}{(s + 1)(s^2 + s + 1)}$$

(a) Apply the transformation of Eq. (7.12a) to obtain a 2-D continuous transfer function $H_{A2}(s_1, s_2)$.

(b) Apply the bilinear transfer transformation of Eq. (7.2) to $H_{A2}(s_1, s_2)$ obtained in part (a).

(c) Verify the stability of the transfer function obtained in part (b).

7.3 Repeat Problem 7.2 with the transformation in Eq. (7.12a) replaced by that in Eq. (7.15) where $a_1 = a_2 = 1$ and $c = 0.1$.

7.4 (a) Show that the transformations in Eqs. (7.12a) and (7.12b) are globally type preserving.

(b) Show that the transformation in Eq. (7.15) is locally type preserving with Ω_I given by

$$\Omega_I = \{(\omega_1, \omega_2) : \omega_1 \geq 0, \omega_2 \geq 0, \omega_1\omega_2 < 1/c\}$$

7.5 Using MATLAB, implement the design procedure described in Example 7.1.

7.6 Design a circularly symmetric lowpass 2-D filter by applying the method of Costa and Venetsanopoulos to the third-order Butterworth lowpass transfer function in Problem 7.2. Use $K = 6$.

7.7 Repeat Problem 7.6 with $H_{A1}(s)$ replaced by

$$H_{A1}(s) = \frac{0.2457}{\prod_{i=1}^{2}(s - p_i)(s - p_i^*)}$$

where

$$p_1, p_1^* = -0.1395 \pm j0.9834$$

$$p_2, p_2^* = -0.3369 \pm j0.4073$$

7.8 Design a circularly symmetric lowpass 2-D filter by applying the method of Goodman to the third-order Butterworth transfer function. Use $K = 6$.

7.9 Repeat Problem 7.8 with $H_{A1}(s)$ given in Problem 7.7.

7.10 Design a lowpass filter with a rectangular passband satisfying the following specifications:

$$\omega_{s1} = \omega_{s2} = 2\pi \text{ rad/s}$$

$$\omega_{p1} = \omega_{p2} = 0.5\pi \text{ rad/s}$$

$$\omega_{a1} = \omega_{a2} = 0.58\pi \text{ rad/s}$$

$$\delta_{p1} \doteq \delta_{p2} = 0.03$$

$$\delta_{a1} = \delta_{a2} = 0.05$$

7.11 Using the method of Costa and Venetsanopoulos design a circularly symmetric lowpass filter satisfying the following specifications:

$$\omega_{s1} = \omega_{s2} = 2\pi \text{ rad/s}$$

$$\Omega_p = 0.5\pi \text{ rad/s}, \ \Omega_a = 0.62\pi \text{ rad/s}$$

$$\delta_p = \delta_a = 0.08$$

8
Design of Recursive Filters by Optimization

8.1 INTRODUCTION

In Chap. 7, several methods for the design of 2-D recursive filters have been described. In these methods, the 2-D discrete transfer function is obtained indirectly by applying transformations to 1-D continuous transfer functions. The designs are readily obtained using closed-form solutions, and, therefore, low approximation error can be achieved with a minimal amount of computational effort. Unfortunately, the application of these methods is restricted to certain types of filter characteristics, for example, for filters with well-defined passbands and stopbands and approximately piecewise-constant amplitude responses.

In this chapter, we examine design methods that involve the application of optimization techniques. These methods are both flexible and versatile and can be used to obtain a variety of filter characteristics. It should be mentioned, however, that the amount of computation required can sometimes be considerable.

In Secs. 8.2 and 8.3, the design is carried out directly in the z domain using standard least pth or minimax algorithms [1,2]. In Sec. 8.4, a stable 2-D digital-filter is obtained from a corresponding two-variable stable analog network, and its parameters are optimized to yield some specified amplitude response. In Sec. 8.5, a method based on the singular-value decomposition is described in which the problem of designing a 2-D filter

230

is broken down into a problem of designing a set of 1-D digital filters [3]. The 1-D filters obtained are designed using standard optimization methods.

8.2 DESIGN BY LEAST pth OPTIMIZATION

The least pth optimization method has been used quite extensively in the past in a variety of applications. In this approach, an objective function in the form of a sum of elemental error functions, each raised to the pth power, is first formulated and is then minimized using any one of the many available unconstrained optimization methods [4,5].

8.2.1 Problem Formulation

Consider the transfer function

$$H(z_1, z_2) = H_0 \prod_{k=1}^{K} \frac{N_k(z_1, z_2)}{D_k(z_1, z_2)} = H_0 \prod_{k=1}^{K} \frac{\sum_{l=0}^{L_{1k}} \sum_{m=0}^{M_{1k}} a_{lm}^{(k)} z_1^{-l} z_2^{-m}}{\sum_{l=0}^{L_{2k}} \sum_{m=0}^{M_{2k}} b_{lm}^{(k)} z_1^{-l} z_2^{-m}} \quad (8.1)$$

where $N_k(z_1, z_2)$ and $D_k(z_1, z_2)$ are polynomials of order equal to or less than 2 and H_0 is a constant, and let

$$\mathbf{x} = [\mathbf{a}^T \, \mathbf{b}^T \, H_0]^T \quad (8.2)$$

where

$$\mathbf{a}^T = [a_{10}^{(1)} \, a_{20}^{(1)} \cdots a_{L_{11}M_{11}}^{(1)} \, a_{10}^{(2)} \, a_{20}^{(2)} \cdots a_{L_{12}M_{12}}^{(2)} \cdots a_{L_{1K}M_{1K}}^{(K)}]$$

and

$$\mathbf{b}^T = [b_{10}^{(1)} \, b_{20}^{(1)} \cdots b_{L_{21}M_{21}}^{(1)} \, b_{10}^{(2)} \, b_{20}^{(2)} \cdots b_{L_{22}M_{22}}^{(2)} \cdots b_{L_{2K}M_{2K}}^{(K)}]$$

are row vectors whose elements are the coefficients of $N_k(z_1, z_2)$ and $D_k(z_1, z_2)$, respectively. The design task at hand amounts to finding a parameter vector $\mathbf{x}$ that minimizes the least pth *objective* function $J(\mathbf{x})$ defined by

$$J(\mathbf{x}) = \sum_{n_1=1}^{N_1} \sum_{n_2=1}^{N_2} [M(n_1, n_2) - M_I(n_1, n_2)]^p \quad (8.3)$$

where

$$M(n_1, n_2) = |H(e^{j\omega_{1n_1} T_1}, e^{j\omega_{2n_2} T_2})|, \quad n_1 = 1, \ldots, N_1, n_2 = 1, \ldots, N_2$$

are samples of the amplitude response of the filter at a set of frequency pairs $\{(\omega_{1n_1}, \omega_{2n_2}) : n_1 = 1, \ldots, N_1, n_2 = 1, \ldots, N_2\}$ with

$$\omega_{1n_1} = \omega_{s_1}(n_1 - 1)/2(N_1 - 1) \quad \text{and} \quad \omega_{1n_2} = \omega_{s_2}(n_2 - 1)/2(N_2 - 1)$$

$M_I(n_1, n_2)$ represents the desired amplitude response at frequencies $(\omega_{1n_1}, \omega_{2n_2})$ and p is an even positive integer.

8.2.2 Quasi-Newton Optimization Algorithms

The design problem described above can be solved by using any one of the standard unconstrained optimization algorithms. A class of such algorithms that have been found to be very versatile, efficient, and robust is the class of quasi-Newton algorithms [4,5]. These are based on the principle that the minimum point $\mathbf{x}^*$ of a quadratic convex function $F(\mathbf{x})$ of N variables can be obtained by applying the correction

$$\boldsymbol{\delta} = -\mathbf{H}^{-1}\mathbf{g}$$

to an arbitrary point $\mathbf{x}$, that is,

$$\mathbf{x}^* = \mathbf{x} + \boldsymbol{\delta}$$

where vector

$$\mathbf{g} = \nabla f(\mathbf{x})$$

and $N \times N$ matrix

$$\mathbf{H} = \begin{bmatrix} \dfrac{\partial^2 f(\mathbf{x})}{\partial x_1^2} & \dfrac{\partial^2 f(\mathbf{x})}{\partial x_1\,\partial x_2} & \cdots & \dfrac{\partial^2 f(\mathbf{x})}{\partial x_1\,\partial x_n} \\[2ex] \dfrac{\partial^2 f(\mathbf{x})}{\partial x_2\,\partial x_1} & \dfrac{\partial^2 f(\mathbf{x})}{\partial x_2^2} & \cdots & \dfrac{\partial^2 f(\mathbf{x})}{\partial x_2\,\partial x_n} \\[2ex] \vdots & \vdots & & \vdots \\[2ex] \dfrac{\partial^2 f(\mathbf{x})}{\partial x_n\,\partial x_1} & \dfrac{\partial^2 f(\mathbf{x})}{\partial x_n\,\partial x_2} & \cdots & \dfrac{\partial^2 f(\mathbf{x})}{\partial x_n^2} \end{bmatrix}$$

are the gradient vector and Hessian matrix of $f(\mathbf{x})$ at point $\mathbf{x}$, respectively.

The basic quasi-Newton algorithm as applied to our design problem is as follows.

Algorithm 1

STEP 1: Input $\mathbf{x}_0$ and ε. Set $\mathbf{S}_0 = \mathbf{I}_N$, where $\mathbf{I}_N$ is the $N \times N$ unity matrix and N is the dimension of $\mathbf{x}$, and set $k = 0$. Compute $g_0 = \nabla J(\mathbf{x}_0)$.

STEP 2: Set $\mathbf{d}_k = -\mathbf{S}_k\mathbf{g}_k$ and find α_k, the value of α that minimizes $J(\mathbf{x}_k + \alpha\mathbf{d}_k)$, using a line search.

STEP 3: Set $\boldsymbol{\delta}_k = \alpha_k\mathbf{d}_k$ and $\mathbf{x}_{k+1} = \mathbf{x}_k + \boldsymbol{\delta}_k$.

STEP 4: If $\|\boldsymbol{\delta}_k\|_2 < \varepsilon$, then output $\mathbf{x}^* = \mathbf{x}_{k+1}$, $J(\mathbf{x}^*) = J(\mathbf{x}_{k+1})$ and stop.

STEP 5: Compute $\mathbf{g}_{k+1} = \nabla J(\mathbf{x}_{k+1})$ and set $\boldsymbol{\gamma}_k = \mathbf{g}_{k+1} - \mathbf{g}_k$.

STEP 6: Compute $\mathbf{S}_{k+1} = \mathbf{S}_k + \mathbf{C}_k$, where $\mathbf{C}_k$ is a suitable matrix correction (see below).

STEP 7: Set $k = k + 1$ and go to Step 2.

Vector x_0 is an initial point and ε is a termination tolerance which depends on the application. In Steps 1 and 5, the gradient vector ∇J is required. From Eq. (8.3), we have

$$\nabla J(x) = \left[\frac{\partial J}{\partial x_1} \cdots \frac{\partial J}{\partial x_N} \right]^T$$

where

$$\frac{\partial J}{\partial x_k} = p \sum_{n_1=1}^{N_1} \sum_{n_2=1}^{N_2} [M(n_1, n_2) - M_I(n_1, n_2)]^{p-1} \frac{\partial M(n_1, n_2)}{\partial x_k}$$

$$= p \sum_{n_1=1}^{N_1} \sum_{n_2=1}^{N_2} [M(n_1, n_2)$$

$$- M_I(n_1, n_2)]^{p-1} M^{-1}(n_1, n_2) \mathrm{Re}\left(\overline{H} \frac{\partial H}{\partial x_k} \right) \qquad (8.4)$$

and $\mathrm{Re}[\overline{H}(\partial H/\partial x_k)]$ is the real part of

$$H(e^{-j\omega_1 n_1 T_1}, e^{-j\omega_2 n_2 T_2}) \frac{\partial H(e^{-j\omega_1 n_1 T_1}, e^{-j\omega_2 n_2 T_2})}{\partial x_k}$$

The correction matrix C_k required in Step 6 can be computed by using either the Davidon–Fletcher–Powell (DFP) formula

$$C_k = \frac{\delta_k \delta_k^T}{\delta_k^T \gamma_k} - \frac{S_k \gamma_k \gamma_k^T S_k}{\gamma_k^T S_k \gamma_k} \qquad (8.5)$$

or the Broyden–Fletcher–Goldfarb–Shanno (BFGS) formula

$$C_k = \left(1 + \frac{\gamma_k^T S_k \gamma_k}{\gamma_k^T \delta_k} \right) \frac{\delta_k \delta_k^T}{\gamma_k^T \delta_k} - \left(\frac{\delta_k \gamma_k^T S_k + S_k \gamma_k \delta_k^T}{\gamma_k^T \delta_k} \right) \qquad (8.6)$$

These formulas have two very important properties. First, for a quadratic convex function $J(x)$, matrix S_N becomes equal to H^{-1}; that is, the preceding algorithm will yield the solution in $N + 1$ iterations. Second, under certain conditions that are easily satisfied in practice, a positive-definite matrix S_k will yield a positive-definite matrix S_{k+1}. This property is often referred to as the "hereditary" property [4].

Initially, in Step 1 of the algorithm, matrix S_0 is set equal to the $N \times N$ unity matrix. Hence vector d_0 is the direction of maximum descent. Minimizing $J(x_0 + \alpha d_0)$ with respect to α in Step 2 will yield a certain reduction in the objective function. Since S_0 is positive definite, matrix S_1 obtained in Step 6 will be positive definite by virtue of the hereditary property. Therefore, the minimization of $J(\mathbf{x}_1 + \alpha\mathbf{d}_1)$ during the second iteration will yield a further reduction in the objective function. Proceeding in the same way, the objective function will progressively be reduced from iteration to iteration and eventually a point in the neighborhood of a local minimum point will be obtained. In such a domain, $J(\mathbf{x})$ can accurately be represented by a quadratic convex approximation and, as a result, $\mathbf{S}_{k+1}$ will become an accurate representation of the inverse Hessian matrix. Therefore, the test in Step 4 will eventually be satisfied and the algorithm will converge.

Quasi-Newton algorithms have a number of important advantages relative to other unconstrained optimization algorithms as follows:

1. The second derivatives of the objective function are not required.
2. Matrix inversion is unnecessary.
3. The hereditary property of the DFP and BFGS updating formulas eliminates the need to check and possibly manipulate matrix $\mathbf{S}_k$.
4. An inexact line search can be used for the minimization of $J(\mathbf{x}_k + \alpha\mathbf{d}_k)$, particularly if the BFGS updating formula is used [4].

As a result, quasi-Newton algorithms are readily applicable to a large range of optimization problems, they are very efficient, and they are not susceptible to numerical ill-conditioning.

The application of Algorithm 1 for the design of 2-D digital filters may result in an unstable transfer function. This problem can be avoided as suggested by Maria and Fahmy [1] by checking the stability of the filter in Step 3, and if any one of the filter sections is found to be unstable, then δ_k is progressively halved until a stable design is obtained. However, the outcome is likely to be a transfer function that has a very small stability margin. Alternatively, the designer may try different initial points $\mathbf{x}_0$ located in the stable domain of the parameter space and choose the best stable design achieved. Design methods that eliminate the problem of stability altogether are described in Secs. 8.4 and 8.5.

Owing to the large number of variables in 2-D digital filters and the large number of sample points needed to construct the objective function, the amount of computation required to complete a design is usually very large. The amount of computation can often be reduced by starting with an approximate design based on some closed-form solution. For example, the design of filters with circular or elliptical passbands and stopbands may start with filters that have square or rectangular passbands and stopbands (see Sec. 7.3) [1].

Example 8.1 Design a circularly symmetric lowpass filter of order (2, 2) with $\omega_{p1} = \omega_{p2} = 0.08\pi$ rad/s and $\omega_{a1} = \omega_{a2} = 0.12\pi$ rad/s assuming that $\omega_{s1} = \omega_{s2} = 2\pi$ rad/s.

Solution.   Let the ideal amplitude response of the filter be

$$M_I(n_1, n_2) = \begin{cases} 1 & \text{for } \sqrt{\omega_{1n_1}^2 + \omega_{2n_2}^2} \leq 0.08\pi \\ 0.5 & \text{for } 0.08\pi < \sqrt{\omega_{1n_1}^2 + \omega_{2n_2}^2} < 0.12\pi \\ 0 & \text{otherwise} \end{cases}$$

where

$$\{\omega_{1n_1}\} = 0, 0.02\pi, 0.04\pi, \ldots, 0.2\pi, 0.4\pi, 0.6\pi, 0.8\pi, \pi$$

$$\{\omega_{1n_2}\} = 0, 0.02\pi, 0.04\pi, \ldots, 0.2\pi, 0.4\pi, 0.6\pi, 0.8\pi, \pi$$

To reduce the amount of computation, a 1-D lowpass filter with pass-band edge $\omega_p = 0.08\pi$ and stopband edge $\omega_a = 0.1\pi$ was designed [1] and the 1-D transfer function

$$H_1(z) = 0.11024 \frac{1 - 1.64382z^{-1} + z^{-2}}{1 - 1.79353z^{-1} + 0.84098z^{-2}}$$

was obtained. Then a 2-D transfer function with a rectangular passband was obtained as

$$H_1(z_1)H_1(z_2)$$

$$= 0.01215 \frac{[1 \; z_1^{-1} \; z_1^{-2}] \begin{bmatrix} 1.0 & -1.64382 & 1.0 \\ -1.64382 & 2.70214 & -1.64382 \\ 1.0 & -1.64382 & 1.0 \end{bmatrix} \begin{bmatrix} 1 \\ z_2^{-1} \\ z_2^{-2} \end{bmatrix}}{[1 \; z_1^{-1} \; z_1^{-2}] \begin{bmatrix} 1.0 & -1.79353 & 0.84098 \\ -1.79353 & 3.21675 & -1.50832 \\ 0.84098 & -1.50832 & 0.70725 \end{bmatrix} \begin{bmatrix} 1 \\ z_2^{-1} \\ z_2^{-2} \end{bmatrix}}$$

Using this transfer function with $p = 4$, the objective function $J(\mathbf{x})$ assumes the value 0.32985. In order to apply Algorithm 1, the gradient of $J(\mathbf{x})$ is required. Since the order of the filter to be designed is $(2, 2)$, the parameters in (8.1) and (8.2) can be specified as $K = 1$, $L_{11} = M_{11} = L_{21} = M_{21} = 2$, and

$$\mathbf{x} = [a_{10} \; a_{20} \; a_{01} \; a_{11} \; a_{21} \; a_{02} \; a_{12} \; a_{22} \; b_{10} \; b_{20} \; b_{01} \; b_{11} \; b_{21} \; b_{02} \; b_{12} \; b_{22} \; H_0]^T$$

The gradient of $J(\mathbf{x})$ can be computed by using the formulas

$$\frac{\partial J}{\partial a_{ij}} = H_0 \frac{z_1^{-i} z_2^{-j}}{D_1(z_1, z_2)} \qquad \text{for } i, j = 0, 1, 2$$

$$\frac{\partial J}{\partial b_{ij}} = -H_0 \frac{N_1(z_1, z_2)}{D_1^2(z_1, z_2)} z_1^{-i} z_2^{-j} \qquad \text{for } i, j = 0, 1, 2$$

and

$$\frac{\partial J}{\partial H_0} = \frac{N_1(z_1, z_2)}{D_1(z_1, z_2)}$$

With $\varepsilon = 5 \times 10^{-4}$, Algorithm 1 converged after 7 iterations and lead to the transfer function

$$H(z_1, z_2)$$

$$= 0.01153 \frac{[1 \; z_1^{-1} \; z_1^{-2}] \begin{bmatrix} 1.0 & -1.64174 & 0.99250 \\ -1.64174 & 2.70580 & -1.64181 \\ 0.99250 & -1.64181 & 1.00813 \end{bmatrix} \begin{bmatrix} 1 \\ z_2^{-1} \\ z_2^{-2} \end{bmatrix}}{[1 \; z_1^{-1} \; z_1^{-2}] \begin{bmatrix} 1.0 & -1.78684 & 0.81687 \\ -1.78684 & 3.23408 & -1.49888 \\ 0.81687 & -1.49888 & 0.70521 \end{bmatrix} \begin{bmatrix} 1 \\ z_2^{-1} \\ z_2^{-2} \end{bmatrix}}$$

for which $J(\mathbf{x}) = 0.07341$. The amplitude response of the filter designed is depicted in Fig. 8.1a and b.

8.3 MINIMAX METHOD

In the design of recursive and nonrecursive filters, minimax optimization methods are often preferred because their application minimizes the maximum of the approximation error. In this section, the design of recursive filters is formulated as a minimax problem. Then two minimax optimization algorithms that have been used quite extensively in the past are described.

8.3.1 Problem Formulation

Consider again the transfer function $H(z_1, z_2)$ given by Eq. (8.1). Since the filter is to have an amplitude response that is approximately circularly symmetric, $H(z_1, z_2)$ has separable denominator [6]. Consequently, the transfer function can be expressed as

$$H_1(z_1, z_2) = H_0(z_1, z_2)^{-1} \prod_{k=1}^{K}$$

$$\frac{z_1 z_2 + (z_1 z_2)^{-1} + \alpha_k(z_1 + z_1^{-1} + z_2 + z_2^{-1}) + z_1^{-1} z_2 + z_1 z_2^{-1} + \beta_k}{(1 + c_k z_1^{-1} + d_k z_1^{-2})(1 + c_k z_2^{-1} + d_k z_2^{-2})}$$

$$(8.7)$$

and if the numerator of $H_1(z_1, z_2)$ is also separable, then the transfer

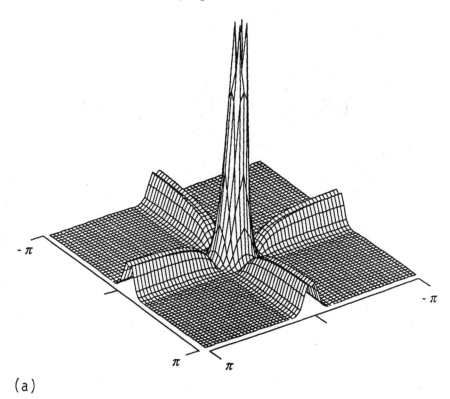

(a)

Figure 8.1 Amplitude response of circularly symmetric lowpass filter (Example 8.1): (a) for $-\pi \leq \omega_1, \omega_2 \leq \pi$; (b) for $-0.08\pi \leq \omega_1, \omega_2 \leq 0.08\pi$.

function assumes the form

$$H_2(z_1, z_2) = H_0 \prod_{k=1}^{K} \frac{(1 + \alpha_k z_1^{-1} + \beta_k z_1^{-2})(1 + \alpha_k z_2^{-1} + \beta_k z_2^{-2})}{(1 + c_k z_1^{-1} + d_k z_1^{-2})(1 + c_k z_2^{-1} + d_k z_2^{-2})} \quad (8.8)$$

The amplitude response of the filter represented by $H_1(z_1, z_2)$ is given by

$$M_1(\mathbf{x}, \omega_1, \omega_2) = \prod_{k=1}^{K} \frac{N_k}{D_k} \quad (8.9)$$

where

$$N_k = |4 \cos \omega_1 T_1 \cos \omega_2 T_2 + 2\alpha_k(\cos \omega_1 T_1 + \cos \omega_2 T_2) + \beta_k|$$

$$D_k = D_{k1} D_{k2}$$

$$D_{ki} = [1 + c_k^2 + d_k^2 + 2d_k(2 \cos^2 \omega_i T_i - 1) + 2c_k(1 + d_k) \cos \omega_i T_i]^{1/2},$$

$$i = 1,2$$

$$\mathbf{x} = [\alpha_1 \, \beta_1 \, c_1 \, d_1 \, \cdots \, \alpha_k \, \beta_k \, c_k \, d_k \, H_0]^T$$

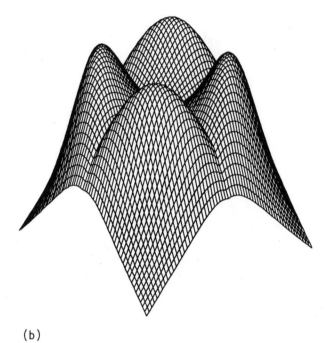

(b)

Figure 8.1 Continued.

and the amplitude response for the filter represented by $H_2(z_1, z_2)$ is given by

$$M_2(\mathbf{x}, \omega_1, \omega_2) = H_0 \prod_{k=1}^{K} \frac{N_{k1}N_{k2}}{D_{k1}D_{k2}} \qquad (8.10)$$

where

$$N_{ki} = [1 + \alpha_k^2 + \beta_k^2 + 2\beta_k(2 \cos^2 \omega_i T_i - 1)$$
$$+ 2\alpha_k(1 + \beta_k)\cos \omega_i T_i]^{1/2}, \qquad i = 1, 2$$

The design problem can be transformed into an unconstrained minimax optimization problem by defining an appropriate objective function. To be specific, consider designing a circularly symmetric lowpass filter with ω_p and ω_a as its passband and stopband edges, respectively, and let

$$M_I(\omega_1, \omega_2) = \begin{cases} 1 & \text{for } \sqrt{\omega_1^2 + \omega_2^2} < \omega_p \\ 0 & \text{for } \sqrt{\omega_1^2 + \omega_2^2} \geq \omega_a \text{ and } |\omega_1| \leq \frac{\omega_{1s}}{2}, |\omega_2| \leq \frac{\omega_{2s}}{2} \end{cases}$$

be the ideal amplitude response. An error function can be defined as

$$E(\mathbf{x}, \omega_1, \omega_2) = M(\mathbf{x}, \omega_1, \omega_2) - M_I(\omega_1, \omega_2) \qquad (8.11)$$

where $M(\mathbf{x}, \omega_1, \omega_2)$ is given by Eq. (8.9) or (8.10) depending on the type of numerator. This error function can be minimized by minimizing the sampled error function $E(\mathbf{x}, \omega_{1i}, \omega_{2i})$ with respect to a set of m sample points $(\omega_{1i}, \omega_{2i}) \in \Omega$ where Ω denotes the union of points in the passband and stopband regions. The preceding error function can be minimized in the minimax sense by finding a vector $\mathbf{x}$ that solves the following optimization problem:

$$\underset{\mathbf{x}}{\text{minimize }} F(\mathbf{x})$$

where

$$F(\mathbf{x}) = \max_{1 \le i \le m} f_i(\mathbf{x}) \qquad (8.12)$$

and

$$f_i(\mathbf{x}) = |E(\mathbf{x}, \omega_{1i}, \omega_{2i})|, \qquad i = 1, 2, \ldots, m$$

8.3.2 Minimax Optimization Algorithms

The problem described in the preceding section can be solved by formulating an objective function $\Phi(\mathbf{x})$ in terms of the L_p norm of the error function and then minimizing $\Phi(\mathbf{x})$ for increasing values of p. Such an objective function can be obtained from Eq. (8.11) as

$$\Phi(\mathbf{x}) = F(\mathbf{x}) \left\{ \sum_{i=1}^{m} \left[\frac{E(\mathbf{x}, \omega_{1i}, \omega_{2i})}{F(\mathbf{x})} \right]^p \right\}^{1/p}$$

where $F(\mathbf{x})$ is given by Eq. (8.12). A minimax algorithm based on $\Phi(\mathbf{x})$ is as follows [7]:

Algorithm 2

STEP 1: Input $\mathbf{x}_0$. Set $k = 1$, $p = 2$, $\mu = 2$, $F_0 = 10^{99}$.

STEP 2: Initialize frequencies ω_{1i}, ω_{2i} for $i = 1, 2, \ldots, m$.

STEP 3: Using point $\tilde{\mathbf{x}}_{k-1}$ as initial point, minimize $\Phi(\mathbf{x})$ with respect to $\mathbf{x}$ to obtain $\tilde{\mathbf{x}}_k$. Set $F_k = F(\tilde{\mathbf{x}}_k)$.

STEP 4: If $|F_{k-1} - F_k| < \varepsilon$, then output $\mathbf{x}^* = \tilde{\mathbf{x}}_k$ and F_k and stop. Else, set $p = \mu p$, $k = k + 1$ and go to Step 3.

The minimization in Step 3 can be carried out using Algorithm 1. The gradient of $F(\mathbf{x})$ is given by [7]

$$\nabla \Phi(\mathbf{x}) = \left\{ \sum_{i=1}^{m} \left[\frac{|E(\mathbf{x}, \omega_{1i}, \omega_{2i})|}{F(\mathbf{x})} \right]^{p} \right\}^{(1/p)-1}$$

$$\times \sum_{i=1}^{m} \left[\frac{|E(\mathbf{x}, \omega_{1i}, \omega_{2i})|}{F(\mathbf{x})} \right]^{p-1} \nabla |E(\mathbf{x}, \omega_{1i}, \omega_{2i})|$$

The preceding algorithm gives excellent results except that it requires a considerable amount of computation. An alternative and much more efficient algorithm is the minimax algorithm described in Charalambous [2] and Charalambous and Antoniou [8]. This algorithm is based on principles developed by Charlambous [9] and involves the minimization of the objective function $\Phi(\mathbf{x}, \xi, \lambda)$ defined by

$$\Phi(\mathbf{x}, \xi, \lambda) = \sum_{i \in I_1} \tfrac{1}{2}\lambda_i [\phi_i(\mathbf{x}, \xi)]^2 + \sum_{i \in I_2} \tfrac{1}{2}[\phi_i(\mathbf{x}, \xi)]^2 \qquad (8.13)$$

where

$$\phi_i(\mathbf{x}, \xi) = f_i(\mathbf{x}) - \xi$$
$$I_1 = \{i : \phi_i(\mathbf{x}, \xi) > 0 \text{ and } \lambda_i > 0\}$$

and

$$I_2 = \{i : \phi_i(\mathbf{x}, \xi) > 0 \text{ and } \lambda_i = 0\}$$

Parameters ξ and λ_i ($i = 1, 2, \ldots, m$) are constants. It can be shown [9] that if (a) the second-order sufficient conditions for a minimum hold at $\mathbf{x}^*$, (b) $\lambda_i = \lambda_i^*$, $i = 1, 2, \ldots, m$, where λ_i^* are the minimax multipliers corresponding to a minimum optimum solution $\mathbf{x}^*$, and (c) $F(\mathbf{x}^*) - \xi$ is sufficiently small, then $\mathbf{x}^*$ is a strong local minimum point of $\Phi(\mathbf{x}, \xi, \lambda)$. The conditions in (a) are usually satisfied in practice. Therefore, a local minimum point $\mathbf{x}^*$ can be found by forcing λ_i to approach λ_i^* ($i = 1, 2, \ldots, m$) and making $F(\mathbf{x}^*) - \xi$ sufficiently small. These two constraints can be simultaneously satisfied by applying the following algorithm.

Algorithm 3

STEP 1:   Set $\xi = 0$ and $\lambda_i = 1$ for $i = 1, 2, \ldots, m$; initialize $\mathbf{x}$.

STEP 2:   Minimize function $\Phi(\mathbf{x}, \xi, \lambda)$ to obtain $\bar{\mathbf{x}}$.

STEP 3:   Set

$$S = \sum_{i \in I_1} \lambda_i \phi_i(\bar{\mathbf{x}}, \xi) + \sum_{i \in I_2} \phi_i(\bar{\mathbf{x}}, \xi)$$

and update λ_i and ξ as

$$\lambda_i = \begin{cases} \lambda_i \phi_i(\tilde{\mathbf{x}}, \xi)/S & \text{if } \phi_i(\tilde{\mathbf{x}}, \xi) \geq 0, \lambda_i > 0 \\ \phi_i(\tilde{\mathbf{x}}, \xi)/S & \text{if } \phi_i(\tilde{\mathbf{x}}, \xi) > 0, \lambda_i = 0 \\ 0 & \text{if } \phi_i(\tilde{\mathbf{x}}, \xi) < 0 \end{cases}$$

$$\xi = \sum_{i=1}^{m} \lambda_i f_i(\tilde{\mathbf{x}})$$

STEP 8: Stop if

$$(F(\tilde{\mathbf{x}}) - \xi)/F(\tilde{\mathbf{x}}) < \varepsilon$$

Otherwise go to Step 2.

The parameter ε is a prescribed termination tolerance. When the algorithm converges, constraints (b) and (c) are satisfied and $\tilde{\mathbf{x}} = \mathbf{x}^*$. The unconstrained minimization required in Step 2 can be performed by using any one of the quasi-Newton algorithms (see Sec. 8.2.2). From Eqs. (8.9)–(8.11) and (8.13) the gradient of function $\Phi(\mathbf{x}, \xi, \lambda)$ with respect to $\mathbf{x}$, which is required in quasi-Newton methods, is obtained as

$$\nabla \Phi(\mathbf{x}, \xi, \lambda) = \sum_{i \in I_1} \lambda_i \phi_i(\mathbf{x}, \xi) \nabla \phi_i(\mathbf{x}, \xi) + \sum_{i \in I_2} \lambda_i \phi_i(\mathbf{x}, \xi) \nabla \phi_i(\mathbf{x}, \xi)$$

where

$$\begin{aligned} \nabla \phi_i(\mathbf{x}, \xi) &= \nabla f_i(\mathbf{x}) \\ &= \nabla |E(\mathbf{x}, \omega_{1i}, \omega_{2i})| \\ &= \nabla |M(\mathbf{x}, \omega_{1i}, \omega_{2i}) - M_I(\omega_{1i}, \omega_{2i})| \\ &= \text{sgn}[E(\mathbf{x}, \omega_{1i}, \omega_{2i})] \nabla M(\mathbf{x}, \omega_{1i}, \omega_{2i}) \end{aligned}$$

with

$$\text{sgn}(E) = \begin{cases} 1 & \text{if } E \geq 0 \\ -1 & \text{if } E < 0 \end{cases}$$

It follows from Eqs. (8.9) and (8.10) that ∇M is equal to either ∇M_1 when transfer function $H_1(z_1, z_2)$ is adopted or ∇M_2 when $H_2(z_1, z_2)$ is adopted, and

$$\nabla M_j = \left[\frac{\partial M_j}{\partial \alpha_1} \frac{\partial M_j}{\partial \beta_1} \frac{\partial M_j}{\partial c_1} \frac{\partial M_j}{\partial d_1} \cdots \frac{\partial M_j}{\partial \alpha_k} \frac{\partial M_j}{\partial \beta_k} \frac{\partial M_j}{\partial c_k} \frac{\partial M_j}{\partial d_k} \frac{\partial M_j}{\partial H_0} \right]^T, \quad j = 1, 2$$

where

$$\frac{\partial M_1}{\partial \alpha_k} = \frac{M_1(\mathbf{x}, \omega_{1i}, \omega_{2i})}{\sigma_k N_{1k} N_{2k}} (\cos \omega_{1i} T_1 + \cos \omega_{2i} T_2)$$

$$\frac{\partial M_1}{\partial \beta_k} = \frac{M_1(\mathbf{x}, \omega_{1i}, \omega_{2i})}{\sigma_k N_{1k} N_{2k}}$$

$$\frac{\partial M_1}{\partial c_k} = -\frac{M_1(\mathbf{x}, \omega_{1i}, \omega_{2i})}{D_{k1}^2 D_{k2}^2} \{D_{k1}^2[c_k + (1 + d_k)\cos \omega_{2i} T_2]$$
$$+ D_{k2}^2[c_k + (1 + d_k)\cos \omega_{1i} T_1]\}$$

$$\frac{\partial M_1}{\partial d_k} = -\frac{M_1(\mathbf{x}, \omega_{1i}, \omega_{2i})}{D_{k1}^2 D_{k2}^2} [D_{k1}^2(-1 + 2\cos^2\omega_{2i} T_2 + c_k \cos \omega_{2i} T_2 + d_k)$$
$$+ D_{k2}^2(-1 + 2\cos^2\omega_{1i} T_1 + c_k \cos \omega_{1i} T_1 + d_k)]$$

$$\frac{\partial M_1}{\partial H_0} = M_1(\mathbf{x}, \omega_{1i}, \omega_{2i})/H_0$$

$$\frac{\partial M_2}{\partial \alpha_k} = \frac{M_2(\mathbf{x}, \omega_{1i}, \omega_{2i})}{N_{k1}^2 N_{k2}^2} \{N_{k1}^2[\alpha_k + (1 + \beta_k)\cos \omega_{2i} T_2]$$
$$+ N_{k2}^2[\alpha_k + (1 + \beta_k)\cos \omega_{1i} T_1]\}$$

$$\frac{\partial M_2}{\partial \beta_k} = \frac{M_2(\mathbf{x}, \omega_{1i}, \omega_{2i})}{N_{k1}^2 N_{k2}^2} [N_{k1}^2(-1 + 2\cos^2\omega_{2i} T_2 + \alpha_k \cos \omega_{2i} T_2 + \beta_k)$$
$$+ N_{k2}^2(-1 + 2\cos^2\omega_{1i} T_1 + \alpha_k \cos \omega_{1i} T_1 + \beta_k)]$$

$$\frac{\partial M_2}{\partial c_k} = -\frac{M_2(\mathbf{x}, \omega_{1i}, \omega_{2i})}{D_{k1}^2 D_{k2}^2} \{D_{k1}^2[c_k + (1 + d_k)\cos \omega_{2i} T_2]$$
$$+ D_{k2}^2[c_k + (1 + d_k)\cos \omega_{1i} T_1]\}$$

$$\frac{\partial M_2}{\partial d_k} = -\frac{M_2(\mathbf{x}, \omega_{1i}, \omega_{2i})}{D_{k1}^2 D_{k2}^2} [D_{k1}^2(-1 + 2\cos^2\omega_{2i} T_2 + c_k \cos \omega_{2i} T_2 + d_k)$$
$$+ D_{k2}^2(-1 + 2\cos^2\omega_{1i} T_1 + c_k \cos \omega_{1i} T_1 + d_k)]$$

$$\frac{\partial M_2}{\partial H_0} = M_2(\mathbf{x}, \omega_{1i}, \omega_{2i})/H_0$$

Example 8.2   Design a 2-D circularly symmetric lowpass filter of order (2, 2) with $\omega_p = 0.08\pi$, $\omega_a = 0.1\pi$ assuming that $\omega_{s1} = \omega_{s2} = 2\pi$ by using Algorithm 3 described in Sec. 8.3.2.

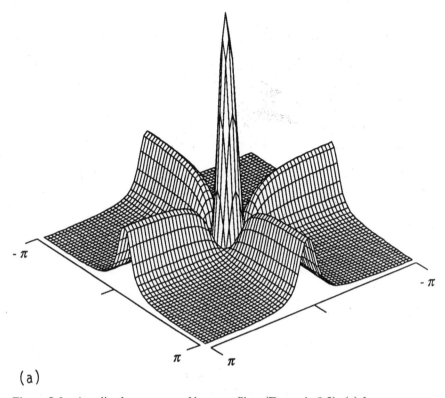

(a)

Figure 8.2 Amplitude response of lowpass filter (Example 8.2): (a) for $-\pi \le \omega_1$, $\omega_2 \le \pi$; (b) for $-0.08\pi \le \omega_1, \omega_2 \le 0.08\pi$.

Solution.   A solution of this problem obtained by Charalambous [2] is as follows. Since the passband and stopband contours are circles, the sample points can be placed on arcs of a set of circles centered at the origin. Five circles with radii $r_1 = 0.3\omega_p$, $r_2 = 0.6\omega_p$, $r_3 = 0.8\omega_p$, $r_4 = 0.9\omega_p$ and $r_5 = \omega_p$ are placed in the passband and the five circles with radii $r_6 = \omega_a$, $r_7 = \omega_a + 0.1(\pi - \omega_a)$, $r_8 = \omega_a + 0.2(\pi - \omega_a)$, $r_9 = \omega_a + 0.55(\pi - \omega_a)$ and $r_{10} = \pi$ are placed in the stopband. For circularly symmetric filters, the amplitude response is uniquely specified by the amplitude response in the sector $[0°, 45°]$. We, therefore, choose six equally spaced sample points on each circle described above between $0°$ and $45°$. These points plus the origin $(\omega_1, \omega_2) = (0, 0)$ form a set of 61 sample points. The transfer function of the filter is assumed to be of the form given by Eq. (8.7). With parameter H_0 fixed as $H_0 = (0.06582)^2$, $\varepsilon = 0.01$, and a starting point given by

$$\alpha_1^{(0)} = -1.514, \quad \beta_1^{(0)} = (\alpha_1^{(0)})^2, \quad c_1^{(0)} = -1.784, \quad d_1^{(0)} = 0.8166$$

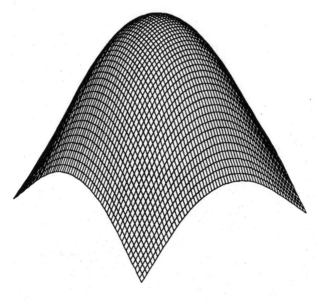

(b)

Figure 8.2 Continued.

Algorithm 3 yields the solution

$$\alpha_1^* = 1.96493, \qquad \beta_1^* = -10.9934, \qquad c_1^* = -1.61564,$$
$$d_1^* = 0.66781, \qquad F(\mathbf{x}^*) = 0.37995$$

The 3-D plot of the amplitude response of the filter designed is shown in Fig. 8.2a and b.

Table 8.1 summarizes the progress of the algorithm. As can be seen, the objective function $F(\mathbf{x})$ decreases and ξ increases with the number of iterations and both tend to the same limit $F(\mathbf{x}^*)$ when the algorithm converges.

The reason for fixing parameter H_0 is to avoid variable dependency among parameters α_1, β_1, and H_0. Such dependency can introduce ill-conditioning, which can, in turn, cause a serious convergence problem for the minimax algorithm.

8.4 DESIGN BASED ON TWO-VARIABLE NETWORK THEORY

8.4.1 Design of Quarter-Plane Filters

As noted in Sec. 8.2.2 (see p. 234), the method based on least pth optimization may lead to unstable designs, and, therefore, special care must be taken to assure the stability of the transfer function obtained. In this sec-

Table 8.1 Progress of Minimax Algorithm (Example 8.2)

Number of iterations	Number of function evaluations	ξ	$F(\mathbf{x})$
1	33	0.27249	0.52010
2	42	0.36653	0.42079
3	55	0.37869	0.41547
4	50	0.37958	0.37995

Source: Ref. 2.

tion, a design method is described in which the problem of stability is eliminated. The method involves the use of two-variable (2-V) strictly Hurwitz polynomials and was proposed by Ramamoorthy and Bruton [10,11]. A 2-V polynomial $b(s_1, s_2)$ is said to be strictly Hurwitz if

$$b(s_1, s_2) \neq 0 \quad \text{for Re } s_1 \geq 0 \text{ and Re } s_2 \geq 0 \quad (8.14)$$

In this approach, a family of 2-V strictly Hurwitz polynomials is obtained by applying network theory [12–14] to the frequency-independent, 2-V lossless network depicted in Fig. 8.3. The network has $1 + N_1 + N_2 + N_r$ ports and ports N_1 and N_2 are terminated in unit capacitors in complex variables s_1 and s_2, respectively, and N_r is terminated in unit resistors. Since the network is lossless and frequency independent, its admittance matrix $\mathbf{Y}$ is a real and skew-symmetric matrix given by

$$\mathbf{Y} = \begin{bmatrix} 0 & y_{12} & y_{13} & \cdots & y_{1N} \\ \hline -y_{12} & 0 & y_{23} & \cdots & y_{2N} \\ -y_{13} & -y_{23} & 0 & \cdots & y_{3N} \\ \vdots & \vdots & \vdots & & \vdots \\ -y_{1N} & -y_{2N} & -y_{3N} & \cdots & 0 \end{bmatrix}$$

$$= \begin{bmatrix} \mathbf{Y}_{11} & \mathbf{Y}_{12} \\ -\mathbf{Y}_{12}^T & \mathbf{Y}_{22} \end{bmatrix}, \quad N = 1 + N_1 + N_2 + N_r \quad (8.15)$$

If we define

$$\hat{\mathbf{Y}}_{22}(s_1, s_2, y_{kl}) = \mathbf{Y}_{22} + \text{diag}\{\overset{N_r}{1 \cdots 1} \ \overset{N_1}{s_1 \cdots s_1} \ \overset{N_2}{s_2 \cdots s_2}\}$$

and

$$\Delta(s_1, s_2, y_{kl}) = \det[\hat{\mathbf{Y}}_{22}(s_1, s_2, y_{kl})] \quad (8.16)$$

where $\text{diag}\{1 \cdots 1 \ s_1 \cdots s_1 \ s_2 \cdots s_2\}$ represents a diagonal matrix in which each of the first N_r elements is unity, each of the next N_1 elements is s_1, and each of the last N_2 elements is s_2, then from network theory, the input admittance at port 1 is given by

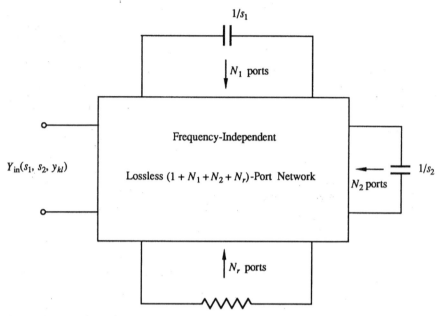

Figure 8.3 A $(1 + N_1 + N_2 + N_r)$-port 2-V lossless network.

$$Y_{in}(s_1, s_2, y_{kl}) = \frac{\mathbf{Y}_{12}\, \text{adj}[\hat{\mathbf{Y}}_{22}(s_1, s_2, y_{kl})]\mathbf{Y}_{12}^T}{\det[\hat{\mathbf{Y}}_{22}(s_1, s_2, y_{kl})]} \equiv \frac{p(s_1, s_2, y_{kl})}{\Delta(s_1, s_2, y_{kl})} \quad (8.17)$$

In addition, $\Delta(s_1, s_2, y_{kl})$ defined by Eq. (8.16) is a strictly Hurwitz polynomial for any set of real values of the $(N - 1)(N - 2)/2$ independent parameters $\{y_{kl} : 1 < k < l \leq N\}$. Table 8.2 shows polynomial $\Delta(s_1, s_2, y_{kl})$ for $N_r = 1$, $(N_1, N_2) = (2, 1)$ and $(N_1, N_2) = (2, 2)$.

Having obtained the parameterized strictly Hurwitz polynomial $\Delta(s_1, s_2, y_{kl})$, the double bilinear transformation can be applied to $\Delta(s_1, s_2, y_{kl})$ to generate a class of stable discrete transfer functions $H(z_1, z_2, y_{kl})$ as

$$H(z_1, z_2, y_{kl}) = \left. \frac{p(s_1, s_2)}{\Delta(s_1, s_2, y_{kl})} \right|_{s_i = (z_i - 1)/(z_i + 1), \ i = 1, 2} \quad (8.18)$$

where $p(s_1, s_2)$ is an arbitrary 2-V polynomial in s_1 and s_2 with degree in each variable not greater than the corresponding degree of the denominator. An optimization procedure can now be used to minimize the objective function

$$J(\mathbf{x}) = \sum_{n_1} \sum_{n_2} [M(n_1, n_2) - M_I(n_1, n_2)]^p, \quad p > 0 \text{ even} \quad (8.19)$$

Table 8.2 2-V Strictly Hurwitz Polynomials

N_1	N_2	N	$\Delta(s_1, s_2, y_{kl})$
1	1	4	$s_1 s_2 + y_{24}^2 s_1 + y_{23}^2 s_2 + y_{34}^2$
2	1	5	$s_1^2 s_2 + y_{25}^2 s_1^2 + (y_{23}^2 + y_{24}^2)s_1 s_2 + (y_{35}^2 + y_{45}^2)s_1 + y_{34}^2 s_2 + (y_{23}y_{45} - y_{24}y_{35} + y_{25}y_{34})^2$
2	2	6	$s_1^2 s_2^2 + (y_{23}^2 + y_{24}^2)s_1 s_2^2 + (y_{25}^2 + y_{26}^2)s_1^2 s_2 + y_{56}^2 s_1^2 + y_{34}^2 s_2^2 + (y_{35}^2 + y_{36}^2 + y_{45}^2 + y_{46}^2)s_1 s_2$ $+ [(y_{23}y_{56} - y_{25}y_{36} + y_{26}y_{35})^2 + (y_{24}y_{56} - y_{25}y_{46} + y_{26}y_{45})^2]s_1 + [(y_{23}y_{45} - y_{24}y_{35} + y_{25}y_{34})^2$ $+ (y_{23}y_{46} - y_{24}y_{36} + y_{26}y_{34})^2]s_2 + (y_{34}y_{56} - y_{35}y_{46} + y_{36}y_{45})^2$

Source: Ref. 10.

where $M(n_1, n_2)$ and $M_I(n_1, n_2)$ are the actual and desired amplitude responses, respectively, of the required filter at frequencies $(\omega_{1n_1}, \omega_{2n_2})$ and $\mathbf{x}$ is the vector consisting of parameters $[y_{kl} : 1 < k < l \leq N]$ and $\{a_{ij} : 0 \leq i \leq N_1, 0 \leq j \leq N_2\}$.

Example 8.3 [10]. By using the preceding approach, design a 2-D circularly symmetric lowpass filter of order $(5, 5)$ with $\omega_p = 0.2\pi$, assuming that $\omega_{s_1} = \omega_{s_2} = 1.2\pi$.

Solution. The continuous transfer function in Eq. (18.8) is assumed to be an all-pole transfer function of the form

$$\frac{p(s_1, s_2)}{\Delta(s_1, s_2, y_{kl})} = \frac{1}{\Delta(s_1, s_2, y_{kl})}$$

Hence the corresponding discrete transfer function can be written as

$$H(z_1, z_2, y_{kl}) = \frac{A(z_1 + 1)^5(z_2 + 1)^5}{\sum_{i=0}^{5} \sum_{j=0}^{5} b_{ij} z_1^i z_2^j} \tag{8.20}$$

where

$$\sum_{i=0}^{5} \sum_{j=0}^{5} b_{ij} z_1^i z_2^j = (z_1 + 1)^5 (z_2 + 1)^5 \Delta(s_1, s_2, y_{kl})|_{s_i = (z_i - 1)/(z_i + 1), \quad i=1, 2}$$

contains $(N - 1)(N - 2)/2 = 36$ parameters.

The desired amplitude response is given by

$$M_I(\omega_{1n_1}, \omega_{2n_2}) = \begin{cases} 1 & \text{for } (\omega_{1n_1}^2 + \omega_{2n_2}^2)^{1/2} \leq 0.2\pi \\ 0 & \text{otherwise} \end{cases}$$

where

$$\omega_{1n_1} = \begin{cases} 0.01\pi n_1 & \text{for } 0 \leq n_1 \leq 20 \\ 0.1\pi n_1 & \text{for } 21 \leq n_1 \leq 24 \end{cases}$$

and

$$\omega_{2n_2} = \omega_{1(24-n_2)} \qquad \text{for } 0 \leq n_2 \leq 24$$

A conventional quasi-Newton algorithm has been applied [5] to minimize the objective function $J(\mathbf{x})$ given in Eq. (8.19) with $p = 2$. The coefficients obtained are listed in Table 8.3. The amplitude response of the lowpass filter obtained is depicted in Fig. 8.4.

Table 8.3 Coefficients of Transfer Function in Eq. (8.20) [A = 0.28627 $\{b_{ij}: 0 \le i \le 5, 0 \le j \le 5\}$]

0.65181E-1	−0.64500E0	0.33632E1	−0.48317E1	0.32176E0	−0.16445E0
−0.79298E0	0.78851E1	−0.25871E2	0.23838E2	0.34048E1	0.34667E1
0.42941E1	−0.28734E2	0.61551E2	−0.29302E2	−0.13249E2	−0.25519E2
−0.63054E1	0.28707E2	−0.33487E2	−0.72275E1	−0.22705E2	0.83011E2
0.79070E0	0.14820E1	−0.74214E1	−0.33313E2	0.13676E3	−0.12843E3
−0.41341E0	0.60739E1	−0.36029E2	0.10147E3	0.14020E3	0.78428E2

Source: Ref.10.

8.4.2 Design of Nonsymmetric Half-Plane Filters

The preceding principles can also be employed for the design of stable nonsymmetric half-plane (NSHP) digital filters [11].

The transfer functions considered here assume the form

$$H(z_1, z_2) = \frac{N(z_1, z_2)}{D_1(z_1, z_2)D_2(z_1, z_2)} \tag{8.21}$$

where

$$N(z_1, z_2) = \sum_{n_1=0}^{N_1} a_{n_10}z_1^{-n_1} + \sum_{n_1=-N_1}^{N_1} \sum_{n_2=1}^{N_2} z_1^{-n_1}z_2^{-n_2} \tag{8.22}$$

$$D_1(z_1, z_2) = \sum_{n_1=0}^{N_1} \sum_{n_2=0}^{N_2} b_{n_1n_2}^{(1)}z_1^{-n_1}z_2^{-n_2} \tag{8.23}$$

$$D_2(z_1, z_2) = \sum_{n_1=-N_1}^{0} \sum_{n_2=0}^{N_2} b_{n_1n_2}^{(2)}z_1^{-n_1}z_2^{-n_2} \tag{8.24}$$

A family of stable polynomials $D_1(z_1, z_2, y_{kl})$ can readily be obtained by applying the double bilinear transformation to the strictly Hurwitz 2-V polynomial $\Delta(s_1, s_2, y_{kl})$ given by Eq. (8.16), that is,

$$D_1(z_1, z_2, y_{kl}) = (z_1 + 1)^{N_1}(z_2 + 1)^{N_2}$$

$$\times \Delta(s_1, s_2, y_{kl})|_{s_i=(z_i-1)/(z_i+1)}, \quad i = 1, 2 \tag{8.25}$$

By expressing polynomial $D_2(z_1, z_2)$ in Eq. (8.24) as

$$D_2(z_1, z_2) = \sum_{n_1=0}^{N_1} \sum_{n_2=0}^{N_2} b_{n_1n_2}^{(2)}z_1^{n_1}z_2^{-n_2}$$

and applying the transformation $z_1 := z_1^{-1}$ (symbol $:=$ stands for *is replaced by*), polynomial $D_2(z_1, z_2)$ is transformed into

$$\hat{D}_2(z_1, z_2) = \sum_{n_1=0}^{N_1} \sum_{n_2=0}^{N_2} b_{n_1n_2}^{(2)}z_1^{-n_1}z_2^{-n_2}$$

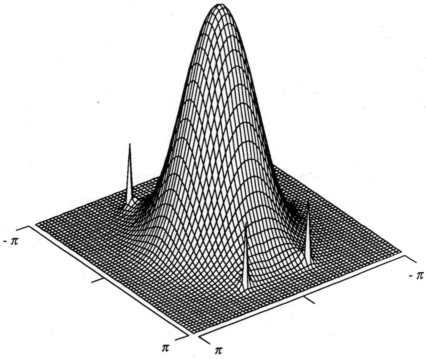

Figure 8.4 Amplitude response of circularly symmetric lowpass filter (Example 8.3).

This polynomial has a region of support in the first quadrant. Consequently, a family of stable polynomials $\hat{D}_2(z_1, z_2, y'_{kl})$ can be obtained as

$$\hat{D}_2(z_1, z_2, y'_{kl}) = (z_1 + 1)^{N_1}(z_2 + 1)^{N_2}\Delta(s_1, s_2, y'_{kl})|_{s_i = (z_i - 1)/(z_i + 1)}, \quad i = 1, 2$$

Since the z-domain transformation $z_1 := z_1^{-1}$ corresponds to the s-domain transformation $s_1 := -s_1$, we conclude that a family of stable 2-D polynomials $D_2(z_1, z_2, y'_{kl})$ can be obtained as

$$D_2(z_1, z_2, y'_{kl}) = (z_1 + 1)^{N_1}(z_2 + 1)^{N_2}$$

$$\times \Delta(-s_1, s_2, y'_{kl})|_{s_i = (z_i - 1)/(z_i + 1)}, \quad i = 1, 2 \quad (8.26)$$

where

$$\Delta(-s_1, s_2, y'_{kl}) = \det[\hat{\mathbf{Y}}_{22}(-s_1, s_2, y'_{kl})] \quad (8.27)$$

In summary, a stable NSHP digital filter with a desired amplitude response $M_I(\omega_1, \omega_2)$ can be designed by carrying out the following steps:

STEP 1:   Obtain polynomial $D_1(z_1, z_2, y_{kl})$ from Eq. (8.25)

STEP 2:   Obtain polynomial $D_2(z_1, z_2, y'_{kl})$ from Eq. (8.26)

STEP 3:   Form the transfer function

$$H(z_1, z_2, y_{kl}, y'_{kl}) = \frac{N(z_1, z_2)}{D_1(z_1, z_2, y_{kl})D_2(z_1, z_2, y'_{kl})} \qquad (8.28)$$

and construct the objective function

$$J(\mathbf{x}) = \sum_{n_1} \sum_{n_2} [M(n_1, n_2) - M_l(n_1, n_2)]^p, \qquad p \text{ even}$$

where $\mathbf{x}$ is the vector consisting of $\{y_{kl} : 1 \leq k < l \leq N\}$, $\{y'_{kl} : 1 < k < l \leq N\}$, $\{a_{n_10} : 0 \leq n_1 \leq N_1\}$ and $\{a_{n_1 n_2} : -N_1 \leq n_1 \leq N_1, 1 < n_2 \leq N_2\}$.

STEP 4:   Find a vector $\mathbf{x}$ that minimizes $J(\mathbf{x})$ and substitute the resulting $\mathbf{x}$ into (8.28) to obtain the required transfer function $H(z_1, z_2)$.

8.5   DESIGN OF RECURSIVE FILTERS USING SINGULAR-VALUE DECOMPOSITION

Another important method for the design of 2-D recursive digital filters is based on the application of the singular-value decomposition (SVD) [3]. The advantages of this design approach include the following: (1) the design can be accomplished by designing a set of 1-D subfilters and, therefore, the many well-established techniques for the design of 1-D filters can be employed, (2) the resulting 2-D filter is stable if the 1-D subfilters employed are stable, (3) the 1-D subfilters form a parallel structure that allows extensive parallel processing. The design can be accomplished in terms of recursive or nonrecursive linear-phase 1-D subfilters.

In this section, the definition and basic properties of the SVD are briefly reviewed. The technique is then applied for the design of recursive zero-phase 2-D filters whose frequency responses are quadrantally symmetric.

8.5.1   The SVD of a Sampled Amplitude Response

Let $A = \{a_{lm}\}$ be a desired amplitude response that is sampled at frequencies $(\omega_{1l}, \omega_{2l}) = (\pi\mu_l/T_1, \pi\upsilon_m/T_2)$ where

$$\mu_l = \frac{l - 1}{L - 1}, \qquad \upsilon_m = \frac{m - 1}{M - 1} \qquad \text{for } 1 \leq l \leq L, 1 \leq m \leq M$$

are normalized frequencies, that is

$$a_{lm} = |H(e^{j\pi\mu_l}, e^{j\pi\upsilon_m})| \qquad (8.29)$$

The SVD of matrix $\mathbf{A}$ gives [15]

$$\mathbf{A} = \sum_{i=1}^{r} \sigma_i \mathbf{u}_i \mathbf{v}_i^T = \sum_{i=1}^{r} \boldsymbol{\phi}_i \boldsymbol{\gamma}_i^T \tag{8.30}$$

where σ_i are the singular values of $\mathbf{A}$ such that $\sigma_1 \geq \sigma_2 \geq \cdots \geq \sigma_r > 0$, r is the rank of $\mathbf{A}$, $\mathbf{u}_i$ and $\mathbf{v}_i$ are the ith eigenvector of $\mathbf{A}\mathbf{A}^T$ and $\mathbf{A}^T\mathbf{A}$, respectively, $\boldsymbol{\phi}_i = \sigma_i^{1/2}\mathbf{u}_i$, $\boldsymbol{\gamma}_i = \sigma_i^{1/2}\mathbf{v}_i$, and $\{\boldsymbol{\phi}_i : 1 \leq i \leq r\}$ and $\{\boldsymbol{\gamma}_i : 1 \leq i \leq r\}$ are sets of orthogonal L-dimensional and M-dimensional vectors, respectively.

An important property of the SVD can be stated as

$$\left\| \mathbf{A} - \sum_{i=1}^{K} \boldsymbol{\phi}_i \boldsymbol{\gamma}_i^T \right\| = \min_{\hat{\boldsymbol{\phi}}_i, \hat{\boldsymbol{\gamma}}_i} \left\| \mathbf{A} - \sum_{i=1}^{K} \hat{\boldsymbol{\phi}}_i \hat{\boldsymbol{\gamma}}_i^T \right\| \qquad \text{for } 1 \leq K \leq r \tag{8.31}$$

where $\hat{\boldsymbol{\phi}}_i \in R^L$, $\hat{\boldsymbol{\gamma}}_i \in R^M$, and norm $\|\mathbf{X}\|$ may be either the Frobenius norm, that is,

$$\|\mathbf{X}\| = \left[\sum_{l=1}^{L} \sum_{m=1}^{M} x_{lm}^2 \right]^{1/2}$$

or the L_2 norm, that is,

$$\|\mathbf{X}\| = \max_{1 \leq i \leq M} \sqrt{\lambda_i}$$

where λ_i is an eigenvalue of $\mathbf{X}^T\mathbf{X}$ and $\mathbf{X} = \{x_{lm}\} \in R^{L \times M}$. The preceding relation shows that for any fixed K ($1 \leq K \leq r$), $\Sigma_{i=1}^{K} \boldsymbol{\phi}_i \boldsymbol{\gamma}_i^T$ is a minimal mean-square-error approximation to $\mathbf{A}$.

Since all the entries of $\mathbf{A}$ are nonnegative, it follows that all the entries of $\boldsymbol{\phi}_1$ and $\boldsymbol{\gamma}_1$ are nonnegative [16,17]. Nevertheless, some elements of $\boldsymbol{\phi}_i$ and $\boldsymbol{\gamma}_i$ for $i \geq 2$ may assume negative values.

8.5.2 Design Approach

In a quadrantally symmetric filter, $H(z_1, z_2)$ has a separable denominator [6]. Therefore, $H(z_1, z_2)$ can be expressed as

$$H(z_1, z_2) = \sum_{i=1}^{K} F_i(z_1) G_i(z_2) \tag{8.32}$$

Now note that Eq. (8.30) can be written as

$$\mathbf{A} = \boldsymbol{\phi}_1 \boldsymbol{\gamma}_1^T + \boldsymbol{\varepsilon}_1 \tag{8.33}$$

where

$$\boldsymbol{\varepsilon}_1 = \sum_{i=2}^{r} \boldsymbol{\phi}_i \boldsymbol{\gamma}_i^T$$

On comparing Eq. (8.33) with Eq. (8.32) and assuming that $K = 1$ and that ϕ_1, γ_1 are sampled versions of the desired amplitude responses for the 1-D filters characterized by $F_1(z_1)$ and $G_1(z_2)$, respectively, a 2-D digital filter can be designed through the following steps:

1. Design 1-D filters F_1 and G_1 characterized by $F_1(z_1)$ and $G_1(z_2)$.
2. Connect filters F_1 and G_1 in cascade.

Step 1 can be carried out by using a quasi-Newton optimization algorithm, such as Algorithm 1 of Sec. 8.2.2 or by using a minimax algorithm, such as Algorithm 2 in Sec. 8.3.2. When filters F_1 and G_1 are designed, we have

$$|F_1(e^{j\pi\mu_l})| \approx \phi_{1l}, \qquad 1 \le l \le L$$

and

$$|G_1(e^{j\pi v_m})| \approx \gamma_{1m}, \qquad 1 \le m \le M$$

where ϕ_{1l} and γ_{1m} denote the lth component of ϕ_1 and mth component of γ_1, respectively. The transfer function of the cascade filter obtained in Step 2 is given by

$$H_1(z_1, z_2) = F_1(z_1)G_1(z_2)$$

where

$$(|H_1(e^{j\pi\mu_l}, e^{j\pi v_m})|) \approx (\phi_{1l}\gamma_{1m}) = \phi_1\gamma_1^T$$

and from Eq. (8.33)

$$\| |A - H_1(e^{j\pi\mu_l}, e^{j\pi v_m})| \| \approx \|A - \phi_1\gamma_1^T\| = \|\epsilon_1\|$$

The approximation error associated with transfer function $H_1(z_1, z_2)$ can be reduced by realizing more of the terms in Eq. (8.30) by means of parallel filter sections. From Eq. (8.30), we can write

$$A = \phi_1\gamma_1^T + \phi_2\gamma_2^T + \epsilon_2 \qquad (8.34)$$

where

$$\epsilon_2 = \sum_{i=3}^{r} \phi_i\gamma_i^T$$

Since ϕ_2 and γ_2 may have some negative components, a careful treatment of the second term in Eq. (8.34) is necessary.

Let ϕ_2^- and γ_2^- be the absolute values of the most negative components of ϕ_2 and γ_2, respectively. If

$$e_\phi = [11 \cdots 1]^T \in R^L \qquad \text{and} \qquad e_\gamma = [11 \cdots 1]^T \in R^M$$

then all components of

$$\tilde{\phi}_2 = \phi_2 + \phi_2^- e_\phi \qquad \text{and} \qquad \tilde{\gamma}_2 = \gamma_2 + \gamma_2^- e_\gamma$$

are nonnegative. Now let us assume that it is possible to design 1-D linear-phase or zero-phase filters characterized by $\tilde{F}_1(z_1)$, $\tilde{G}_1(z_2)$, $\tilde{F}_2(z_1)$, and $\tilde{G}_2(z_2)$ such that

$$\tilde{F}_i(e^{j\pi\mu_l}) = |\tilde{F}_i(e^{j\pi\mu_l})| e^{j\alpha_1\mu_l}, \qquad 1 \le l \le L, \, i = 1, 2$$

$$\tilde{G}_i(e^{j\pi\upsilon_m}) = |\tilde{G}_i(e^{j\pi\upsilon_m})| e^{j\alpha_2\upsilon_m}, \qquad 1 \le m \le M, \, i = 1, 2$$

where

$$|\tilde{F}_1(e^{j\pi\mu_l})| \approx \tilde{\phi}_{1l}$$

$$|\tilde{F}_2(e^{j\pi\mu_l})| \approx \tilde{\phi}_{2l}$$

$$|\tilde{G}_1(e^{j\pi\upsilon_m})| \approx \tilde{\gamma}_{1m}$$

$$|\tilde{G}_2(e^{j\pi\upsilon_m})| \approx \tilde{\gamma}_{2m}$$

Here $\tilde{\phi}_{2l}$ and $\tilde{\gamma}_{2m}$ are the lth component of $\tilde{\phi}_2$ and mth component of $\tilde{\gamma}_2$, respectively, and α_1, α_2 are constants that are equal to zero if zero-phase filters are to be employed. Let

$$\alpha_1 = -\pi n_1, \qquad \alpha_2 = -\pi n_2 \qquad \text{with integers } n_1, n_2 \ge 0 \qquad (8.35)$$

and define

$$F_2(z_1) = \tilde{F}_2(z_1) - \phi_2^- z_1^{-n_1}$$

$$G_2(z_2) = \tilde{G}_2(z_2) - \gamma_2^- z_2^{-n_1}$$

Under these circumstances, it follows that

$$F_2(e^{j\pi\mu_l}) = [\tilde{F}_2(e^{j\pi\mu_l}) - \phi_2^-] e^{-j\pi n_1\mu_l} \approx \phi_{2l} e^{j\pi\mu_l}$$

and

$$G_2(e^{j\pi\upsilon_m}) = [\tilde{G}_2(e^{j\pi\upsilon_m}) - \gamma_2^-] e^{-j\pi n_2\upsilon_m} \approx \gamma_{2l} e^{j\pi\upsilon_m}$$

Moreover, if we form

$$H_1(z_1, z_2) = F_1(z_1)G_1(z_2) + F_2(z_1)G_2(z_2)$$

then

$$|H_2(e^{j\pi\mu_l}, e^{j\pi\upsilon_m})| = |F_1(e^{j\pi\mu_l})G_1(e^{j\pi\upsilon_m}) + F_2(e^{j\pi\mu_l})G_2(e^{j\pi\upsilon_m})|$$

$$\approx |\phi_{1l}\gamma_{1m} + \phi_{2l}\gamma_{2m}|$$

which in conjunction with Eq. (8.34) implies that

$$\|\mathbf{A} - |H_2(e^{j\pi\mu_l}, e^{j\pi\nu_m})|\,\| \approx \|\mathbf{A} - |\boldsymbol{\phi}_1\boldsymbol{\gamma}_1^T + \boldsymbol{\phi}_2\boldsymbol{\gamma}_2^T|\,\|$$

$$\leq \|\mathbf{A} - (\boldsymbol{\phi}_1\boldsymbol{\gamma}_1^T + \boldsymbol{\phi}_2\boldsymbol{\gamma}_2^T)\| = \|\boldsymbol{\varepsilon}_2\|$$

$$= \min_{\hat{\boldsymbol{\phi}}_i, \hat{\boldsymbol{\gamma}}_i}\|\mathbf{A} - (\hat{\boldsymbol{\phi}}_1\hat{\boldsymbol{\gamma}}_1^T + \hat{\boldsymbol{\phi}}_2\hat{\boldsymbol{\gamma}}_2^T)\| \qquad (8.36)$$

Evidently, through the preceding technique it is possible to realize the second term in Eq. (8.34) by means of a parallel subfilter, thereby reducing the approximation error from $\boldsymbol{\varepsilon}_1$ to $\boldsymbol{\varepsilon}_2$. According to Eq. (8.36), the two-section 2-D filter obtained has an amplitude response that is a minimal mean-square-error approximation to the desired amplitude response.

Since $F_1(z_1)G_1(z_2)$ corresponds to the largest singular value σ_1, the subfilter characterized by $F_1(z_1)G_1(z_1)$ is said to be the main section of the 2-D filter. On the other hand, $|F_2(e^{j\pi\mu_l})G_2(e^{j\pi\nu_m})|$ represents a correction to the amplitude response and the subfilter characterized by $F_2(z_1)G_2(z_2)$ is said to represent a correction section.

Through the use of data $\boldsymbol{\phi}_i$ and $\boldsymbol{\gamma}_i$ ($i = 3, \ldots , K, K \leq r$) given in Eq. (8.30), vectors $\hat{\boldsymbol{\phi}}_i$ and $\hat{\boldsymbol{\gamma}}_i$ can be found, and correction sections characterized by $F_i(z_1)G_i(z_2)$ can then be designed in a similar manner. When K sections are designed, $H_K(z_1, z_2)$ can be formed as

$$H_K(z_1, z_2) = \sum_{i=1}^{K} F_i(z_1)G_i(z_2)$$

and from Eq. (8.36) we have

$$\|\mathbf{A} - |H_K(e^{j\pi\mu_l}, e^{j\pi\nu_m})|\,\| \approx \left\|\mathbf{A} - \left|\sum_{i=1}^{K} \boldsymbol{\phi}_i\boldsymbol{\gamma}_i^T\right|\right\|$$

$$\leq \|\boldsymbol{\varepsilon}_K\| = \min_{\hat{\boldsymbol{\phi}}_i, \hat{\boldsymbol{\gamma}}_i}\left\|\mathbf{A} - \sum_{i=1}^{K} \boldsymbol{\phi}_i\boldsymbol{\gamma}_i^T\right\|$$

In effect, a 2-D digital filter comprising K sections is obtained whose amplitude response is a minimal mean-square-error approximation to the desired amplitude response.

The method leads to an asymptotically stable 2-D filter, provided that all 1-D subfilters employed are stable. This requirement is easily satisfied in practice.

The general structure of the 2-D filter obtained is illustrated in Fig. 8.5, where the various 1-D subfilters may be either linear-phase or zero-phase filters, as was shown earlier. The structure obtained is a parallel arrangement of cascade low-order sections, and, consequently, the traditional advantages associated with parallel and/or cascade structures apply. These

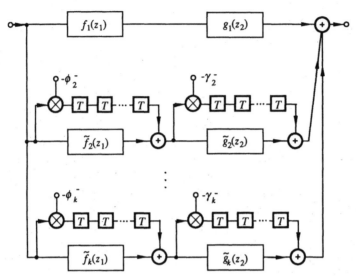

Figure 8.5   General structure of 2-D filter.

include low sensitivity to coefficient quantization, efficient computation due to parallel processing, and a relatively low number of required multipliers.

If linear-phase subfilters are to be employed, the equalities in Eq. (8.35) must be satisfied. This implies that the subfilters must have constant group delays. Causal subfilters of this class can be designed as nonrecursive filters by using the weighted-Chebyshev approximation method [18,19] or by using the Fourier series method along with the Kaiser window function [20]. In such a case, the 2-D filter obtained can be used in real-time applications. The application for the SVD for the design of 2-D nonrecursive filters will be studied in detail in Chap. 9.

If a record of the data to be processed is available, the processing can be carried out in nonreal time. In such a case, the subfilter in Fig. 8.5 can be designed as zero-phase recursive filters. The resulting structure is depicted in Fig. 8.6, where $\overline{F}_i(z_1)$ and $\overline{F}_i(z_1^{-1})$[$\overline{G}_i(z_2)$ and $\overline{G}_i(z_2^{-1})$] contribute equally to the amplitude response of the 2-D filter. The design can be completed by assuming that the desired amplitude response for subfilters $\overline{F}_1$, $\overline{G}_1$, $\overline{F}_i$, and $\overline{G}_i$ for $i = 2, \ldots, K$ are $\phi_1^{1/2}$, $\gamma_1^{1/2}$, $\phi_i^{1/2}$, and $\gamma_i^{1/2}$ for $i = 2, \ldots, K$, respectively.

When a circularly symmetric 2-D filter is required, the design work can be reduced significantly. Matrix **A** defined in Eq. (8.29) is symmetric and, therefore, Eq. (8.30) becomes

$$\mathbf{A} = \sum_{i=1}^{r} s_i \phi_i \phi_i^T \tag{8.37}$$

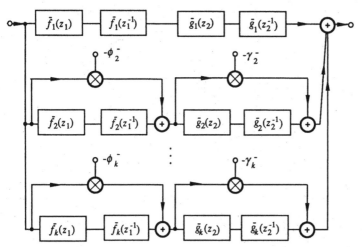

Figure 8.6 Structure using zero-phase recursive filters.

where $s_1 = 1$ and $s_i = \pm 1$ for $2 \le i \le r$. This implies that each parallel section requires only one 1-D subfilter to be designed and as a consequence the design work is reduced by 50 percent.

8.5.3 Error Compensation

When the main section and the correction sections are designed by using an optimization method, approximation errors inevitably occur that will accumulate and manifest themselves as the overall approximation error in the design. The accumulation of error can to a certain extent be reduced by using the following compensation technique.

When the design of the main section is completed, define an error matrix

$$\mathbf{E}_1 = \mathbf{A} - F_1(e^{j\pi\mu_l})G_1(e^{j\pi\upsilon_m})$$

and then perform SVD on $\mathbf{E}_1$ to obtain

$$\mathbf{E}_1 = s_{22}\boldsymbol{\phi}_{22}\boldsymbol{\gamma}_{22}^T + \cdots + s_{r2}\boldsymbol{\phi}_{r2}\boldsymbol{\gamma}_{r2}^T \tag{8.38}$$

Data $\boldsymbol{\phi}_{22}$ and $\boldsymbol{\gamma}_{22}$ in Eq. (8.38) can be used to deduce $F_2(z_1)$ and $G_2(z_2)$ as in Sec. 8.5.2; thus, the first correction section can be designed. Now form error matrix $\mathbf{E}_2$ as

$$\begin{aligned}
\mathbf{E}_2 &= \mathbf{E}_1 - s_{22}F_2(e^{j\pi\mu_l})G_2(e^{j\pi\upsilon_m}) \\
&= \mathbf{A} - [F_1(e^{j\pi\mu_l})G_1(e^{j\pi\upsilon_m}) + s_{22}F_2(e^{j\pi\mu_l})G_2(e^{j\pi\upsilon_m})]
\end{aligned}$$

and perform SVD on $\mathbf{E}_2$ to obtain

$$\mathbf{E}_2 = s_{33}\boldsymbol{\phi}_{33}\boldsymbol{\gamma}_{33}^T + \cdots + s_{r3}\boldsymbol{\phi}_{r3}\boldsymbol{\gamma}_{r3}^T$$

As before, data $\boldsymbol{\phi}_{33}$ and $\boldsymbol{\gamma}_{33}$ can be used to design the second correction section. The procedure is continued until the norm of the error matrix becomes sufficiently small for the application at hand.

Example 8.4   Design a circularly symmetric, zero-phase 2-D filter specified by

$$|H(e^{j\omega_1 T_1}, e^{j\omega_2 T_2})| = \begin{cases} 1 & \text{for } \sqrt{\omega_1^2 + \omega_2^2} \leq 0.35\pi \\ 0 & \text{for } \sqrt{\omega_1^2 + \omega_2^2} \geq 0.65\pi \end{cases}$$

assuming that $\omega_{s1} = \omega_{s2} = 2\pi$.

Solution.   By taking $L = M = 21$ and assuming that the amplitude response varies linearly with the radius in the transition band, the sampled amplitude response is given by a 21×21 matrix as

$$\mathbf{A} = \begin{bmatrix} \mathbf{A}_1 & \mathbf{0} \\ \mathbf{0} & \mathbf{0} \end{bmatrix} \tag{8.39}$$

where

$$\mathbf{A}_1 = \begin{bmatrix} 1 & 1 & \cdots & 1 & 0.75 & 0.5 & 0.25 \\ 1 & 1 & \cdots & 0.75 & 0.5 & 0.25 & 0 \\ \vdots & \vdots & & & & & \vdots \\ 1 & 0.75 & & & & & \\ 0.75 & 0.5 & & & & & \\ 0.5 & 0.25 & & & & & \\ 0.25 & 0 & \cdots & \cdots & \cdots & \cdots & 0 \end{bmatrix}_{12 \times 12}$$

The ideal amplitude response of the 2-D filter is shown in Fig. 8.7.

It is worth noting that although the vector $\boldsymbol{\phi}_1(=\boldsymbol{\gamma}_1)$ obtained from the SVD of $\mathbf{A}$ is a typical sampled amplitude response for a 1-D lowpass filter, the data given by the SVD error matrices $\mathbf{E}_1$ and $\mathbf{E}_2$ lead to the necessity of designing 1-D filters with arbitrary amplitude response. For example, given $\mathbf{A}$ as in Eq. (8.39), the square root of $\boldsymbol{\phi}_1$ is obtained as [3]

$$\boldsymbol{\phi}_1^{1/2} = [1.0415\ 1.0263\ 1.0263\ 1.0005\ 0.9625\ 0.9120\ 0.8300\ 0.7075$$
$$0.5514\ 0.3705\ 0.1866\ 0 \ldots 0]^T$$

Now if a sixth-order approximation is obtained for transfer function $\overline{F}_1(z_1)$,

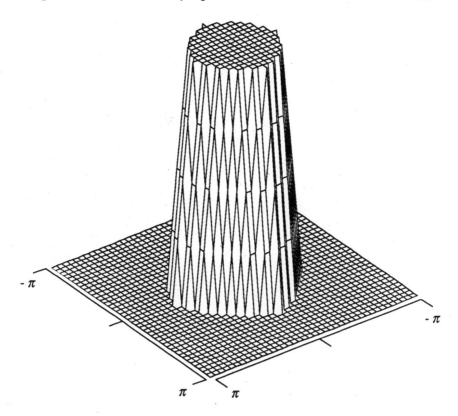

Figure 8.7 Ideal amplitude response of circularly symmetric lowpass filter (Example 8.4).

the SVD of E_1 gives

$$\tilde{\phi}_2^{1/2} = [1.0556 \ 0.9588 \ 0.9615 \ 0.8114 \ 0.8200 \ 0.6211 \ 0.3735 \ 0.0 \ 0.2030$$
$$0.4668 \ 0.6436 \ 0.7795 \ 0.8368 \ 0.8376 \ 0.8376 \ 0.8376 \ 0.8376$$
$$0.8376 \ 0.8376 \ 0.8376 \ 0.8376]^T$$

which represents an irregular amplitude response.

By using the procedure in Sec. 8.5.2 along with the error compensation technique in Sec. 8.5.3 and a 1-D optimization algorithm based on Algorithm 3, a 2-D zero-phase filter comprising the main section and two cor-

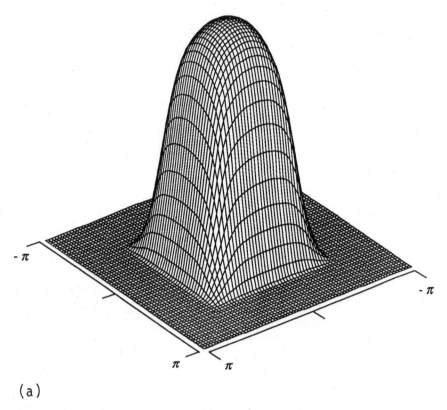

(a)

Figure 8.8 Amplitude response of (a) main section; (b) main section plus one correction; (c) main section plus two corrections (Example 8.4).

rection sections has been designed. The filter coefficients obtained with sixth-order 1-D transfer functions for the various subfilters are given in Table 8.4. The amplitude responses of (a) the main section, (b) the main section plus one correction section, and (c) the main section plus two correction sections are depicted in Fig. 8.8a–c.

Design of 1-D digital filters by using optimization methods can sometimes yield unstable filters. This problem can be eliminated by replacing poles outside the unit circle of the z plane by their reciprocals and simultaneously adjusting the multiplier constant to compensate for the change in gain (see Sec. 7.4 of Antoniou [20]).

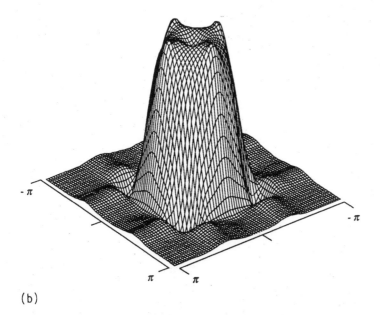

(b)

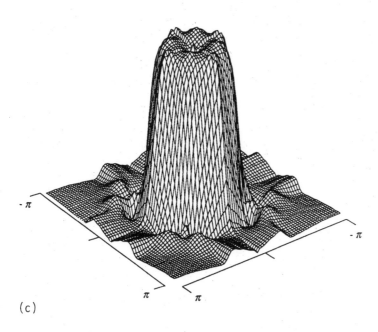

(c)

Table 8.4 Design Based on Sixth-Order Subfilters (Example 8.4)

Main section

$$\bar{F}_1(z) = 0.07283 \frac{(z^2 + 0.6656z + 0.9946)(z^2 + 1.8510z + 0.9733)(z^2 + 1.0838z + 0.8708)}{(z^2 + 0.1693z + 0.7407)(z^2 + 0.2463z - 0.0669)(z^2 - 0.3816z + 0.3328)}$$

First correction section

$$\bar{F}_2(z) = 0.1308 \frac{(z^2 + 4.6145z - 0.9949)(z^2 - 0.7341z + 0.4189)(z^2 - 0.8685z + 0.9822)}{(z^2 - 0.2721z - 0.1582)(z^2 - 0.0433z + 0.3971)(z^2 - 1.0524z + 0.6393)}$$

$$\phi_2^- = 0.7016$$

Second correction section

$$\bar{F}_3(z) = 0.4917 \frac{(z^2 + 1.7809z + 1.0320)(z^2 - 0.9564z - 0.0545)(z^2 - 1.3422z + 0.9450)}{(z^2 - 0.7131z + 0.5857)(z^2 - 1.6959z + 0.9429)(z^2 - 1.4170z + 0.4219)}$$

$$\phi_3^- = 0.3405$$

REFERENCES

1. G. A. Maria and M. M. Fahmy, An l_p design technique for two-dimensional digital recursive filters, *IEEE Trans. Acoust., Speech, Signal Process.*, vol. ASSP-22, pp. 15–21, Feb. 1974.

2. C. Charalambous, Design of 2-dimensional circularly-symmetric digital filters, *IEE Proc.*, vol. 129, pt. G, pp. 47–54, April 1982.

3. A. Antoniou and W.-S. Lu, Design of two-dimensional digital filters by using the singular value decomposition, *IEEE Trans. Circuits Syst.*, vol. CAS-34, pp. 1191–1198, Oct. 1987.

4. R. Fletcher, *Practical Methods of Optimization*, 2nd ed., Chichester: Wiley, 1987.

5. D. G. Luenberger, *Linear and Nonlinear Programming*, 2nd ed., Reading, Mass.: Addison-Wesley, 1984.

6. P. K. Rajan and M. N. S. Swamy, Quadrantal symmetry associated with two-dimensional digital transfer functions, *IEEE Trans. Circuits Syst.*, vol. CAS-29, pp. 340–343, June 1983.

7. C. Charalambous, A unified review of optimization, *IEEE Trans. Microwave Theory and Techniques*, vol. MTT-22, pp. 289–300, March 1974.

8. C. Charalambous and A. Antoniou, Equalisation of recursive digital filters, *IEE Proc.*, vol. 127, pt. G, pp. 219–225, Oct. 1980.

9. C. Charalambous, Acceleration of the least pth algorithm for minimax optimization with engineering applications, *Math Program.*, vol. 17, pp. 270–297, 1979.

10. P. A. Ramamoorthy and L. T. Bruton, Design of stable two-dimensional analog and digital filters with applications in image processing, *Circuit Theory Appl.*, vol. 7, pp. 229–245, 1979.

11. P. A. Ramamoorthy and L. T. Bruton, Design of stable two-dimensional recursive filters, *Topics in Applied Physics*, vol. 42, T. S. Huang, ed., pp. 41–83, New York: Springer Verlag, 1981.

12. H. Ozaki and T. Kasami, Positive real function of several variables and their applications to variable networks, *IRE Trans. Circuit Theory*, vol. CT-7, pp. 251–260, 1960.

13. H. G. Ansel, On certain two-variable generalizations of circuit theory, with applications to networks of transmission lines and lumped reactances, *IEEE Trans. Circuit Theory*, vol. CT-11, pp. 214–223, 1964.

14. T. Koga, Synthesis of finite passive networks with prescribed two-variable reactance matrices, *IEEE Trans. Circuit Theory*, vol. CT-13, pp. 31–52, 1966.

15. G. W. Stewart, *Introduction to Matrix Computations*, New York: Academic Press, 1973.

16. R. E. Twogood and S. K. Mita, Computer-aided design of separable two-dimensional digital filters, *IEEE Trans. Acoust., Speech, Signal Processing*, vol. ASSP-25, pp. 165–169, Feb. 1977.

17. P. Lancaster and M. Tismenetsky, *The Theory of Matrices*, 2nd ed., New York: Academic Press, 1985.

18. J. H. McClellan, T. W. Parks and L. R. Rabiner, A computer program for designing optimum FIR linear phase digital filters, *IEEE Trans. Audio Electroacoust.*, vol. AU-21, pp. 506–526, Dec. 1973.

19. A. Antoniou, New improved method for the design of weighted-Chebyshev, nonrecursive, digital filters, *IEEE Trans. Circuits Syst.*, vol. CAS-30, pp. 740–750, Oct. 1983.

20. A. Antoniou, *Digital Filters: Analysis and Design*, New York: McGraw-Hill, 1979 (2nd ed. in press).

PROBLEMS

8.1 Verify Eq. (8.4).

8.2 Consider Algorithm 1 with $\mathbf{C}_k$ obtained by using the DFP or BFGS formulas. Show that if $\mathbf{S}_k$ is positive definite, then

$$J(\mathbf{x}_k + \alpha \mathbf{d}_k) < J(\mathbf{x}_k)$$

for any $\alpha > 0$. (In other words, the hereditary property of the DFP and BFGS formulas assures the descent property of the algorithm.)

8.3 Design a circularly symmetric highpass filter of order $(2, 2)$ with $\omega_{p_1} = \omega_{p_2} = 0.12$ rad/s, $\omega_{a_1} = \omega_{a_2} = 0.08\pi$ rad/s, and $\omega_{s_1} = \omega_{s_2} = 2\pi$ rad/s by the least pth optimization.

8.4 Verify Eqs. (8.9) and (8.10).

8.5 Design a circularly symmetric highpass filter of order $(2, 2)$ with the same specifications as in Problem 8.3 by using Algorithm 2.

8.6 Repeat Problem 8.5 by using Algorithm 3.

8.7 Design a circularly symmetric highpass quarter-plane filter of order $(5, 5)$ with $\omega_a = 0.3\pi$ rad/s and $\omega_{s_1} = \omega_{s_2} = 1.2\pi$ rad/s.

8.8 Design a circularly symmetric highpass nonsymmetric half-plane filter of order $(5, 5)$ with $\omega_a = 0.3\pi$ rad/s and $\omega_{s_1} = \omega_{s_2} = 1.2\pi$ rad/s.

8.9 By applying the SVD method to the idealized amplitude response

$$|H_I(e^{j\omega_1 T}, e^{j\omega_2 T})| = \begin{cases} 0 & \sqrt{(\omega_1 T)^2 + (\omega_2 T)^2} \leq 0.3\pi \\ 1 & \text{otherwise} \end{cases}$$

where $T = 1$ s, obtain a 2-D transfer function of the form given in Eq. (8.32) with $K = 3$. The orders of the 1-D transfer functions in Eq. (8.32) should be equal to or less than 6.

8.10 Repeat Problem 8.9 with $K = 6$. Compare the design results with those obtained in Problem 8.9

8.11 Repeat Problem 8.9 with the error compensation method described in Sec. 8.5.3. Compare the design results with those obtained in Problem 8.9.

9

Design of Nonrecursive Filters by Optimization

9.1 INTRODUCTION

In Chap. 8, several optimization methods that can be used for the design of recursive filters have been described in detail. With some modifications, these methods can also be applied for the design of nonrecursive digital filters [1–5].

The most serious problem in applying optimization methods for the design of nonrecursive filters relates to the fact that these filters have low selectivity. Consequently, even a moderately demanding application would require a high-order filter that would, in turn, entail a large number of filter coefficients. The magnitude of this problem can to some extent be reduced by reducing the number of independent filter coefficients. As was shown in Chap. 6, a linear phase response implies that the impulse response of the filter satisfies a set of symmetry conditions [see Eq. (6.9)], and, as a result, the number of independent filter coefficients is reduced by half. If, in addition, the amplitude response of the filter has certain types of symmetry, then the number of independent filter coefficients can be reduced further.

This chapter begins with a study of some general symmetry properties of 2-D nonrecursive filters. Then a minimax optimization method for the design of linear-phase nonrecursive filters due to Charalambous [3] is described. The second half of the chapter deals with application of the SVD method of Sec. 8.5 for the design of linear-phase nonrecursive filters with arbitrary amplitude responses.

9.2 MINIMAX DESIGN OF LINEAR-PHASE NONRECURSIVE FILTERS

The minimax method of Sec. 8.3 yields some fairly good recursive designs, as was demonstrated in [2] of Chap. 8. With some modifications that address the basic differences between recursive and nonrecursive filters, the minimax method of Sec. 8.3 can readily be applied for the design of nonrecursive filters.

9.2.1 Symmetry Properties

Consider a nonrecursive 2-D filter with a transfer function given by

$$H(z_1, z_2) = \sum_{n_1=0}^{N_1-1} \sum_{n_2=0}^{N_2-1} h(n_1 T_1, n_2 T_2) z_1^{-n_1} z_2^{-n_2}$$

The filter has a linear phase response if its impulse response satisfies Eq. (6.9), as was shown in Sec. 6.2.1. The frequency response of such a filter is given by Eq. (6.10).

Many practically useful digital filters possess certain symmetry properties in the frequency domain. For instance, circularly symmetric lowpass, bandpass, and highpass filters have octagonal symmetry. That is, the amplitude response of a filter of this class satisfies the relations

$$|H(e^{j\omega_1 T_1}, e^{j\omega_2 T_2})| = |H(e^{-j\omega_1 T_1}, e^{j\omega_2 T_2})| = |H(e^{j\omega_1 T_1}, e^{-j\omega_2 T_2})| \quad (9.1)$$

and

$$|H(e^{j\omega_1 T_1}, e^{j\omega_2 T_2})| = |H(e^{j\omega_2 T_2}, e^{j\omega_1 T_1})| \quad (9.2)$$

Note that Eq. (9.2) implies that $N_1 = N_2$ and

$$a(n_1, n_2) = a(n_2, n_1) \quad (9.3)$$

where

$$a(0, 0) = h\left[\left(\frac{N_1 - 1}{2}\right) T_1, \left(\frac{N_2 - 1}{2}\right) T_2\right]$$

$$a(0, n_2) = 2h\left[\left(\frac{N_1 - 1}{2}\right) T_1, \left(\frac{N_2 - 1}{2} - n_2\right) T_2\right]$$

$$a(n_1, 0) = 2h\left[\left(\frac{N_1 - 1}{2} - n_1\right) T_1, \left(\frac{N_2 - 1}{2}\right) T_2\right]$$

$$a(n_1, n_2) = 4h\left[\left(\frac{N_1 - 1}{2} - n_1\right) T_1, \left(\frac{N_2 - 1}{2} - n_2\right) T_2\right]$$

$$\text{for } 1 \le n_1 \le \frac{N_1 - 1}{2}, \, 1 \le n_2 \le \frac{N_2 - 1}{2}$$

Another important class of filters is the class of fan filters which have quadrantal symmetry. The amplitude response of these filters satisfies Eq. (9.1), which implies that

$$h(n_1, n_2) = h(n_1, N_2 - 1 - n_2) = h(N_1 - 1 - n_1, n_2)$$

$$\text{for } 0 \le n_1 \le \frac{N_1 - 1}{2}, \, 0 \le n_2 \le \frac{N_2 - 1}{2} \qquad (9.4)$$

For a digital filter with octagonal symmetry and $N_1 = N_2 = N$ with N odd, the frequency response can be written as [3]

$$H(e^{j\omega_1 T_1}, e^{j\omega_2 T_2}) = M(\omega_1, \omega_2)e^{-j(N-1)(\omega_1 T_1 + \omega_2 T_2)/2} \qquad (9.5)$$

where

$$M(\omega_1, \omega_2) = \sum_{n_2=0}^{(N-3)/2} \sum_{n_1=n_2+1}^{(N-1)/2} a(n_1, n_2)[\cos(n_1\omega_1 T_1) \cos(n_2\omega_2 T_2)$$

$$+ \cos(n_2\omega_1 T_1) \cos(n_1\omega_2 T_2)]$$

$$+ \sum_{n=0}^{(N-1)/2} a(n, n) \cos(n\omega_1 T_1) \cos(n\omega_2 T_2) \qquad (9.6)$$

On the other hand, for a digital filter with quadrantal symmetry with N_1 and N_2 odd, the frequency response can be written as

$$H(e^{j\omega_1 T_1}, e^{j\omega_2 T_2}) = M(\omega_1, \omega_2)e^{-j[(N_1-1)\omega_1 T_1 + (N_2-1)\omega_2 T_2]/2}$$

where

$$M(\omega_1, \omega_2) = \sum_{n_1=0}^{(N_1-1)/2} \sum_{n_2=0}^{(N_2-1)/2} a(n_1, n_2) \cos(n_1\omega_1 T_1) \cos(n_2\omega_2 T_2) \qquad (9.7)$$

9.2.2 Minimax Optimization Algorithm

The design of nonrecursive filters can be transformed into an unconstrained minimax optimization problem by defining the objective function as

$$E(\mathbf{x}, \omega_1, \omega_2) = M(\mathbf{x}, \omega_1, \omega_2) - M_I(\omega_1, \omega_2)$$

where $M(\mathbf{x}, \omega_1, \omega_2)$ is the actual amplitude response of the filter to be designed and $M_I(\omega_1, \omega_2)$ represents the ideal amplitude response. The former quantity is given by Eq. (9.6) or (9.7) depending upon the symmetry properties of the filter. As in Sec. 8.3, the preceding error function can be minimized in the minimax sense by finding a vector $\mathbf{x}$ that solves the optimization problem

$$\underset{\mathbf{x}}{\text{minimize}} \; F(\mathbf{x})$$

where

$$F(\mathbf{x}) = \max_{1 \le i \le m} f_i(\mathbf{x})$$

and

$$f_i(\mathbf{x}) = |E(\mathbf{x}, \omega_{1i}, \omega_{2i})|, \qquad i = 1, \dots, m$$

Although Algorithm 3 of Sec. 8.3.2 is directly applicable to the present design problem, Charalambous [3] has found that better results can be achieved by using the conjugate direction method of Powell [6] for the unconstrained minimization required in Step 2 instead of a quasi-Newton method. Conjugate direction methods are preferred because they are more efficient than other methods when the number of independent variables is large and the objective function is a positive-definite quadratic function with respect to $\mathbf{x}$. Function $\Phi(\mathbf{x}, \xi, \lambda)$ can be shown to have these properties by using Eqs. (9.6) and (9.7). An excellent analysis of the properties of conjugate direction methods can be found in Fletcher [7, pp. 63–69].

Example 9.1 Design a 2-D nonrecursive circularly symmetric lowpass filter with $\omega_p = 0.4\pi$, $\omega_a = 0.6\pi$, and $\omega_{s1} = \omega_{s2} = \omega_s = 2\pi$ by using the optimization algorithm described in Sec. 9.2.2.

Solution. A solution of this problem obtained by Charalambous [3] is as follows. Since the frequency response of the filter to be designed is circularly symmetric, it satisfies Eqs. (9.1) and (9.2). Consequently, sampling points need to be chosen only in the [0–45°] sector of the (ω_1, ω_2) plane. Furthermore, it is appropriate to choose the sampling points on arcs of circles encircling the origin. If n_p and n_c are the numbers of circles in the passband and stopband regions, respectively, then the radii of the circles may be determined as

$$r_i = \omega_p \cos\left[(n_p - i)\frac{\pi}{2n_p - 1}\right], \qquad i = 1, 2, \dots, n_p$$

$$r_{n_p + n_c - i} = \omega_a + \left\{1 - \cos\left[(n_c - i - 1)\frac{\pi}{2n_c - 3}\right]\right\}(1 - \omega_a),$$

$$i = 1, 2, \dots, n_c - 2$$

$$r_{n_p + 1} = \omega_a$$

$$r_{n_p + n_c} = 1$$

With $\omega_p = 0.4\pi$, $\omega_a = 0.6\pi$, $n_p = 8$ and $n_a = 9$, the radii of the circles described above can be calculated as

$$0.04181\pi, \ 0.1236\pi, \ 0.2\pi, \ 0.2677\pi, \ 0.3236\pi, \ 0.3654\pi, \ 0.3913\pi, \ 0.4\pi$$

for the passband region, and

$$0.6\pi, \ 0.6087\pi, \ 0.6346\pi, \ 0.6764\pi, \ 0.7323\pi, \ 0.8\pi, \ 0.8763\pi, \ 0.9582\pi, \ \pi$$

for the stopband region. On each circle in the sector between $0°$ and $45°$, seven equally spaced sample points can be chosen, and the sample points (ω_1, ω_2) can be obtained as

$$(\omega_{11}, \omega_{22}) = (0, 0)$$

$$(\omega_{1k}, \omega_{2k}) = (r_i \cos \gamma_j, \ r_i \sin \gamma_j),$$

where

$$\gamma_j = \frac{\pi(j-1)}{24}$$

for $i = 1, 2, \ldots, 17, j = 1, 2, \ldots, 7$, and $k = 2, \ldots, 120$. As can be seen from Fig. 9.1, additional sample points are needed to cover the points whose distance to the origin is larger than π. On each of the circles with radius

$$\left\{ 1 + \left[\tan(i-1)\frac{\pi}{24} \right]^2 \right\}^{1/2} \pi$$

for $i = 2, 3, \ldots, 7, (8-i)$ equally spaced points can be chosen in the sector between $(i-1)180°/24$ and $45°$. With these additional 21 points and the origin included, there are a total of 141 sample points over the $[0-45°]$ sector of the (ω_1, ω_2) plane. By defining

$$\mathbf{x} = \begin{bmatrix} a(0, 0) \\ a(0, 1) \\ \vdots \\ a\left(0, \dfrac{N-1}{2}\right) \\ \hline \vdots \\ \hline a(i, i) \\ \vdots \\ a\left(i, \dfrac{N-1}{2}\right) \\ \hline \vdots \\ \hline a\left(\dfrac{N-1}{2}, \dfrac{N-1}{2}\right) \end{bmatrix} \begin{matrix} \left.\vphantom{\begin{matrix}1\\1\\1\\1\end{matrix}}\right\} \dfrac{N+1}{2} \\ \\ \left.\vphantom{\begin{matrix}1\\1\\1\\1\end{matrix}}\right\} \dfrac{N-2i+1}{2} \\ \\ \end{matrix}$$

and

$$\mathbf{c}(\omega_1, \omega_2)$$

$$= \begin{bmatrix} 1 \\ \cos \omega_1 + \cos \omega_2 \\ \vdots \\ \cos[(N-1)\omega_1/2] + \cos[(N-1)\omega_2/2] \\ \hline \vdots \\ \hline \cos i\omega_1 \cos i\omega_2 \\ \vdots \\ \cos i\omega_1 \cos[(N-1)\omega_1/2] + \cos[(N-1)\omega_2/2] \cos i\omega_2 \\ \hline \vdots \\ \hline \cos[(N-1)\omega_1/2] \cos[(N-1)\omega_2/2] \end{bmatrix} \left.\begin{matrix} \\ \\ \\ \\ \end{matrix}\right\}\dfrac{N+1}{2} \quad \left.\begin{matrix} \\ \\ \\ \\ \end{matrix}\right\}\dfrac{N-2i+1}{2}$$

Equation (9.6) can be written as

$$M(\omega_1, \omega_2, \mathbf{x}) = \mathbf{c}^T(\omega_1, \omega_2)\mathbf{x}$$

With a given total number of circles n_t, the number of circles in the passband n_p and the number of circles in the stopband n_a should be chosen as

$$n_p = \text{int}\left[\frac{\omega_p n_t}{\omega_p + (\pi - \omega_a)} + 0.5\right]$$

$$n_a = n_t - n_p$$

where $\text{int}[\nu]$ denotes the largest integer not greater than ν.

Some results obtained using a starting point $\mathbf{x}(0) = \mathbf{0}$ and $\varepsilon = 10^{-2}$ are summarized in Table 9.1.

The amplitude response of a 17×17 design is depicted in Fig. 9.1b.

9.3 DESIGN OF LINEAR-PHASE NONRECURSIVE FILTERS USING SVD

In the design of recursive filters by means of the SVD (see Sec. 8.5), high selectivity can be achieved by using low-order recursive designs for the parallel 1-D subfilters. However, zero phase is required for each subfilter. This necessitates data transpositions at the inputs and outputs of subfilters, and, as a result, the usefulness of these designs is limited to nonreal-time applications where the delay introduced in the processing is unimportant.

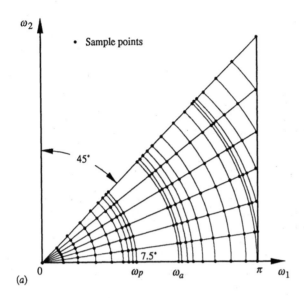

(a)

Figure 9.1 Design of circularly symmetric lowpass filter (Example 9.1): (a) sample points of filter; (b) amplitude response achieved using a filter of order 17 × 17. (From [3].)

In this section, it is shown that the SVD of the sampled frequency response of a 2-D digital filter with real coefficients possesses a special structure: every singular vector is either mirror-image symmetric or anti-symmetric about its midpoint. Consequently, the SVD method can be applied along with 1-D nonrecursive-filter techniques for the design of linear-phase 2-D filters with arbitrary amplitude responses that are symmetrical with respect to the origin of the (ω_1, ω_2) plane.

Table 9.1 Results for the Filter in Example 9.1

Passband edge	Stopband edge	Order	Maximum error
		5 × 5	0.26780
		7 × 7	0.12643
0.4	0.6	9 × 9	0.11433
		13 × 13	0.05023
		17 × 17	0.02282

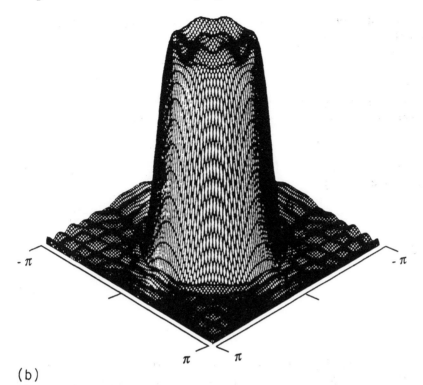

(b)

Figure 9.1 Continued

9.3.1 Symmetry Properties of the SVD of a Sampled Frequency Response

A 2-D nonrecursive digital filter with support in the rectangle defined by $-N_i/2 \le n_i \le N_i/2$, $i = 1, 2$, can be characterized by the transfer function

$$H(z_1, z_2) = \sum_{n_1 = -N_1/2}^{N_1/2} \sum_{n_2 = -N_2/2}^{N_2/2} h(n_1, n_2) z_1^{-n_1} z_2^{-n_2}$$

where $h(n_1, n_2)$ is the impulse response. If $h(n_1, n_2)$ is real and

$$h(n_1, n_2) = h(-n_1, -n_2)$$

then the frequency response of the filter given by

$$H(e^{j\omega_1 T_1}, e^{j\omega_2 T_2}) = \sum_{n_1 = -N_1/2}^{N_1/2} \sum_{n_2 = -N_2/2}^{N_2/2} h(n_1, n_2) e^{-j\omega_1 n_1 T_1} e^{-j\omega_2 n_2 T_2}$$

$$= X(\omega_1, \omega_2) \tag{9.8}$$

is a real function that is symmetrical with respect to the origin of the (ω_1, ω_2) plane such that

$$X(\omega_1, \omega_2) = X(-\omega_1, -\omega_2) \tag{9.9}$$

where $-\pi \leq \omega_1, \omega_2 \leq \pi$.

Now assume that matrix $\mathbf{A} = \{a_{lm}\}$ represents a desired arbitrary frequency response such that Eq. (9.9) is satisfied, that is,

$$X(\pi\mu_l, \pi\upsilon_m) = X(-\pi\mu_l, -\pi\upsilon_m) = a_{lm}$$

where $1 \leq l \leq L$ and $1 \leq m \leq M$. The quantities μ_l, υ_m are normalized frequencies such that

$$\mu_l = -1 + 2\left(\frac{l-1}{L-1}\right), \qquad \upsilon_m = -1 + 2\left(\frac{m-1}{M-1}\right)$$

and $-1 \leq \mu_l \leq 1$, $-1 \leq \upsilon_m \leq 1$. The SVD of $\mathbf{A}$ gives

$$\mathbf{A} = \sum_{i=1}^{r} \sigma_i \mathbf{u}_i \mathbf{v}_i^T \tag{9.10a}$$

$$= \sum_{i=1}^{r} \tilde{\mathbf{u}}_i \tilde{\mathbf{v}}_i^T \tag{9.10b}$$

where σ_i are the singular values of $\mathbf{A}$ such that $\sigma_1 \geq \sigma_2 \geq \ldots \geq \sigma_r$, $\mathbf{u}_i$ is the ith eigenvector of $\mathbf{A}\mathbf{A}^T$ associated with the ith eigenvalue σ_i^2, $\mathbf{v}_i$ is the ith eigenvector of $\mathbf{A}^T\mathbf{A}$ associated with the ith eigenvalue σ_i^2, r is the rank of $\mathbf{A}$, and $\tilde{\mathbf{u}}_i = \sigma_i^{1/2}\mathbf{u}_i$, $\tilde{\mathbf{v}}_i = \sigma_i^{1/2}\mathbf{v}_i$. A very important property of nonrecursive filters can now be stated in terms of the following theorem.

Theorem 9.1 *If the frequency response of a nonrecursive filter satisfies Eq. (9.9), then vectors $\mathbf{u}_i$ and $\mathbf{v}_i$ in the SVD of Eq. (9.10b) are either mirror-image symmetric or antisymmetric simultaneously for $j = 1, 2, \ldots, r$.*

Proof. Let $\tilde{\mathbf{H}} = \{\tilde{h}_{i,j}: 1 \leq i, j \leq 2N\}$ be a $2N \times 2N$ arbitrary matrix with real elements such that

$$\tilde{h}_{i,j} = \tilde{h}_{2N+1-i, 2N+1-j}$$

and, for the sake of simplicity, assume that the matrix is square and of even dimension and has distinct singular values. If matrices $\hat{\mathbf{I}}$, $\bar{\mathbf{I}}$, and $\mathbf{H}$ are defined by

$$\hat{\mathbf{I}} = \begin{bmatrix} 0 & \cdots & 0 & 1 \\ 0 & \cdots & 1 & 0 \\ \vdots & \vdots & \vdots & \vdots \\ 1 & \cdots & 0 & 0 \end{bmatrix}, \qquad \bar{\mathbf{I}} = \begin{bmatrix} \mathbf{I}_N & 0 \\ 0 & \hat{\mathbf{I}}_N \end{bmatrix}, \qquad \mathbf{H} = \bar{\mathbf{I}}\tilde{\mathbf{H}}\bar{\mathbf{I}} = \begin{bmatrix} \mathbf{H}_1 & \mathbf{H}_2 \\ \mathbf{H}_2 & \mathbf{H}_1 \end{bmatrix}$$

respectively, where the dimensions of $\hat{\mathbf{I}}_N$, $\mathbf{I}_N$, $\mathbf{H}_1$, and $\mathbf{H}_2$ are $N \times N$, and the dimensions of $\tilde{\mathbf{I}}$ and $\mathbf{H}$ are $2N \times 2N$, then matrix $\tilde{\mathbf{H}}$ can be decomposed as

$$\tilde{\mathbf{H}} = \tilde{\mathbf{I}}\mathbf{H}\tilde{\mathbf{I}} = \mathbf{U}\Sigma\mathbf{V}^T = [\mathbf{u}_1 \cdots \mathbf{u}_i \cdots \mathbf{u}_{2N}]\Sigma[\mathbf{v}_1 \cdots \mathbf{v}_i \cdots \mathbf{v}_{2N}]^T \quad (9.11)$$

where $\mathbf{u}_i$ and $\mathbf{v}_i$ are eigenvectors of $\tilde{\mathbf{H}}\tilde{\mathbf{H}}^T$ and $\tilde{\mathbf{H}}^T\tilde{\mathbf{H}}$, respectively. Assume that $\mathbf{u}_i$ is a normalized eigenvector of $\tilde{\mathbf{H}}\tilde{\mathbf{H}}^T$, that is, there exists a σ_i such that

$$\tilde{\mathbf{H}}\tilde{\mathbf{H}}^T\mathbf{u}_i = \sigma_i\mathbf{u}_i \quad (9.12)$$

with $\|\mathbf{u}_i\| = 1$. Substituting Eq. (9.11) into Eq. (9.12), we have

$$\tilde{\mathbf{I}}\mathbf{H}\tilde{\mathbf{I}}(\tilde{\mathbf{I}}\mathbf{H}\tilde{\mathbf{I}})^T\mathbf{u}_i = (\tilde{\mathbf{I}}\mathbf{H}\mathbf{H}^T\tilde{\mathbf{I}})\mathbf{u}_i = \sigma_i\mathbf{u}_i$$

and

$$\mathbf{H}\mathbf{H}^T\tilde{\mathbf{I}}\mathbf{u}_i = \sigma_i\tilde{\mathbf{I}}\mathbf{u}_i \quad (9.13)$$

If we let

$$\mathbf{u}_i = \begin{bmatrix} \mathbf{u}_{i1} \\ \mathbf{u}_{i2} \end{bmatrix}$$

then

$$\tilde{\mathbf{I}}\mathbf{u}_i = \begin{bmatrix} \mathbf{u}_{i1} \\ \hat{\mathbf{I}}\mathbf{u}_{i2} \end{bmatrix} \equiv \begin{bmatrix} \mathbf{x}_{i1} \\ \mathbf{x}_{i2} \end{bmatrix}$$

and, therefore, Eq. (9.13) becomes

$$\mathbf{H}\mathbf{H}^T\begin{bmatrix} \mathbf{x}_{i1} \\ \mathbf{x}_{i2} \end{bmatrix} = \sigma_i\begin{bmatrix} \mathbf{x}_{i1} \\ \mathbf{x}_{i2} \end{bmatrix} \quad (9.14)$$

We can now write

$$\mathbf{H}\mathbf{H}^T = \begin{bmatrix} \mathbf{H}_1 & \mathbf{H}_2 \\ \mathbf{H}_2 & \mathbf{H}_1 \end{bmatrix}\begin{bmatrix} \mathbf{H}_1^T & \mathbf{H}_2^T \\ \mathbf{H}_2^T & \mathbf{H}_1^T \end{bmatrix}$$

$$= \begin{bmatrix} \mathbf{H}_1\mathbf{H}_1^T + \mathbf{H}_2\mathbf{H}_2^T & \mathbf{H}_1\mathbf{H}_2^T + \mathbf{H}_2\mathbf{H}_1^T \\ \mathbf{H}_2\mathbf{H}_1^T + \mathbf{H}_1\mathbf{H}_2^T & \mathbf{H}_2\mathbf{H}_2^T + \mathbf{H}_1\mathbf{H}_1^T \end{bmatrix} \equiv \begin{bmatrix} \mathbf{A} & \mathbf{B} \\ \mathbf{B} & \mathbf{A} \end{bmatrix}$$

where $\mathbf{A}$ is a positive semidefinite and $\mathbf{B}$ is a symmetric matrix. Therefore, Eq. (9.13) can be expressed as

$$\begin{bmatrix} \mathbf{A} & \mathbf{B} \\ \mathbf{B} & \mathbf{A} \end{bmatrix}\begin{bmatrix} \mathbf{x}_{i1} \\ \mathbf{x}_{i2} \end{bmatrix} = \sigma_i\begin{bmatrix} \mathbf{x}_{i1} \\ \mathbf{x}_{i2} \end{bmatrix} \quad (9.15)$$

from which two equations can be obtained as

$$\mathbf{A}\mathbf{x}_{i1} + \mathbf{B}\mathbf{x}_{i2} = \sigma_i\mathbf{x}_{i1} \tag{9.16a}$$

$$\mathbf{B}\mathbf{x}_{i1} + \mathbf{A}\mathbf{x}_{i2} = \sigma_i\mathbf{x}_{i2} \tag{9.16b}$$

By writing Eq. (9.16) in another matrix notation, we have

$$\begin{bmatrix} \mathbf{A} & \mathbf{B} \\ \mathbf{B} & \mathbf{A} \end{bmatrix}\begin{bmatrix} \mathbf{x}_{i2} \\ \mathbf{x}_{i1} \end{bmatrix} = \sigma_i\begin{bmatrix} \mathbf{x}_{i2} \\ \mathbf{x}_{i1} \end{bmatrix} \tag{9.17}$$

On comparing Eq. (9.17) with Eq. (9.15), we note that both vectors

$$\begin{bmatrix} \mathbf{x}_{i1} \\ \mathbf{x}_{i2} \end{bmatrix} \quad \text{and} \quad \begin{bmatrix} \mathbf{x}_{i2} \\ \mathbf{x}_{i1} \end{bmatrix}$$

are eigenvectors of matrix $\mathbf{H}\mathbf{H}^T$ associated with the same eigenvalue σ_i. Therefore, the two vectors must be linearly dependent, that is, they must satisfy the relation

$$\begin{bmatrix} \mathbf{x}_{i1} \\ \mathbf{x}_{i2} \end{bmatrix} = \pm\begin{bmatrix} \mathbf{x}_{i2} \\ \mathbf{x}_{i1} \end{bmatrix}$$

which implies that

$$\mathbf{x}_{i1} = \mathbf{x}_{i2} \quad \text{or} \quad \mathbf{x}_{i1} = -\mathbf{x}_{i2}$$

If $\mathbf{x}_{i1} = \mathbf{x}_{i2} \equiv \mathbf{x}_i$, we can write

$$\tilde{\mathbf{I}}\mathbf{u}_i = \begin{bmatrix} \mathbf{x}_i \\ \mathbf{x}_i \end{bmatrix}$$

and so

$$\mathbf{u}_i = \begin{bmatrix} \mathbf{x}_i \\ \hat{\mathbf{I}}\mathbf{x}_i \end{bmatrix}$$

which means that $\mathbf{u}_i$ is mirror-image symmetric. On the other hand, if $\mathbf{x}_{i1} = -\mathbf{x}_{i2} \equiv \mathbf{x}_i$, we have

$$\mathbf{u}_i = \begin{bmatrix} \mathbf{x}_i \\ -\hat{\mathbf{I}}\mathbf{x}_i \end{bmatrix}$$

which implies that $\mathbf{u}_i$ is mirror-image antisymmetric.

Now from Eq. (9.11)

$$\mathbf{V} = \hat{\mathbf{I}}\mathbf{H}^T\tilde{\mathbf{I}}(\mathbf{U}^T)^{-1}\mathbf{\Sigma}^{-1}$$

and since $\tilde{\mathbf{I}}$ is symmetric and $\mathbf{U}$ is orthogonal, that is, $(\mathbf{U}^T)^{-1} = \mathbf{U}$, matrix $\mathbf{V}$ can be expressed as

$$\mathbf{V} = \tilde{\mathbf{I}}\mathbf{H}^T\tilde{\mathbf{I}}\mathbf{U}\mathbf{\Sigma}^{-1} \qquad (9.18)$$

If

$$\mathbf{u}_i = \begin{bmatrix} \mathbf{x}_i \\ \hat{\mathbf{I}}\mathbf{x}_i \end{bmatrix}$$

then Eq. (9.18) implies that

$$\mathbf{v}_i = \sigma_i^{-1}\tilde{\mathbf{I}}\mathbf{H}^T\tilde{\mathbf{I}}\mathbf{u}_i = \sigma_i^{-1}\begin{bmatrix} (\mathbf{H}_1 + \mathbf{H}_2)^T\mathbf{x}_i \\ \hat{\mathbf{I}}(\mathbf{H}_1 + \mathbf{H}_2)^T\mathbf{x}_i \end{bmatrix} = \begin{bmatrix} \mathbf{y}_i \\ \hat{\mathbf{I}}\mathbf{y}_i \end{bmatrix}$$

where $\mathbf{y}_i = \sigma_i^{-1}(\mathbf{H}_1 + \mathbf{H}_2)^T\mathbf{x}_i$. If

$$\mathbf{u}_i = \begin{bmatrix} \mathbf{x}_i \\ -\hat{\mathbf{I}}\mathbf{x}_i \end{bmatrix}$$

then Eq. (9.18) implies that

$$\mathbf{v}_i = \sigma_i^{-1}\tilde{\mathbf{I}}\mathbf{H}^T\tilde{\mathbf{I}}\mathbf{u}_i = \sigma_i^{-1}\begin{bmatrix} (\mathbf{H}_2 - \mathbf{H}_1)^T\mathbf{x}_i \\ -\hat{\mathbf{I}}(\mathbf{H}_2 - \mathbf{H}_1)^T\mathbf{x}_i \end{bmatrix} = \begin{bmatrix} \mathbf{y}_i \\ -\hat{\mathbf{I}}\mathbf{y}_i \end{bmatrix}$$

where $\mathbf{y}_i = \sigma_i^{-1}(\mathbf{H}_1 - \mathbf{H}_2)^T\mathbf{x}_i$. This shows that the two vectors $\mathbf{u}_i$ and $\mathbf{v}_i$ have the same symmetry properties simultaneously, that is, they are both either mirror-image symmetric or antisymmetric. ■

9.3.2 Design Approach

A 2-D nonrecursive filter having an arbitrary amplitude response satisfying Eq. (9.9) can readily be designed by using a parallel arrangement of K 2-D nonrecursive sections each comprising two 1-D subfilters in cascade, as will now be demonstrated. Such an arrangement can be represented by the transfer function

$$H(z_1, z_2) = \sum_{i=1}^{K} F_i(z_1)G_i(z_2) \qquad (9.19)$$

where $F_i(z_1)$ and $G_i(z_2)$ are the transfer functions of two cascaded 1-D subfilters. If these subfilters are nonrecursive filters with support in the rectangle defined by $-N_i/2 \leq n_i \leq N_i/2$, $i = 1, 2$, we have

$$F_i(z_1) = \sum_{n_1=-N_1/2}^{N_1/2} f_i(n_1)z_1^{-n_1} \qquad (9.20)$$

and

$$G_i(z_2) = \sum_{n_2=-N_2/2}^{N_2/2} g_i(n_2)z_2^{-n_2} \tag{9.21}$$

and if $F_i(z_1)$ and $G_i(z_2)$ are assumed to represent zero-phase or $\pi/2$-phase filters, then their frequency responses are given by

$$F_i(e^{j\omega_1 T_1}) = \sum_{n_1=-N_1/2}^{N_1/2} f_i(n_1)e^{-j\omega_1 n_1 T_1}$$

$$= \Phi_i(\omega_1)e^{j\theta_i} \tag{9.22}$$

$$G_i(e^{j\omega_2 T_2}) = \sum_{n_2=-N_2/2}^{N_2/2} g_i(n_2)e^{-j\omega_2 n_2 T_2}$$

$$= \Gamma_i(\omega_2)e^{j\theta_i} \tag{9.23}$$

If $f_i(n_1)$ and $g_i(n_2)$ are mirror-image symmetric, then $\theta_i = 0$ in Eqs. (9.22) and (9.23) and $\Phi_i(\omega_1)$ and $\Gamma_i(\omega_2)$ are real functions that are even with respect to ω_1 and ω_2, respectively; if $f_i(n_1)$ and $g_i(n_2)$ are mirror-image antisymmetric, then $\theta_i = \pi/2$ and $\Phi_i(\omega_1)$ and $\Gamma_i(\omega_2)$ are real functions that are odd with respect to ω_1 and ω_2, respectively. Under these circumstances, a zero-phase 2-D filter is obtained whose frequency response is given by Eqs. (9.19)–(9.23) as

$$H(e^{j\omega_1 T_1}, e^{j\omega_2 T_2}) = \sum_{i=1}^{K} F_i(e^{j\omega_1 T_1})G_i(e^{j\omega_2 T_2})$$

$$= \sum_{i=1}^{K} \pm \Phi_i(\omega_1)\Gamma_i(\omega_2) \tag{9.24}$$

where the plus sign corresponds to $\theta_i = 0$, and the minus sign corresponds to $\theta_i = \pi/2$. On comparing Eq. (9.24) with Eq. (9.8), we obtain

$$X(\omega_1, \omega_2) = \sum_{i=1}^{K} \pm \Phi_i(\omega_1)\Gamma_i(\omega_2) \tag{9.25}$$

On comparing Eq. (9.25) with Eq. (9.10b), $\tilde{u}_i$ and $\tilde{v}_i$ may be taken to be sampled versions of the frequency responses $\Phi_i(\omega_1)$ and $\Gamma_i(\omega_2)$, respectively. By designing the 1-D nonrecursive filters characterized by $F_i(z_1)$ and $G_i(z_2)$ $(1 \le i \le K, 1 \le K \le r)$ as zero-phase or $\pi/2$-phase filters and then interconnecting the filters obtained as in Fig. 8.5, a zero-phase 2-D filter is obtained whose amplitude response is a minimal mean-square-error approximation to the desired amplitude response (see Sec. 8.5.1). The impulse response of the resulting filter is given by

$$h(n_1, n_2) = \sum_{i=1}^{K} f_i(n_1)g_i(n_2) \qquad (9.26)$$

A 2-D causal linear-phase filter can readily be obtained by shifting the impulse response by $N_1/2$ and $N_2/2$ samples with respect to the n_1 axis and n_2 axis, respectively. This can be accomplished by multiplying $F_i(z_1)$ and $G_i(z_2)$ by $z_1^{-N_1/2}$ and $z_2^{-N_2/2}$, respectively.

As $K \to r$, the approximation error in the design of the 2-D filter is reduced to that introduced in the design of the 1-D nonrecursive filter but the number of multiplications required in the realization becomes very large. Hence K is, in practice, chosen to be as small as possible without increasing the approximation error beyond a certain specified upper bound.

The design of the 2-D filter can be completed by using any one of the standard methods for the design of 1-D nonrecursive filters. Using the Fourier series methods in conjunction with window techniques [8] (see Sec. 6.3.1), designs can be obtained very quickly with a small amount of computational effort. These designs are not optimal although the approximation error can be made arbitrarily small by increasing the order of the 1-D filters used. On the other hand, by using methods based on the Remez algorithm [9–12], it may be possible to obtain optimal designs although a large amount of computation would be required to complete the design.

An algorithm implementing the preceding design procedure is as follows:

STEP 1: Specify the desired amplitude response and thereby obtain the corresponding sampled amplitude response matrix **A**.

STEP 2: Decompose matrix **A** using Eq. (9.10b) to get $\tilde{\mathbf{u}}_i$ and $\tilde{\mathbf{v}}_i$, where $1 \leq i \leq r$.

STEP 3: Obtain K ($1 \leq K \leq r$) 2-D nonrecursive filters by designing either two 1-D zero-phase or two $\pi/2$-phase nonrecursive filters characterized by transfer functions $F_i(z_1)$ and $G_i(z_2)$ assuming sampled frequency responses $\tilde{\mathbf{u}}_i$ and $\tilde{\mathbf{v}}_i$, respectively.

STEP 4: Obtain the impulse response of the resulting zero-phase 2-D filter through Eq. (9.26).

STEP 5: Multiply the resulting 2-D zero-phase transfer function by $z_1^{-N_1/2}$ and $z_2^{-N_2/2}$ to obtain a causal linear-phase 2-D filter.

Example 9.2 Design a 2-D lowpass nonrecursive filter with a rotated elliptical passband. The desired amplitude response, shown in Fig. 9.2a,

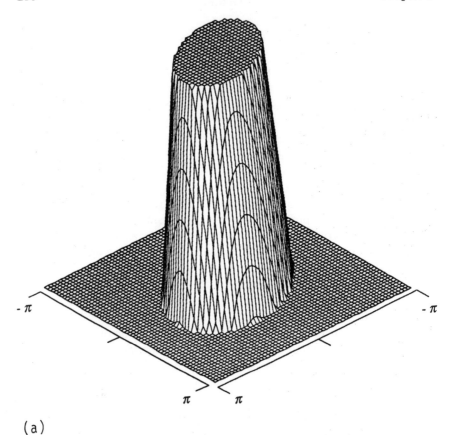

$- \pi$ $- \pi$

π π

(a)

Figure 9.2 Design of lowpass nonrecursive filter with rotated elliptical passband (Example 9.2): (a) ideal amplitude response; (b) amplitude response achieved with $K = 12$.

is specified by

$$|H(e^{j\omega_1 T_1}, e^{j\omega_2 T_2})| = \begin{cases} 1 & \text{for } (\tilde{\omega}_1/\omega_{p_1})^2 + (\tilde{\omega}_2/\omega_{p_2})^2 \leq 1 \\ 0 & \text{for } (\tilde{\omega}_1/\omega_{a_1})^2 + (\tilde{\omega}_2/\omega_{a_2})^2 > 1 \end{cases}$$

where

$$\tilde{\omega}_1 = \omega_1 \cos \alpha + \omega_2 \sin \alpha$$
$$\tilde{\omega}_2 = -\omega_1 \sin \alpha + \omega_2 \cos \alpha$$

and

$$\alpha = \pi/6, \quad \omega_{p_1} = 0.32\pi, \quad \omega_{p_2} = 0.52\pi, \quad \omega_{a_1} = 0.48\pi, \quad \omega_{a_2} = 0.68\pi$$

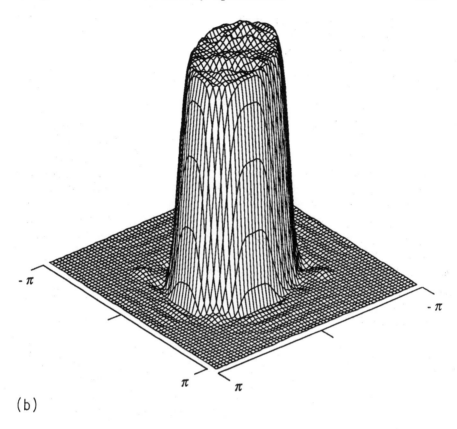

(b)

Solution. The corresponding sampled amplitude response $\mathbf{A} = |H(e^{j\pi\mu_l}, e^{j\pi\nu_m})|$ can be expressed as

$$|H(e^{j\pi\mu_l}, e^{j\pi\nu_m})| = \begin{cases} 1 & \text{for } (\bar{\mu}_l/\omega_{c_1})^2 + (\bar{\nu}_m/\omega_{c_2})^2 \leq 1/\pi^2 \\ 0 & \text{otherwise} \end{cases}$$

where

$$\bar{\mu}_l = \mu_l \cos \alpha + \nu_m \sin \alpha$$

$$\bar{\nu}_m = -\mu_l \sin \alpha + \nu_m \cos \alpha$$

$$\omega_{c_1} = \frac{1}{2}(\omega_{a_1} + \omega_{p_1}) = 0.4\pi, \qquad \omega_{c_2} = \frac{1}{2}(\omega_{a_2} + \omega_{p_2}) = 0.6\pi$$

The 1-D nonrecursive filters can be designed by using the Fourier series method along with the Kaiser window function. As may be expected, the

Table 9.2 Maximum Approximation
Errors in Example 9.2

K	Passband	Stopband
12	0.0393	0.0225
18	0.0124	0.0113
25	0.0117	0.0102

higher the order of the 1-D filters, the lower the approximation error. By trial and error, it has been found that a value of 29 for both N_1 and N_2 gives satisfactory results.

There were 25 nonzero singular values resulting from the SVD of matrix **A**. The resulting amplitude response of the lowpass 2-D nonrecursive filter for $K = 12$ is shown in Fig. 9.2b. The maximum passband and stopband errors for $K = 12$, 18, and 25 are given in Table 9.2.

REFERENCES

1. Y. Kamp and J. P. Thiran, Chebyshev approximations for two-dimensional nonrecursive digital filters, *IEEE Trans. Circuits Syst.*, vol. CAS-22, pp. 208–218, March 1975.
2. D. B. Harris and R. M. Mersereau, A comparison of algorithms for minimax design of two-dimensional linear-phase FIR digital filters, *IEEE Trans. Acoust., Speech, Signal Process.*, vol. ASSP-25, pp. 492–500, Dec. 1977.
3. C. Charalambous, The performance of an algorithm for minimax design of two-dimensional linear phase FIR digital filters, *IEEE Trans. Circuits Syst.*, vol. CAS-32, pp. 1016–1028, Oct. 1985.
4. W.-S. Lu, H.-P. Wang, and A. Antoniou, Design of two-dimensional FIR digital filters by using the singular value decomposition, *IEEE Trans. Circuits Syst.*, vol. CAS-37, pp. 35–46, Jan. 1990.
5. W.-S. Lu, H.-P. Wang, and A. Antoniou, Design of two-dimensional digital filters using singular value decomposition and balanced approximation method, *IEEE Trans. Signal Processing*, vol.39, pp. 2253–2262, Oct. 1991.
6. M. J. D. Powell, "Restart procedures for the conjugate gradient method", *Math. Program.*, vol. 12, pp. 241–254, 1977.
7. R. Fletcher, *Practical Methods of Optimization*, 2nd ed., Chichester: Wiley, 1987.
8. A. Antoniou, *Digital Filters: Analysis and Design*, New York: McGraw-Hill, 1979 (2nd ed. in press).
9. T. W. Parks and J. H. McClellan, Chebyshev approximation for nonrecursive digital filters with linear phase, *IEEE Trans. Circuit Theory*, vol. CT-19, pp. 189–194, March 1972.

10. J. H. McClellan, T. W. Parks, and L. R. Rabiner, A computer program for designing optimum FIR linear phase digital filters, *IEEE Trans. Audio Electroacoust.*, vol. AU-21, pp. 506–526, Dec. 1973.

11. L. R. Rabiner, J. H. McClellan and T. W. Parks, FIR digital filter design techniques using weighted Chebyshev approximation, *Proc. IEEE*, vol. 63, pp. 595–610, April 1975.

12. D. J. Shpak and A. Antoniou, A generalized Reméz method for the design of FIR digital filters, *IEEE Trans. Circuits Syst.*, vol. CAS-37, pp. 161–174, Feb. 1990.

PROBLEMS

9.1 Verify Eqs. (9.4)–(9.7).

9.2 Design a 2-D circularly symmetric nonrecursive highpass filter of order 7×7 with $\omega_p = 0.6\pi$, $\omega_a = 0.4\pi$, $\omega_{s1} = \omega_{s2} = \omega_s = 2\pi$ by using the optimization algorithm described in Sec. 9.2.2.

9.3 Repeat Problem 9.2 for a filter of the same type of order 13×13.

9.4 (a) Using MATLAB, generate a sampled amplitude response matrix **A** for the filter specified in Example 9.2.
(b) Perform the SVD of **A** and show that Theorem 9.1 applies.

9.5 (a) Design zero-phase nonrecursive 1-D filters of order 28 for the first 12 pairs of vectors ($\tilde{u}_i$, $\tilde{v}_i$, $1 \le i \le 12$) of the matrix **A** obtained in Problem 9.4a.
(b) Find the coefficient matrix of the transfer function obtained.

9.6 The number of parallel sections obtained in Problem 9.5 can be further reduced by carrying out the following steps:
(a) Compute the SVD of the coefficient matrix obtained in Problem 9.5(b).
(b) Approximate the coefficient matrix by neglecting its insignificant singular values, say, those that are larger than $0.05\sigma_1$.

10
Realization

10.1 INTRODUCTION

When a 2-D transfer function that satisfies the required specifications is generated, its realization is often required. As was stated in the introduction to Chap. 6, realization is the process of converting the transfer function or some other characterization of the filter into a network.

For a given characterization, several networks or structures can usually be obtained. The various possible structures can vary significantly in their sensitivity to coefficient quantization, their computational complexity, and the level of output noise produced by the quantization of products. The computational complexity of a filter structure is a measure of the amount of computation required in the implementation of the filter and depends on the number of arithmetic operations required per output sample.

There are two general approaches to the solution of the realization problem: direct and indirect. In direct methods, the transfer function is put in some form that allows the identification of an interconnected set of elemental subnetworks. In indirect realizations, on the other hand, an analog-filter network is converted into a topologically related digital-filter network.

In this chapter, some direct and indirect realization methods are considered in detail. The chapter begins with methods that are suitable for the realization of nonrecursive filters of the type that can be designed using the

McClellan transformation [1–4]. Then methods based on matrix decompositions are presented [5–7]. General direct methods that are applicable for the realization of arbitrary recursive transfer functions [8,9] are considered in Sec. 10.4. Indirect methods whereby 2-D filter structures are obtained by applying the wave characterization to analog-filter networks [10] are considered in Sec. 10.6. These structures have some interesting properties and advantages such as low sensitivity to coefficient quantization.

10.2 NONRECURSIVE REALIZATIONS BASED ON THE McCLELLAN TRANSFORMATION

One of the important methods for the design of nonrecursive filters is that based on the McClellan transformation method described in Sec. 6.4.1. In this section, we consider a number of structures that are suitable for the realization of transfer functions of the type that can be obtained by using the McClellan transformation.

A 1-D nonrecursive filter of odd length can be represented by the transfer function

$$H_1(z) = \sum_{n=-(N-1))/2}^{(N-1)/2} h_1(n)z^{-n} \tag{10.1}$$

and if the impulse response of the filter is symmetrical such that $h_1(n) = h_1(-n)$ for $n = 1, 2, \ldots, (N-1)/2$, the frequency response of the filter can be expressed as

$$H_1(e^{j\omega T}) = \sum_{n=0}^{(N-1)/2} a_1(n)[P_1(e^{j\omega T})]^n \tag{10.2}$$

where $a_1(n)$ for $n = 0, 1, 2, \ldots, (N-1)/2$ are constants, and

$$P_1(e^{j\omega T}) = \cos \omega T$$

[see Eq. (6.50)]. Now if we apply the McClellan transformation

$$P_1(e^{j\omega T}) = \cos \omega T = A \cos \omega_1 T_1 + B \cos \omega_2 T_2$$
$$+ C \cos \omega_1 T_1 \cos \omega_2 T_2 + D$$
$$= F(e^{j\omega_1 T_1}, e^{j\omega_2 T_2}) \tag{10.3}$$

to $H_1(e^{j\omega T})$ given by Eq. (10.2), we obtain

$$H_2(e^{j\omega_1 T_1}, e^{j\omega_2 T_2}) = \sum_{n=0}^{(N-1)/2} a_1(n)[F(e^{j\omega_1 T_1}, e^{j\omega_2 T_2})]^n \tag{10.4}$$

Therefore, with $e^{j\omega_1 T_1}$ and $e^{j\omega_2 T_2}$ replaced by z_1 and z_2, respectively, we have

$$H_2(z_1, z_2) = \sum_{n=0}^{(N-1)/2} a_1(n)F^n \qquad (10.5)$$

where

$$F \equiv F(z_1, z_2) = \frac{1}{2} A(z_1 + z_1^{-1}) + \frac{1}{2} B(z_2 + z_2^{-1})$$

$$+ \frac{1}{4} C(z_1 + z_1^{-1})(z_2 + z_2^{-1}) + D$$

Several direct realizations of $H_2(z_1, z_2)$ can readily be deduced at this point by replacing z^{-1} by F in standard nonrecursive structures [11,12]. Two possible structures reported in [3] are depicted in Fig. 10.1a and b. Alternatively, the transfer function in Eq. (10.5) can be factored as

$$H_2(z_1, z_2) = \prod_{i=1}^{M_1} \{b_i(0) + b_i(1)F + b_i(2)F^2\} \times \prod_{j=1}^{M_2} \{c_j(0) + c_j(1)F\}$$

This yields the cascade structure of Fig. 10.2 [3].

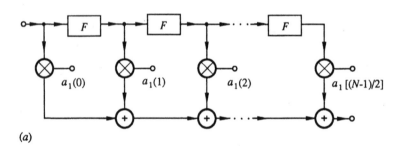

(a)

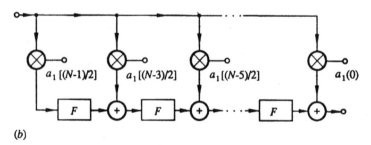

(b)

Figure 10.1 Realization of nonrecursive filters: (a) direct form; (b) transpose direct form.

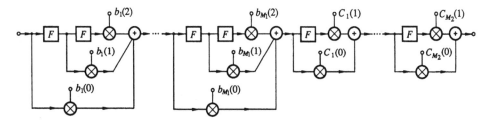

Figure 10.2 Cascade realization of nonrecursive filters.

The direct realizations of Fig. 10.1a and b require a minimal number of multiplications, but, as is the case in direct realizations in general, the sensitivity to coefficient quantization and the output noise due to product quantization can be high. The cascade realization offers low sensitivity and roundoff noise but, unfortunately, the computational complexity is increased since the required number of multiplications is increased.

An alternative realization method that yields structures with low sensitivity and roundoff noise in addition to low computational complexity is one proposed by McClellan and Chan [4]. The steps involved in this method are as follows.

The frequency response of Eq. (10.2) can be expressed as

$$H_1(e^{j\omega T}) = h_1(0) + 2 \sum_{n=1}^{(N-1)/2} h_1(n)C_n[P_1(e^{j\omega T})] \qquad (10.6)$$

where $C_n(x)$ is the nth-order Chebyshev polynomial in x [see Eq. (6.50)]. On applying the McClellan transformation to Eq. (10.5), we obtain

$$H_2(e^{j\omega_1 T_1}, e^{j\omega_2 T_2}) = h_1(0) + 2 \sum_{n=1}^{(N-1)/2} h_1(n) C_n[F(e^{j\omega_1 T_1}, e^{j\omega_2 T_2})]$$

Therefore

$$H_2(z_1, z_2) = h_1(0) + 2 \sum_{n=1}^{(N-1)/2} h_1(n) C_n(F)$$

where

$$C_n(F) = 2FC_{n-1}(F) - C_{n-2}(F) \qquad \text{for } n \geq 2$$
$$C_1(F) = F$$

and

$$C_0(F) = 1$$

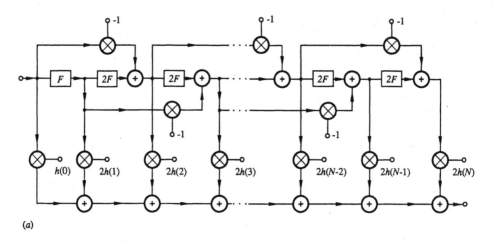

(a)

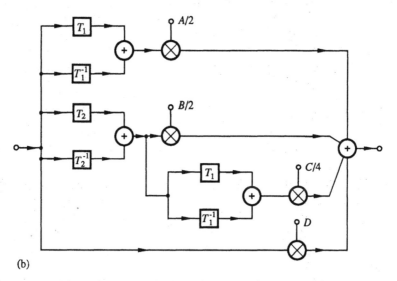

(b)

Figure 10.3 (a) Chebyshev realization of nonrecursive filters; (b) a realization of $F(z_1, z_2)$.

(see Sec. 6.4.3). A realization of $H_2(z_1, z_2)$, commonly referred to as the Chebyshev realization, is depicted in Fig. 10.3a. A noncausal realization of $F(z_1, z_2)$ is depicted in Fig. 10.3b. As can be seen, a total of six additions and four multiplications are required per sample for the realization of $F(z_1, z_2)$. Hence, it follows from Fig. 10.3 that the Chevyshev realization requires $9N - 2$ additions and $5N + 1$ multiplications per sample.

10.3 NONRECURSIVE REALIZATIONS BASED ON MATRIX DECOMPOSITIONS

The transfer function of a nonrecursive 2-D filter can be expressed as a quadratic form where the matrix involved is actually an array of the filter coefficients. By applying matrix decompositions such as the Jordan decomposition, the LU decomposition (LUD), the singular-value decomposition (SVD), etc., a number of useful realization schemes can be obtained [5,6,7].

10.3.1 LUD Realization

Consider a nonrecursive 2-D filter with a transfer function given by

$$H(z_1, z_2) = \sum_{n_1=0}^{N_1-1} \sum_{n_2=0}^{N_2-1} h(n_1, n_2)z_1^{-n_1} z_2^{-n_2} \tag{10.7}$$

If we let $\mathbf{z}_i = [1 \quad z_i^{-1} \ldots z_i^{-(N_i-1)}]^T$ for $i = 1, 2$, then $H(z_1, z_2)$ can be written as

$$H(z_1, z_2) = \mathbf{z}_1^T \mathbf{C} \mathbf{z}_2 \tag{10.8}$$

where $\mathbf{C}$ is the $N_1 \times N_2$ coefficient matrix given by

$$\mathbf{C} = \begin{bmatrix} h(0, 0) & h(0, 1) & \cdots & h(0, N_2 - 1) \\ h(1, 0) & h(1, 1) & \cdots & h(1, N_2 - 1) \\ \vdots & \vdots & & \vdots \\ h(N_1 - 1, 0) & h(N_1 - 1, 1) & \cdots & h(N_1 - 1, N_2 - 1) \end{bmatrix}$$

If r_c is the rank of matrix $\mathbf{C}$, then

$$r_c \leq \min(N_1, N_2)$$

and the LU decomposition of matrix $\mathbf{C}$ [13] is given by

$$\mathbf{C} = \mathbf{LU}$$

where $\mathbf{L} \in \mathbf{R}^{N_1 \times r_c}$ and $\mathbf{U} \in \mathbf{R}^{r_c \times N_1}$ are a lower- and an upper-triangular matrix, respectively, that is,

$$\mathbf{L} = \begin{bmatrix} * & 0 & \cdots & 0 \\ * & * & \cdots & 0 \\ \vdots & \vdots & \ddots & \vdots \\ \vdots & \vdots & & * \\ \vdots & \vdots & & \vdots \\ * & * & \cdots & * \end{bmatrix} \quad \text{and} \quad \mathbf{U} = \begin{bmatrix} * & * & \cdots & \cdots & * \\ 0 & * & \cdots & \cdots & * \\ \vdots & \vdots & \ddots & & \vdots \\ 0 & 0 & \cdots & * & \cdots & * \end{bmatrix}$$

Consequently, $H(z_1, z_2)$ can be expressed as

$$H(z_1, z_2) = \sum_{i=1}^{r_c} L_i(z_1)U_i(z_2) \tag{10.9}$$

where

$$L_i(z_1) = \sum_{k=i}^{N_1} l_{ki}z_1^{-(k-1)} \tag{10.10}$$

$$U_i(z_2) = \sum_{k=i}^{N_2} u_{ik}z_2^{-(k-1)} \tag{10.11}$$

From Eqs. (10.9)–(10.11), it follows that $H(z_1, z_2)$ can be realized by first realizing r_c pairs of 1-D nonrecursive transfer functions $L_i(z_1)$ and $U_i(z_2)$ and then connecting pairs of cascaded 1-D subfilters in parallel as depicted in Fig. 10.4. From Eqs. (10.10) and (10.11) and Fig. 10.4, it can readily be shown that the numbers of additions and multiplications required by the LUD realization are $r_c(N_1 + N_2 - r_c) - 1$ and $r_c(N_1 + N_2 - r_c + 1)$, respectively.

If the phase response of the filter is linear, then from Eq. (6.6) the coefficient matrix $\mathbf{C}$ can be expressed as

$$\mathbf{C} = \begin{bmatrix} \mathbf{I}_{(N_1+1)/2} & \mathbf{0} \\ \hat{\mathbf{I}}_1 & \mathbf{I}_{(N_2-1)/2} \end{bmatrix} \begin{bmatrix} \tilde{\mathbf{C}} & \mathbf{0} \\ \mathbf{0} & \mathbf{0} \end{bmatrix} \begin{bmatrix} \mathbf{I}_{(N_2+1)/2} & \hat{\mathbf{I}}_2 \\ \mathbf{0} & \mathbf{I}_{(N_1-1)/2} \end{bmatrix} \tag{10.12}$$

where $\tilde{\mathbf{C}} \in \mathbf{R}^{(N_1-1)/2 \times (N_2+1)/2}$, $\hat{\mathbf{I}}_1 \in \mathbf{R}^{(N_2-1)/2 \times (N_1+1)/2}$, and $\hat{\mathbf{I}}_2 \in \mathbf{R}^{(N_2+1)/2 \times (N_1-1)/2}$ with

$$\hat{\mathbf{I}}_1 = \begin{bmatrix} 0 & 0 & \cdots & 1 & 0 \\ 0 & 0 & \cdots & 0 & 0 \\ \vdots & \vdots & & \vdots & \vdots \\ 0 & 1 & \cdots & 0 & 0 \\ 1 & 0 & \cdots & 0 & 0 \end{bmatrix}$$

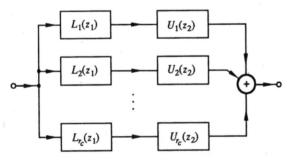

Figure 10.4 LUD realization of nonrecursive filters.

and

$$
\hat{I}_2 = \begin{bmatrix} 0 & 0 & \cdots & 0 & 1 \\ 0 & 0 & \cdots & 1 & 0 \\ \vdots & \vdots & & \vdots & \vdots \\ 1 & 0 & \cdots & 0 & 0 \\ 0 & 0 & \cdots & 0 & 0 \end{bmatrix}
$$

If the LUD of matrix $\tilde{C}$ is given by $\tilde{C} = \tilde{L}\tilde{U}$, then Eq. (10.12) implies that

$$
\begin{aligned}
C &= \begin{bmatrix} \tilde{L} & 0 \\ \hat{I}_1\tilde{L} & 0 \end{bmatrix} \begin{bmatrix} \tilde{U} & \tilde{U}\hat{I}_2 \\ 0 & 0 \end{bmatrix} \\
&= \begin{bmatrix} \tilde{L} \\ \hat{I}_1\tilde{L} \end{bmatrix} [\tilde{U} \quad \tilde{U}\hat{I}_2] \\
&= L^*U^*
\end{aligned}
$$

where $L^* \in R^{N_1 \times r_c}$ and $U^* \in R^{r_c \times N_2}$ assume the forms

$$
L^* = \begin{bmatrix} * & 0 & & 0 \\ * & * & & 0 \\ \vdots & \vdots & \ddots & \vdots \\ & & & * \\ \vdots & \vdots & & \vdots \\ & & & * \\ & & \ddots & \vdots \\ * & * & & 0 \\ * & 0 & & 0 \end{bmatrix}
$$

and

$$
U^* = \begin{bmatrix} * & * & \cdots\cdots\cdots\cdots\cdots\cdots\cdots\cdots & * & * \\ 0 & * & \cdots\cdots\cdots\cdots\cdots\cdots\cdots & * & 0 \\ \vdots & \vdots & \ddots & \vdots & \vdots \\ 0 & 0 & \cdots & * & \cdots & * & \cdots & 0 & 0 \end{bmatrix}
$$

respectively. As in Eq. (10.9), the transfer function $H(z_1, z_2)$ can be written as

$$
H(z_1, z_2) = \sum_{i=1}^{r_c} L_i^*(z_1)U_i^*(z_2) \tag{10.13}
$$

where

$$
L_i^*(z_1) = \sum_{k=i}^{N_1-i+1} l_{ki}^* z_1^{-(k-1)} \tag{10.14}
$$

$$U_i^*(z_2) = \sum_{k=i}^{N_2-i+1} u_{i,k}^* z_2^{-(k-1)} \qquad (10.15)$$

and can be realized using the parallel arrangement of Fig. 10.4. From Eqs. (10.13)–(10.15), the numbers of additions and multiplications required in this special LUD realization are $r_c(N_1 + N_2 - 2r_c + 1) - 1$ and $r_c(N_1 + N_2 - 2r_c + 2)$, respectively.

10.3.2 SVD–LUD Realization

The LUD realization described above requires r_c 2-D subfilters where r_c is the rank of coefficient matrix $\mathbf{C}$. As will be demonstrated in what follows, if some of the singular values of $\mathbf{C}$ are relatively small, a slightly modified version of transfer function $H(z_1, z_2)$ can be obtained that leads to a more economical realization.

The SVD of matrix $\mathbf{C}$ in Eq. (10.8) can be expressed as

$$\mathbf{C} = \mathbf{U}\mathbf{S}\mathbf{V}^T = \sum_{i=1}^{r_c} \sigma_i \mathbf{u}_i \mathbf{v}_i^T$$

where $\mathbf{U} = [\mathbf{u}_1 \cdots \mathbf{u}_{N_1}]$ and $\mathbf{V} = [\mathbf{v}_1 \cdots \mathbf{v}_{N_2}]$ are orthogonal matrices and

$$\mathbf{S} = \begin{bmatrix} \sigma_1 & & & \vdots & \\ & \ddots & & \vdots & \mathbf{0} \\ & & \sigma_{r_c} & \vdots & \\ \hdashline & \mathbf{0} & & \vdots & \mathbf{0} \end{bmatrix}$$

with $\sigma_1 \geq \sigma_2 \geq \ldots \geq \sigma_{r_c} > 0$. If $\sigma_{K+1} \ll \sigma_K$ for some $K < r_c$, then $\mathbf{C}$ can be approximated by

$$\mathbf{C}_K = \sum_{i=1}^{K} \sigma_i \mathbf{u}_i \mathbf{v}_i^T \qquad (10.16)$$

and hence the transfer function

$$H_K(z_1, z_2) = \mathbf{z}_1^T \mathbf{C}_K \mathbf{z}_2 \qquad (10.17)$$

can be obtained.

The L_2 error of the preceding approximation with respect to $H(z_1, z_2)$ can be expressed as

$$e_2 = \left[\oint_{|z_2|=1} \oint_{|z_1|=1} |H(z_1, z_2) - H_K(z_1, z_2)|^2 \frac{dz_1}{z_1} \frac{dz_2}{z_2} \right]^{1/2}$$

From Eqs. (10.8) and (10.16), we have

$$|H(z_1, z_2) - H_k(z_1, z_2)|^2 = |\mathbf{z}_1^T(\mathbf{C} - \mathbf{C}_K)\mathbf{z}_2|^2$$
$$= \mathbf{z}_1^T(\mathbf{C} - \mathbf{C}_K)\mathbf{z}_2\bar{\mathbf{z}}_2^T(\mathbf{C} - \mathbf{C}_K)^T\bar{\mathbf{z}}_1$$
$$= \text{tr}[(\mathbf{C} - \mathbf{C}_K)^T\bar{\mathbf{z}}_1\mathbf{z}_1^T(\mathbf{C} - \mathbf{C}_K)\mathbf{z}_2\bar{\mathbf{z}}_2^T]$$

where $\text{tr}(\cdot)$ is the trace of $(\cdot)$. Hence

$$e_2^2 = \text{tr}\left[(\mathbf{C} - \mathbf{C}_k)^T \oint_{|z_1|=1} \bar{\mathbf{z}}_1\mathbf{z}_1^T \frac{dz_1}{z_1} (\mathbf{C} - \mathbf{C}_K) \oint_{|z_2|=1} \mathbf{z}_2\bar{\mathbf{z}}_2^T \frac{dz_2}{z_2}\right]$$
$$= \text{tr}[(\mathbf{C} - \mathbf{C}_K)^T(\mathbf{C} - \mathbf{C}_K)]$$
$$= \sum_{i=K+1}^{r_c} \sigma_i^2$$

i.e.

$$e_2 = \left(\sum_{i=K+1}^{r_c} \sigma_i^2\right)^{1/2} \tag{10.18}$$

Therefore, if singular values $\sigma_{K+1}, \ldots, \sigma_{r_c}$ of coefficient matrix $\mathbf{C}$ are sufficiently small, then transfer function $H_K(z_1, z_2)$ represents a good approximation of $H(z_1, z_2)$.

The realization of $H_K(z_1, z_2)$ can be obtained by noting that matrix $\mathbf{C}_K$ can be expressed as

$$\mathbf{C}_K = \mathbf{U}\mathbf{S}_K\mathbf{V}^T$$

with

$$\mathbf{S}_K = \left[\begin{array}{ccc|c} \sigma_1 & & & \\ & \ddots & & \mathbf{0} \\ & & \sigma_K & \\ \hline & \mathbf{0} & & \mathbf{0} \end{array}\right]$$

which implies that rank $(\mathbf{C}_K) = K$. Consequently, the LU decomposition of $\mathbf{C}_K$ assumes the form

$$\mathbf{C}_K = \mathbf{L}_K\mathbf{U}_K$$

where $\mathbf{L}_K$ is an $N_1 \times K$ lower-triangular matrix and $\mathbf{U}_K$ is a $K \times N_2$ upper-triangular matrix. Hence $H_K(z_1, z_2)$ can be written as

$$H_K(z_1, z_2) = \sum_{i=1}^{K} \hat{L}_i(z_1)\hat{U}_i(z_2) \tag{10.19}$$

where

$$\hat{L}_{ci}(z_1) = \sum_{k=i}^{N_1} \hat{l}_{ki} z_1^{-(k-1)} \tag{10.20}$$

$$\hat{U}_{ci}(z_2) = \sum_{k=i}^{N_2} \hat{u}_{ik} z_2^{-(k-1)} \tag{10.21}$$

Clearly, Eqs. (10.19)–(10.21) indicate that the filter can be realized using the parallel configuration of Fig. 10.4 where the number of 1-D subfilters has been reduced from r_c to K. Therefore, the numbers of additions and multiplications are $K(N_1 + N_2 - K) - 1$ and $K(N_1 + N_2 - K + 1)$, respectively.

10.4   DIRECT REALIZATION OF RECURSIVE FILTERS

As for the 1-D case, the realization of recursive 2-D digital filters can be achieved by directly manipulating the transfer function involved. Consider the transfer function

$$H(z_1, z_2) = \frac{\overline{N}(z_1^{-1}, z_2^{-1})}{\overline{D}(z_1^{-1}, z_2^{-1})} = \frac{\sum_{i=0}^{N_1} \sum_{j=0}^{N_2} a_{ij} z_1^{-i} z_2^{-j}}{1 + \sum_{i=0}^{N_1} \sum_{j=0}^{N_2} b_{ij} z_1^{-i} z_2^{-j}} \tag{10.22}$$

where $b_{00} = 0$. Both polynomials $\overline{N}(z_1^{-1}, z_2^{-1})$ and $\overline{D}(z_1^{-1}, z_2^{-1})$ can be expressed as polynomials in z_1^{-1}, whose coefficients are polynomials in z_2^{-1} (or the other way around). Hence $H(z_1, z_2)$ may be written as

$$H(z_1, z_2) = \frac{\sum_{i=0}^{N_1} A_i(z_2^{-1}) z_1^{-i}}{1 + \sum_{i=0}^{N_1} B_i(z_2^{-1}) z_1^{-i}} \tag{10.23}$$

where

$$A_i(z_2^{-1}) = \sum_{j=0}^{N_2} a_{ij} z_2^{-j}$$

$$B_i(z_2^{-1}) = \sum_{j=0}^{N_2} b_{ij} z_2^{-j}$$

A direct realization of $H(z_1, z_2)$ can be obtained as shown in Fig. 10.5 by replacing the numerator and denominator multipliers in the standard 1-D canonic realization [12] by 1-D subfilters characterized by the polynomials $A_i(z_2^{-1})$ and $B_i(z_2^{-1})$. The realization of $H(z_1, z_2)$ is then completed by realizing $A_i(z_2^{-1})$ and $B_i(z_2^{-1})$ as nonrecursive filters. The scheme requires $N_1 + 2N_2$ delay elements. A direct realization of a 2-D transfer function of order (3, 2) is depicted in Fig. 10.6.

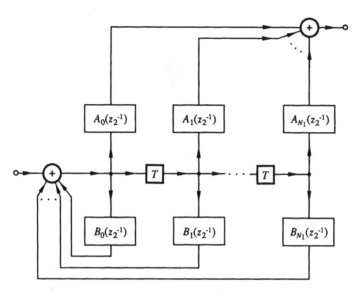

Figure 10.5 Direct realization of $H(z_1, z_2)$.

A realization of $H(z_1, z_2)$ of order $(N_1 + N_2)$ is said to be canonic with respect to the delay elements if the total number of delay elements needed in the realization is $N_1 + N_2$. Although such realizations can readily be obtained for 1-D digital filters, in the case of 2-D filters they may not exist [8,14,15]. However, if the numerator or denominator of the transfer function is separable, then a canonic realization always exists. As examples, canonic realizations for the transfer functions

$$H(z_1, z_2) = \frac{\sum_{i=0}^{3} \sum_{j=0}^{2} a_{ij} z_1^{-i} z_2^{-j}}{(\alpha_0 + \alpha_1 z_1^{-1} + \alpha_2 z_1^{-2} + \alpha_3 z_1^{-3})(\beta_0 + \beta_1 z_2^{-1} + \beta_2 z_2^{-2})}$$

and

$$H(z_1, z_2) = \frac{(\alpha_0 + \alpha_1 z_1^{-1} + \alpha_2 z_1^{-2} + \alpha_3 z_1^{-3})(\beta_0 + \beta_1 z_2^{-1} + \beta_2 z_2^{-2})}{\sum_{i=0}^{3} \sum_{j=0}^{2} b_{ij} z_1^{-i} z_2^{-j}}$$

can be obtained as shown in Figs. 10.7 and 10.8, respectively.

Direct realization methods applicable to M-D digital filters with $M > 2$ can be found in [16].

10.5 LUD REALIZATION OF RECURSIVE FILTERS

As was demonstrated in Sec. 10.3, matrix decompositions such as the LUD and SVD can be used to obtain useful realization schemes for nonrecursive

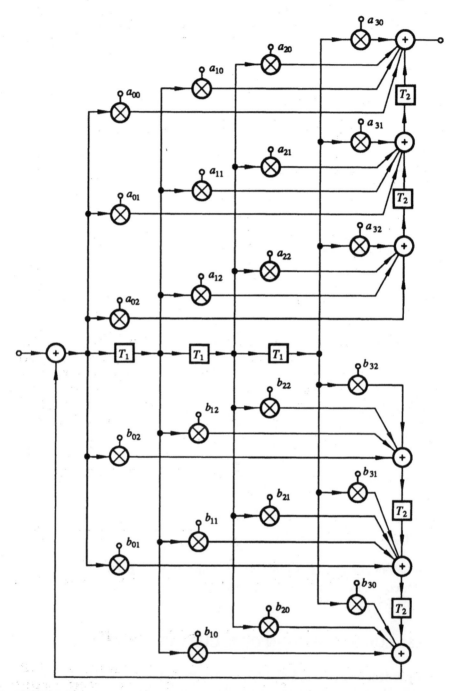

Figure 10.6   Direct realization of a 2-D transfer function of order (3, 2).

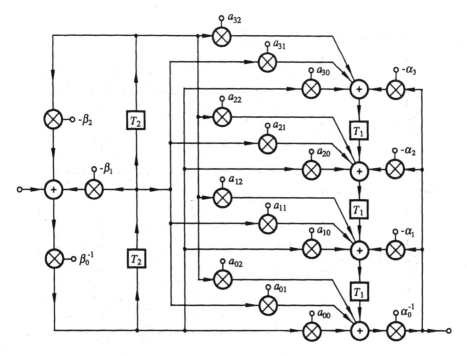

Figure 10.7 Canonic realization of a 2-D filter with separable denominator.

filters. By representing a recursive filter as an interconnection of nonrecursive filters, the methods of Sec.10.3 can readily be extended to the realization of recursive filters [5].

By rewriting $H(z_1, z_2)$ as

$$H(z_1, z_2) = \frac{\overline{N}(z_1^{-1}, z_2^{-1})}{1 + \tilde{D}(z_1^{-1}, z_2^{-1})} \tag{10.24}$$

where

$$\overline{N}(z_1^{-1}, z_2^{-1}) = \sum_{i=0}^{N_1} \sum_{j=0}^{N_2} a_{ij} z_1^{-i} z_2^{-j} \tag{10.25}$$

$$\tilde{D}(z_1^{-1}, z_2^{-1}) = \sum_{i=0}^{N_1} \sum_{j=0}^{N_2} b_{ij} z_1^{-i} z_2^{-j}, \qquad b_{00} = 0 \tag{10.26}$$

a 2-D recursive filter can be realized using the two-block feedback scheme shown in Fig. 10.9a or that in Fig. 10.9b. From Eqs. (10.25) and (10.26), the two blocks of each realization in Fig. 10.9 can be treated as nonrecursive 2-D filters, and can be realized by the LUD method. Specifically, the 2-D

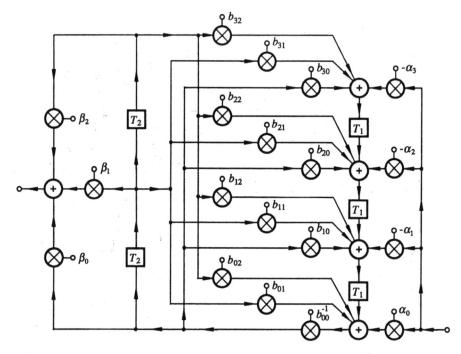

Figure 10.8 Canonic realization of a 2-D filter with separable numerator.

polynomials $\overline{N}(z_1^{-1}, z_2^{-1})$ and $\tilde{D}(z_1^{-1}, z_2^{-1})$ can be written in matrix form as

$$\overline{N}(z_1^{-1}, z_2^{-1}) = \mathbf{z}_1^T \mathbf{A} \mathbf{z}_2$$

and

$$\tilde{D}(z_1^{-1}, z_2^{-1}) = \mathbf{z}_1^T \mathbf{B} \mathbf{z}_2$$

respectively, where $\mathbf{A}, \mathbf{B} \in R^{N_1 \times N_2}$. If rank $(\mathbf{A}) = r_a$ and rank $(\mathbf{B}) = r_b$, then the LU decompositions of $\mathbf{A}$ and $\mathbf{B}$ give

$$\mathbf{A} = \mathbf{L}_A \mathbf{U}_A, \quad \mathbf{L}_A \in R^{N_1 \times r_a}, \quad \mathbf{U}_A \in R^{r_a \times N_2}$$

and

$$\mathbf{B} = \mathbf{L}_B \mathbf{U}_B, \quad \mathbf{L}_B \in R^{N_1 \times r_b}, \quad \mathbf{U}_B \in R^{r_b \times N_2}$$

respectively. From Sec.10.3, it immediately follows that

$$\overline{N}(z_1^{-1}, z_2^{-1}) = \sum_{i=1}^{r_a} L_{ai}(z_1) U_{ai}(z_2) \qquad (10.27)$$

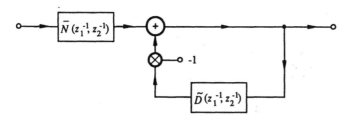

(a)

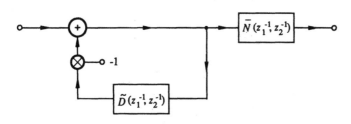

(b)

Figure 10.9 Two feedback schemes for the realization of $H(z_1, z_2)$.

with

$$L_{ai}(z_1) = \sum_{k=i}^{N_1} l_{ki}^{(a)} z_1^{-(k-1)}$$

$$U_{ai}(z_2) = \sum_{k=i}^{N_2} u_{ik}^{(a)} z_2^{-(k-1)}$$

and

$$\tilde{D}(z_1^{-1}, z_2^{-1}) = \sum_{i=1}^{r_b} L_{bi}(z_1)U_{bi}(z_2) \tag{10.28}$$

with

$$L_{bi}(z_1) = \sum_{k=i}^{N_1} l_{ki}^{(b)} z_1^{-(k-1)}$$

$$U_{bi}(z_2) = \sum_{k=i}^{N_2} u_{ik}^{(b)} z_2^{-(k-1)}$$

Figure 10.9 in conjunction with Eqs. (10.27) and (10.28) implies that $r_a +$

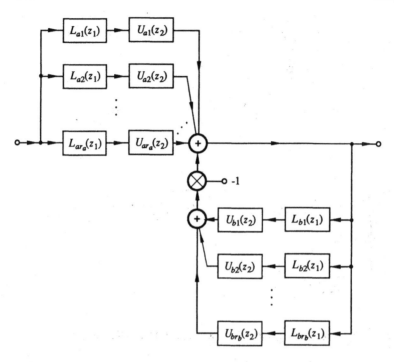

Figure 10.10   Realization of a recursive filter using LUD.

r_b filters are needed to realize $\overline{N}(z_1^{-1}, z_2^{-1})$ and $\tilde{D}(z_1^{-1}, z_2^{-1})$, and each of these filters is composed of two cascaded 1-D nonrecursive subfilters as in Fig. 10.4. The complete realization scheme is depicted in Fig. 10.10

10.6   INDIRECT REALIZATION OF RECURSIVE FILTERS

A number of different indirect 1-D realizations can be obtained by converting analog-filter networks into corresponding digital-filter networks through the application of network-theoretic concepts in conjunction with some simple transformations. By applying the wave network characterization in conjunction with the bilinear transformation of Eq. (7.2) to equally terminated LC filters, the class of 1-D wave digital filters proposed by Fettweis [17] and later developed further by Sedlemeyer and Fettweis [18] can be obtained. Similarly, by transforming an analog network configuration using generalized-immittance converters (GICs) [19], the class of 1-D GIC digital filters reported in [20] can be obtained. These realizations have several attractive properties such as low sensitivity to coefficient quantization and low output roundoff noise; further, they can be designed to be free from limit-cycle oscillations.

In Sec. 10.6.1, we show that the methodology of Fettweis can be applied for the realization of 2-D circularly symmetric lowpass wave digital filters. Then, in Sec. 10.6.2, we show that the 1-D GIC approach can be extended to include the realization of arbitrary 2-D transfer functions. Finally, the realization of circularly symmetric 2-D GIC digital filters is presented in Sec. 10.6.3.

10.6.1 Circularly Symmetric Wave Digital Filters

In Sec. 7.4, transformations that generate rotation in the amplitude response of an analog 1-D filter were used to solve the approximation problem in the design of 2-D recursive filters having piecewise-constant amplitude responses with circular symmetry. These transformations can also be applied in solving the realization problem for this class of filters. The steps involved are as follows:

STEP 1: Obtain an analog equally terminated LC ladder realization of the lowpass transfer function $H_{A1}(s)$ given by Eq. (7.25) using one of the classical synthesis methods [21–24].

STEP 2: For each rotation angle identified by Eq. (7.26), apply the transformation $s = g_1(s_1, s_2)$ of Eq. (7.12a) to the analog realization obtained in Step 1 to obtain a set of 2-D analog realizations.

STEP 3: For each 2-D analog realization obtained in Step 2, apply the methodology in [17]–[18] to obtain a set of 2-D digital realizations.

STEP 4: Cascade the 2-D digital realizations obtained in Step 3 to yield the required 2-D wave digital filter.

The application of the transformation of Eq. (7.12a) to an impedance sL_0 yields

$$sL_0|_{s=g_1(s_1,\ s_2)} = s_1 L_1 + s_2 L_2 \qquad (10.29)$$

where

$$L_1 = -L_0 \sin \beta_k \qquad \text{and} \qquad L_2 = L_0 \cos \beta_k$$

Similarly, for an admittance sC_0, we have

$$sC_0|_{s=g_1(s_1,\ s_2)} = s_1 C_1 + s_2 C_2 \qquad (10.30)$$

where

$$C_1 = -C_0 \sin \beta_k \qquad \text{and} \qquad C_2 = C_0 \cos \beta_k$$

Therefore, the 2-D analog filters required in Step 2 can readily be obtained by replacing each inductor in the 1-D analog filter by a pair of series inductors and each capacitor by a pair of parallel capacitors, as depicted in Fig. 10.11.

Realizations obtained by applying the transformation in Eq. (7.12a) have a nonessential singularity of the second kind on the bicircle U_2 (see p. 206) and can be unstable. This problem can be overcome by using the transformation $s = g_4(s_1, s_2)$ of Eq. (7.43) instead of that in Eq. (7.12a) (see Sec. 7.4.3). With this transformation, an impedance sL_0 becomes

$$sL_0|_{s=g_4(s_1, s_2)} = \frac{s_1L_1(1/s_2C_1)}{s_1L_1 + (1/s_2C_1)} + \frac{s_2L_2(1/s_1C_2)}{s_2L_2 + (1/s_1C_2)} \quad (10.31)$$

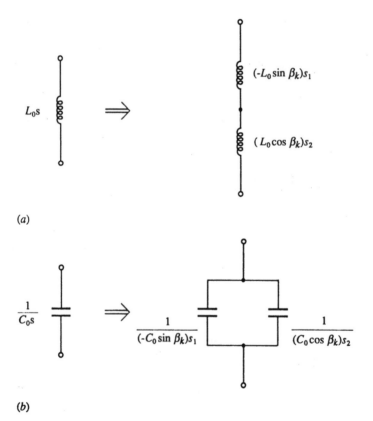

(a)

(b)

Figure 10.11 (a) 2-D analog network for inductance; (b) 2-D analog network for capacitance.

where

$$L_1 = L_0 \cos \beta_k, \qquad C_1 = c/L_0 \cos \beta_k$$
$$L_2 = L_0 \sin \beta_k, \qquad C_2 = c/L_0 \sin \beta_k$$

Similarly, an admittance sC_0 becomes

$$sC_0|_{s=g_4(s_1, s_2)} = \frac{s_1 C_3(1/s_2 L_3)}{s_1 C_3 + (1/s_2 L_3)} + \frac{s_2 C_4(1/s_1 L_4)}{s_2 C_4 + (1/s_1 L_4)} \qquad (10.32)$$

where

$$C_3 = C_0 \cos \beta_k, \qquad L_3 = c/L_0 \cos \beta_k$$
$$C_4 = C_0 \sin \beta_k, \qquad L_4 = c/L_0 \sin \beta_k$$

Hence the 2-D analog filters can readily be obtained by replacing each inductor in the 1-D analog filter by two parallel resonant circuits in series and each capacitor by two series resonant circuits in parallel as depicted in Fig. 10.12 [25].

10.6.2 Realization of 2-D GIC Digital Filters

The concept of the current-conversion GIC (CGIC) has been used extensively in the past for the realization of 1-D digital as well as analog filters [19,20,26]. In this section, we show that the CGIC can also be used for the realization of 2-D digital filters. Three general steps are involved as follows:

STEP 1: The 2-D counterpart of the 1-D CGIC is defined.

STEP 2: A general configuration comprising 2-D CGICs and conductances that realizes arbitrary 2-D continuous transfer functions is obtained.

STEP 3: The 2-D analog configuration obtained in Step 2 is transformed into a corresponding 2-D digital configuration.

A 2-D GIC is defined as a two-port analog network that can be represented as shown in Fig. 10.13 where

$$V_1 = V_2, \qquad I_1 = -h(s_1, s_2)I_2 \qquad (10.33)$$

The function $h(s_1, s_2)$ is said to be the admittance-conversion function (ACF) of the CGIC. A 2-D digital realization of the CGIC can be obtained by applying the wave characterization [17,27]:

$$A_v = V_v + I_v/G_v \qquad (10.34a)$$
$$B_v = V_v - I_v/G_v \qquad (10.34b)$$

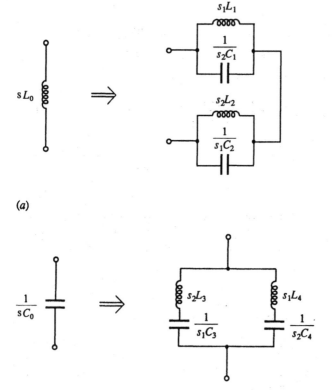

(a)

(b)

Figure 10.12 (a) 2-D analog network for inductance; (b) 2-D analog network for capacitance.

for $v = 1, 2$, to the network of Fig. 10.13. In Eq. (10.34) A_v, B_v, and G_v represent, respectively, the incident and reflected wave quantities and the port conductance of the vth port. By solving Eqs. (10.33)–(10.34) for B_1 and B_2 and then applying the double bilinear transformation of Eq. (7.14), we obtain

$$B_1 = A_2 + (A_1 - A_2)F(z_1, z_2) \qquad (10.35a)$$

$$B_2 = A_1 + (A_1 - A_2)F(z_1, z_2) \qquad (10.35b)$$

where

$$F(z_1, z_2) = \frac{G_1 - G_2 h(z_1, z_2)}{G_1 + G_2 h(z_1, z_2)} \qquad (10.36)$$

and

$$h(z_1, z_2) = h(s_1, s_2)|_{s_r = (1-z_r)/(1+z_r),\ r=1,\ 2} \qquad (10.37)$$

Consider now the two special cases

$$h(s_1, s_2) = s_r \qquad (10.38a)$$

and

$$h(s_1, s_2) = 1/s_r \qquad (10.38b)$$

for $r = 1, 2$. If we assume that

$$G_1 = G_2 \qquad (10.39)$$

then from Eqs. (10.36), (10.37), and (10.39) we obtain

$$F(z_1, z_2) = z_r$$

and

$$F(z_1, z_2) = -z_r$$

for Eqs. (10.38a) and (10.38b), respectively. The digital realization of the 2D CGICs for the preceding two cases and their symbolic representations are depicted in Fig. 10.14a and b, respectively. These will be referred to as GIC adaptors.

Let us consider a 2-D analog reference network comprising n 2-D CGICs and $n + 1$ conductances connected as shown in Fig. 10.15. The transfer function realized by the network can be written as

$$H_A(s_1, s_2) = \frac{V_0}{V_i} = \frac{N_A(s_1, s_2)}{D_A(s_1, s_2)} \qquad (10.40)$$

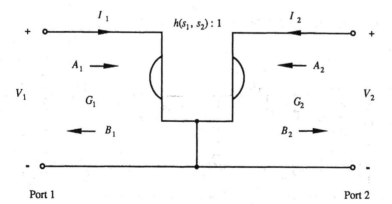

Port 1   Port 2

Figure 10.13 2-D digital realization of the CGIC.

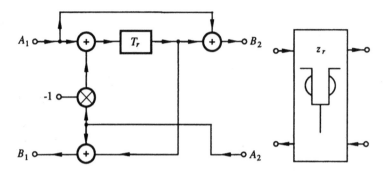

(a)

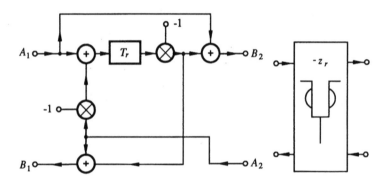

(b)

Figure 10.14 GIC adaptor: (a) for $h(s_1, s_2) = s_r$; (b) for $h(s_1, s_2) = 1/s_r$.

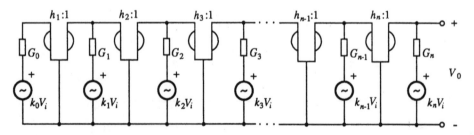

Figure 10.15 2-D analog reference network.

where

$$N_A(s_1, s_2) = k_0 G_0 + k_1 G_1 h_1 + \cdots + k_n G_n h_1 h_2 \cdots h_n$$

and

$$D_A(s_1, s_2) = G_0 + G_1 h_1 + \cdots + G_n h_1 h_2 \cdots h_n$$

with

$$h_i \equiv h_i(s_1, s_2)$$

By letting

$$G_r = b_r \quad \text{and} \quad k_r = a_r/b_r \quad \text{for } r = 0, 1, \ldots, n \quad (10.41)$$

in Eq. (10.40), the 2-D continuous transfer function

$$H_A(s_1, s_2) = \frac{N_A(s_1, s_2)}{D_A(s_1, s_2)} = \frac{a_0 + a_1 h_1 + \cdots + a_n h_1 h_2 \cdots h_n}{b_0 + b_1 h_1 + \cdots + b_n h_1 h_2 \cdots h_n} \quad (10.42)$$

is realized.

A reference network that realizes a 2-D transfer function of a specific order can be obtained by applying the following step-by-step procedure:

STEP 1: Choose the ACFs such that $N_A(s_1, s_2)$ and $D_A(s_1, s_2)$ in Eq. (10.40) represent general polynomials of the same order as the transfer function. This can be achieved by constructing a 2-D polynomial map containing the terms that are present in the polynomial. The polynomial maps corresponding to first-, second-, and third-order transfer functions are given in Fig. 10.16a, b, and c, respectively. Each box in the map represents a particular term $s_1^i s_2^j$ that is present in the polynomial. The values of the variables i, j are indicated in the maps as shown in Fig. 10.16. The map is constructed in such a way that any two adjacent boxes containing the terms $s_1^p s_2^q$ and $s_1^i s_2^j$ satisfy the condition

$$|p - i| + |q - j| = 1 \quad (10.43)$$

The element in each box of the map (except the first box) is assumed to represent a product of ACFs in the polynomials $N_A(s_1, s_2)$ and $D_A(s_1, s_2)$. The element in the first box (denoted as box I) is assumed to be $s_1^0 s_2^0 = 1$.

STEP 2: Draw a path in the polynomial map satisfying the following conditions:

1. It starts from the first box corresponding to the term $s_1^0 s_2^0$.
2. It passes through only adjacent boxes that satisfy Eq. (10.43).
3. It covers all the boxes.
4. No box is covered more than once.

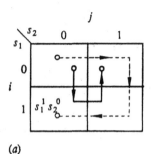

(a)

Figure 10.16 Polynomial maps: (a) first-order transfer function; (b) second-order transfer function; (c) third-order transfer function.

For the polynomial maps given in Fig. 10.16a, b, and c, two possible paths are illustrated in each case.

STEP 3: Obtain the required ACFs as

$$h_1 = h_{II}, \quad h_2 = h_{III}/h_{II}, \quad h_3 = h_{IV}/h_{III}, \ldots, \text{etc.}$$

where h_I, h_{II}, h_{III}, etc., are the elements represented by the boxes covered by the path sequentially. These elements are related to ACFs $h_1, h_2, \ldots,$ h_n as $h_I = 1$, $h_{II} = h_1$, $h_{III} = h_1 h_2$, $h_{IV} = h_1 h_2 h_3, \ldots,$ etc.

STEP 4: Connect n 2-D CGICs as shown in Fig. 10.15 having the h values obtained in Step 3 to yield the required reference network. The h values obtained using the paths corresponding to the solid lines in the polynomial maps shown in Fig. 10.16a, b, and c are given in Table 10.1.

A 2-D wave digital-filter structure can be derived from a reference configuration by using the wave characterization of Eq. (10.34) and assuming that the ACFs of the CGICs are as given in Eq. (10.38). For example, the digital-filter structure corresponding to Fig. 10.15 is derived as shown in Fig. 10.17 by using the GIC adaptors and the known digital equivalents for voltage sources, resistors, and parallel interconnections. The digital structure shown in Fig. 10.17 comprises n GIC adaptors and $n - 1$ parallel adaptors. The assignment of port conductances is done such that port 2 of the first $n - 2$ parallel adaptors is made reflection free [18]. In a three-port parallel adaptor, the port conductance corresponding to the reflection-free port is equal to the sum of the other two port conductances. From Eqs. (10.41) and (10.42) and the formula for the multiplier constants given in [18], it can be shown that the values of the multipliers

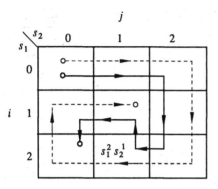

(b)

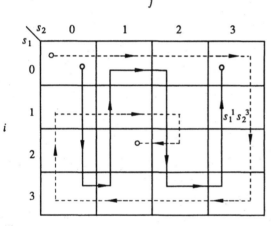

(c)

in Fig. 10.17 are given by

$$m_v = \frac{\sum_{p=0}^{v-1} b_p}{\sum_{p=0}^{v} b_p} \quad \text{for} \quad v = 1, \ldots, n-2 \quad (10.44a)$$

$$m_{n-1} = \frac{2 \sum_{p=0}^{n-2} b_p}{\sum_{p=0}^{n} b_p} \quad (10.44b)$$

$$m_n = \frac{2b_n}{\sum_{p=0}^{n} b_p} \quad (10.44c)$$

Table 10.1 Values of ACFs Obtained from Fig. 10.16a–c Using Solid-Line Path

$h(s_1, s_2)$	First-order (Fig. 10.16a)	Second-order (Fig. 10.16b)	Third-order (Fig. 10.16c)
h_1	s_1	s_2	s_1
h_2	s_2	s_2	s_1
h_3	$1/s_1$	s_1	s_1
h_4		s_1	s_2
h_5		$1/s_2$	$1/s_1$
h_6		$1/s_1$	$1/s_1$
h_7		$1/s_2$	$1/s_1$
h_8		s_1	s_2
h_9			s_1
h_{10}			s_1
h_{11}			s_1
h_{12}			s_2
h_{13}			$1/s_1$
h_{14}			$1/s_1$
h_{15}			$1/s_1$

and

$$k_r = a_r/b_r \quad \text{for} \quad r = 0, 1, \ldots, n \qquad (10.44d)$$

The digital transfer function realized by Fig. 10.17 is related to that of the reference configuration of Fig. 10.15 as

$$H_D(z_1, z_2) = \frac{B_0}{A_i} = 2H_A(s_1, s_2)\big|_{s_i = (1 - z_i)/(1 + z_i)} \quad \text{for } i = 1, 2$$

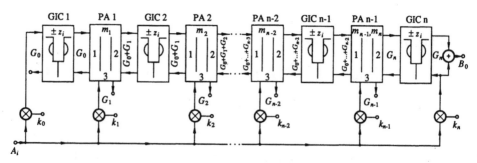

Figure 10.17 2-D digital filter corresponding Fig. 10.15.

10.6.3 Realization of 2-D Circularly Symmetric GIC Digital Filters

The concept of GIC can also be used to realize circularly symmetric lowpass filters [29]. The steps involved are described below. For the sake of simplicity, it is assumed that the 2-D filter to be realized has been designed by using the transformation method described in Sec. 7.4.1.

STEP 1: Decompose the 1-D continuous lowpass transfer function as

$$H_A(s) = K_0 \left(\prod_{j=1}^{M} \frac{s + a_j}{s + b_j} \right) \left(\prod_{j=1}^{N} \frac{s^2 + d_j s + f_j}{s^2 + g_j s + e_j} \right) \qquad (10.45)$$

STEP 2: For each rotation angle identified by Eq. (7.26), apply the transformation $s = g(s_1, s_2)$ where

$$g(s_1, s_2) = c_1 s_1 + c_2 s_2 \qquad (10.46)$$

with

$$c_1 = -\sin \beta_k, \qquad c_2 = \cos \beta_k$$

to obtain a set of 2-D analog transfer functions.

STEP 3: For each 2-D analog transfer function obtained in Step 2, apply the methodology in Sec. 10.6.2 to obtain a set of 2-D digital realizations.

STEP 4: Cascade the 2-D digital realizations obtained in Step 3 to obtain the required 2-D GIC digital filter.

The first-order digital filter corresponding to the jth first-order section in Eq. (10.45) has the transfer function

$$H_{D_1}(z_1, z_2) = \frac{N_{D_1}(z_1, z_2)}{D_{D_1}(z_1, z_2)} \qquad (10.47)$$

where

$$N_{D_1}(z_1, z_2) = k_0 m_1 (1 - z_1 + z_2 - z_1 z_2) + k_1 (2 - m_1 - m_2)$$
$$\times (1 + z_1 + z_2 + z_1 z_2) + k_2 m_2 (1 + z_1 - z_2 - z_1 z_2)$$
$$D_{D_1}(z_1, z_2) = 1 + (1 - m_1) z_1 + (1 - m_2) z_2 + (1 - m_1 - m_2) z_1 z_2$$
$$k_0 = 1, \qquad k_1 = a_1/b_1, \qquad k_2 = 1$$

and

$$m_1 = \frac{2c_1}{c_1 + b_1 + c_2}, \qquad m_2 = \frac{2c_2}{c_1 + b_1 + c_2}$$

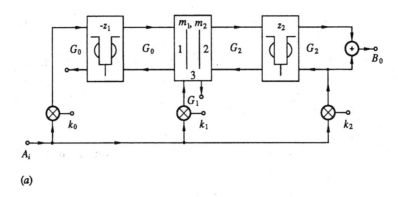

(a)

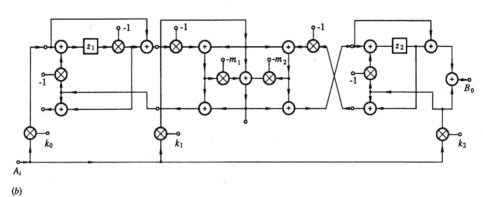

(b)

Figure 10.18 (a) 2-D first-order rotated digital filter; (b) corresponding filter structure.

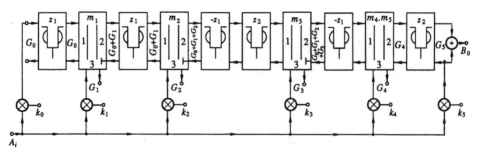

Figure 10.19 2-D second-order rotated digital filter.

Two GIC adaptors and one parallel adaptor are required to realize the transfer function in Eq. (10.47), as shown in Fig. 10.18a. The corresponding digital filter structure is shown in Fig. 10.18b.

The second-order digital filter corresponding to the jth section in Eq. (10.45) can be realized by using six GIC and four parallel adaptors, as shown in Fig. 10.19. The values of multiplier constants $m_1, \ldots, m_5$ of the parallel adaptors are given by

$$m_1 = \frac{e_1}{e_1 + g_1 c_1}$$

$$m_2 = \frac{e_1 + g_1 c_1}{e_1 + g_1 c_1 + c_1^2}$$

$$m_3 = \frac{e_1 + g_1 c_1 + c_1^2}{e_1 + g_1 c_1 + c_1^2 + 2c_1 c_2}$$

$$m_4 = \frac{2(e_1 + g_1 c_1 + 2c_1 c_2)}{e_1 + g_1 c_1 + c_1^2 + 2c_1 c_2 + g_1 c_2 + c_2^2}$$

$$m_5 = \frac{2c_2^2}{e_1 + g_1 c_1 + c_1^2 + 2c_1 c_2 + g_1 c_2 + c_2^2}$$

and the values of multiplier constants k_j are given by

$$k_0 = f_1/e_1, \quad k_1 = d_1/g_1, \quad k_2 = 1, \quad k_3 = 1, \quad k_4 = d_1/g_1, \quad k_5 = 1.$$

REFERENCES

1. J. H. McClellan, The design of two-dimensional filters by transformations, *Proc. 7th Annual Princeton Conf. Information Sciences and Systems,* pp. 247–251, 1973.
2. R. M. Mersereau, W. F. G. Mecklenbräuker, and T. F. Quatieri, Jr., McClellan transformations for two-dimensional digital filtering: I—Design, *IEEE Trans. Circuits Syst.,* vol. CAS-23, pp. 405–414, July 1976.
3. W. F. G. Mecklenbräuker and R. M. Mersereau, McClellan transformations for two-dimensional digital filtering: II—Implementation, *IEEE Trans. Circuits Syst.,* vol. CAS-23, pp. 414–422, July 1976.
4. J. H. McClellan and D. K. S. Chan, A 2-D FIR filter structure derived from the Chebyshev recursion, *IEEE Trans. Circuits Syst.,* vol. CAS-24, pp. 372–378, July 1977.
5. A. N. Venetsanopoulos and C. L. Nikias, Realization of two-dimensional digital filters by LU decomposition of their transfer function, *Proc. IEEE Int. Conf. Acoust., Speech, Signal Process.,* pp. 20.4.1–20.4.4, March 1984.

6. A. N. Venetsanopoulos and B. G. Mertzios, A decomposition theorem and its implications to the design and realization of two-dimensional filters, *IEEE Trans. Acoust., Speech, Signal Process.*, vol. ASSP-33, pp. 1562–1575, Dec. 1985.

7. W.-S. Lu, H.-P. Wang, and A. Antoniou, Design of two-dimensional FIR digital filters by using the singular value decomposition, *IEEE Trans. Circuits Syst.*, vol. CAS-37, pp. 35–46, Jan. 1990.

8. S. Y. Kung, B. C. Levy, M. Morf, and T. Kailath, New results in 2-D systems theory, Part II: 2-D state space model—Realization and the notions of controllability, observability, and minimality, *Proc. IEEE,* vol. 65, pp. 945–961, 1977.

9. R. Eising, Realization and stabilization of 2-D systems, *IEEE Trans. Automat. Contr.,* vol. AC-23, pp. 793–799, Oct. 1978.

10. C. Eswaran, T. Venkateswarlu, and A. Antoniou, Realization of multidimensional GIC digital filters, *IEEE Trans. Circuits Syst.,* vol. CAS-37, pp. 685–694, June 1990.

11. A. Antoniou, Realization of digital filters, *IEEE Trans. Audio Electroacoust.,* vol. AU-20, pp. 95–97, March 1972.

12. A. Antoniou, *Digital Filters: Analysis and Design,* New York: McGraw-Hill, 1979 (2nd ed. in press).

13. G. Strang, *Linear Algebra and Its Applications,* 2nd ed., New York: Academic Press, 1980.

14. E. D. Sontag, On first-order equations for multidimensional filters, *IEEE Trans. Acoust., Speech, Signal Process.,* vol. ASSP-26, pp. 480–482, Oct. 1978.

15. S. H. Zak, E. B. Lee, and W.-S. Lu, Realization of 2-D filters and time-delay systems, *IEEE Trans. Circuits Syst.,* vol. CAS-33, pp. 1241–1244, Dec. 1986.

16. T. Venkateswarlu, C. Eswaran, and A. Antoniou, Realization of multidimensional digital transfer functions, *Multidimensional Systems and Signal Processing,* vol. 1, pp. 179–198, June 1990.

17. A. Fettweis, Digital filter structures related to classical filter networks, *Arch. Elektron. Uebertrag.,* vol. 25, pp. 79–89, 1971.

18. A. Sedlmeyer and A. Fettweis, Realization of digital filters with true ladder configuration, *Int. J. Circuit Theory,* vol. CT-18, pp. 314–316, March 1973.

19. A. Antoniou, Novel RC-active-network synthesis using generalized-immittance converters, *IEEE Trans. Circuit Theory,* vol. CT-17, pp. 212–217, May 1970.

20. A. Antoniou and M. G. Rezk, Digital-filter synthesis using concept of generalized-immittance convertor, *IEE J. Electron. Circuits Syst.,* vol. 1, pp. 207–216, Nov. 1977.

21. J. K. Skwirzynski, *Design Theory and Data for Electrical Filters,* London: Van Nostrand, 1965.

22. R. Saal, *The Design of Filters Using the Catalogue of Normalized Low-Pass Filters,* Backnang: Telefunken AG, 1966.

23. A. I. Zverev, *Handbook of Filter Synthesis,* New York: Wiley, 1967.

24. L. Weinberg, *Network Analysis and Synthesis,* New York: McGraw-Hill, 1962.

25. G. V. Mendonça, *Design, Realization and Implementation of 2-D Circularly Symmetric Pseudo-Rotated Digital Filters,* Ph.D. thesis, Concordia University, 1984.

26. A. Antoniou, Realization of gyrators using operational amplifiers, and their use in RC-active-network synthesis, *IEE Proc.,* vol. 116, pp. 1838–1850, Nov. 1969.

27. A. Fettweis, Some principles of designing digital filters imitating classical filter structures, *IEEE Trans. Circuits Syst.,* vol. CAS-18, pp. 314–316, March 1971.

28. A. Fettweis and K. Meerkötter, On adaptors for wave digital filters, *IEEE Trans. Acoust., Speech, Signal Process.,* vol. ASSP-23, pp. 516–525, Dec. 1975.

29. T. Venkateswarlu, C. Eswaran, and A. Antoniou, Design of circularly symmetric two- and three-dimensional digital filters using the concept of the GIC, *IEE Proc.,* Vol. 138, pt.G, pp. 523–529, Aug. 1991.

PROBLEMS

10.1 Verify Eq. (10.5).

10.2 Realize the filter designed in Example 6.5 of Sec. 6.4.3 by using the method of McClellan and Chan.

10.3 Obtain the SVD–LUD realization of the 2-D filter designed in Problem 9.2.

10.4 Obtain the SVD–LUD realization of the 2-D filter designed in Problem 9.5b.

10.5 Extend the canonic realization depicted in Fig. 10.7 to the general 2-D transfer function given by

$$H(z_1, z_2) = \frac{\sum_{i=0}^{N_1} \sum_{j=0}^{N_2} a_{ij} z_1^{-i} z_2^{-j}}{\left(\sum_{i=0}^{N_1} \alpha_i z_1^{-i}\right)\left(\sum_{j=0}^{N_2} \beta_j z_2^{-j}\right)}$$

10.6 Extend the canonic realization shown in Fig. 10.8 to the general 2-D transfer function given by

$$H(z_1, z_2) = \frac{\left(\sum_{i=0}^{N_1} \alpha_i z_1^{-i}\right)\left(\sum_{j=0}^{N_2} \beta_j z_2^{-j}\right)}{\sum_{i=0}^{N_1} \sum_{j=0}^{N_2} b_{ij} z_1^{-i} z_2^{-j}}$$

10.7 Obtain a direct realization for each of the following transfer functions:

(a) $H(z_1, z_2) = \dfrac{1}{z_1 z_2 + 0.2 z_2 + 0.5}$.

(b) $H(z_1, z_2) = \dfrac{z_1 z_2 + z_2 - 0.5}{z_1 z_2 - 0.5 z_1 + 0.8 z_2 - 0.4}.$

(c) $H(z_1, z_2) = \dfrac{z_1 z_2 + 3 z_1 - z_2 - 3}{z_1 z_2 + 0.8 z_1 + 0.9 z_2 + 0.8}.$

10.8 Obtain the LUD realization of the filter described in Sec. 4.5.2, p. 89.

10.9 Obtain the LUD realization of the state-space filter given in Problem 2.5.

10.10 Obtain a 2-D GIC digital-filter structure for the circularly symmetric lowpass filter designed in Problem 7.6.

10.11 Obtain a 2-D GIC digital-filter structure for the filter designed in Problem 7.7.

11
Finite Wordlength Effects

11.1 INTRODUCTION

When a 2-D digital filter is implemented in terms of either software on a general-purpose computer or dedicated hardware, numbers representing transfer-function coefficients and signals are stored and manipulated in registers of finite length. Transfer-function coefficients are calculated to a high degree of precision during the solution of the approximation problem, and, consequently, they must be quantized before they can be accommodated in registers of finite length. Similarly, products generated when signals are multiplied by coefficients are always too long to fit in the registers and must again be quantized. The quantization of coefficients and products gives rise to quantization errors that propagate through the filter and appear at the output as noise. Product quantization can lead to other problems as well, such as the generation of spurious parasitic oscillations known as limit cycles. Although the effects of coefficient and product quantization are insignificant if a general-purpose computer is to be used for the implementation, owing to the high precision of the hardware, particular attention must be paid to these effects when fixed-point arithmetic or specialized hardware with reduced wordlength is to be employed. In such applications, the design process cannot be considered complete until the effects of quantization are studied in detail.

This chapter deals with the effects of finite wordlength in 2-D digital filters. In the next section, the quantization errors introduced by the use of finite wordlengths for the representation of coefficients and signals are examined and methods for their computation and characterization [1–3] are considered. In Sec. 11.3, a method for the minimization of output roundoff noise in 1-D state-space structures due to Mullis and Roberts [4] based on a state-space noise model proposed by Hwang [5] is extended to the case of 2-D digital filters [6]. In Sec. 11.4, two types of parasitic oscillations, namely, quantization and overflow limit cycles, are examined and conditions for their elimination are delineated.

11.2 QUANTIZATION ERRORS AND THEIR COMPUTATION

As in 1-D digital filters, the use of finite wordlength for the representation of numbers introduces three types of quantization errors, as follows:

1. Coefficient-quantization errors
2. Product-quantization errors
3. Input-quantization errors

Coefficient-quantization errors are due to the representation of the filter coefficients by a finite number of digits. Product-quantization errors are due to the quantization of signals after multiplication. Input-quantization errors occur when the signal to be processed is continuous and must be sampled and encoded through an A/D converter prior to processing.

Quantization of numbers can be carried out by rounding or truncation, and the magnitude of the error introduced tends to depend on the type of arithmetic (i.e., fixed point or floating point) and on the number representation (i.e. sign-magnitude, one's complement, and two's complement).

Throughout the following analysis, it is assumed that quantization errors are uniformly distributed over the quantization interval and that they are statistically independent from error to error, from sample to sample, and with respect to the input signal (see [11], Chap. 11, for further details). Under these circumstances, the mean $\bar{e}$ and variance σ_e^2 of a quantization error e for fixed-point arithmetic are given in Table 11.1 [1]. The 2-D transfer function is assumed to be of the form

$$H(z_1, z_2) = \frac{N(z_1, z_2)}{D(z_1, z_2)} \tag{11.1}$$

where

$$N(z_1, z_2) = \sum_{i=0}^{N_1} \sum_{j=0}^{N_2} a_{ij} z_1^{-i} z_2^{-j}$$

Table 11.1 Mean and Variance of Quantization Error e

Number representation	Quantity	Rounding	Truncation
Sign-magnitude	$\bar{e}$	0	0
	σ_e^2	$E_0^2/12$	$E_0^2/3$
One's complement	$\bar{e}$	0	0
	σ_e^2	$E_0^2/12$	$E_0^2/3$
Two's complement	$\bar{e}$	0	$E_0/2$
	σ_e^2	$E_0^2/12$	$E_0^2/12$

Source: Ref. 1.

and

$$D(z_1, z_2) = 1 + \sum_{i=0}^{N_1} \sum_{j=0}^{N_2} b_{ij} z_1^{-i} z_2^{-j}, \qquad b_{00} = 0$$

11.2.1 Coefficient Quantization

Let $\bar{a}_{ij}$ and $\bar{b}_{ij}$ be the quantized values of a_{ij} and b_{ij} and denote the quantization errors introduced by

$$\mu_{ij} = a_{ij} - \bar{a}_{ij}$$

and

$$\eta_{ij} = b_{ij} - \bar{b}_{ij}$$

respectively. Also let $\bar{y}(n_1, n_2)$ and $y(n_1, n_2)$ be the output of a 2-D digital filter with and without coefficient quantization, respectively. From Eq. (1.8a), we have

$$\bar{y}(n_1, n_2) = \sum_{i=0}^{N_1} \sum_{j=0}^{N_2} \bar{a}_{ij} x(n_1 - i, n_2 - j) - \sum_{i=0}^{N_1} \sum_{j=0}^{N_2} \bar{b}_{ij} \bar{y}(n_1 - i, n_2 - j)$$

and hence the output error produced is given by

$$f(n_1, n_2) = y(n_1, n_2) - \bar{y}(n_1, n_2)$$

$$= u(n_1, n_2) - \sum_{i=0}^{N_1} \sum_{j=0}^{N_2} b_{ij} f(n_1 - i, n_2 - j) \qquad (11.2)$$

where second-order terms have been neglected in the second equation and

$$u(n_1, n_2) = \sum_{i=0}^{N_1} \sum_{j=0}^{N_2} \mu_{ij} x(n_1 - i, n_2 - j)$$

$$- \sum_{i=0}^{N_1} \sum_{j=0}^{N_2} \eta_{ij} y(n_1 - i, n_2 - j) \quad (11.3)$$

Combining the 2-D z transforms of Eqs. (11.2) and (11.3) leads to the z-domain description of the output error given by

$$F(z_1, z_2) = \frac{V(z_1, z_2)}{D(z_1, z_2)} X(z_1, z_2) - \frac{G(z_1, z_2)}{D(z_1, z_2)} Y(z_1, z_2) \quad (11.4)$$

where

$$V(z_1, z_2) = \sum_{i=0}^{N_1} \sum_{j=0}^{N_2} \mu_{ij} z_1^{-i} z_2^{-j}$$

$$G(z_1, z_2) = \sum_{i=0}^{N_1} \sum_{j=0}^{N_2} \eta_{ij} z_1^{-i} z_2^{-j}$$

Equation (11.4) can be used to derive formulas for the mean and variance of the output error. Such formulas for the case of fixed-point arithmetic are given in Table 11.2 [1] where σ_x^2 and σ_y^2 are the variance of the input and output signals, respectively.

Table 11.2 Mean and Variance of the Output Error Due to Coefficient Quantization

Number representation	Quantity	Rounding	Truncation
Sign-magnitude	$\bar{f}$	0	0
	σ_f^2	$\dfrac{E_0^2}{12\Delta}(k\sigma_x^2 + l\sigma_y^2)$	$\dfrac{E_0^2}{3\Delta}(k\sigma_x^2 + l\sigma_y^2)$
One's complement	$\bar{f}$	0	0
	σ_f^2	$\dfrac{E_0^2}{12\Delta}(k\sigma_x^2 + l\sigma_y^2)$	$\dfrac{E_0^2}{3\Delta}(k\sigma_x^2 + l\sigma_y^2)$
Two's complement	$\bar{f}$	0	0
	σ_f^2	$\dfrac{E_0^2}{12\Delta}(k\sigma_x^2 + l\sigma_y^2)$	$\dfrac{E_0^2}{12\Delta}(k\sigma_x^2 + l\sigma_y^2)$

$$\sigma_y^2 = \sigma_x^2 \left(\sum_{i=0}^{N_1} \sum_{j=0}^{N_2} a_{ij}^2 / \Delta \right), \quad \Delta = \sum_{i=0}^{N_1} \sum_{j=0}^{N_2} b_{ij}^2$$

Source: Ref. 1.

11.2.2 Product Quantization

Let $\bar{x}(n_1, n_2)$ for $n_1 \geq 0$, $n_2 \geq 0$ be a quantized input signal. Assuming infinite-precision arithmetic, the output of the filter for $n_1 \geq 0$, $n_2 \geq 0$ can be obtained from Eq. (1.8a) as

$$y(n_1, n_2) = \sum_{i=0}^{N_1} \sum_{j=0}^{N_2} a_{ij}\bar{x}(n_1 - i, n_2 - j) - \sum_{i=0}^{N_1} \sum_{j=0}^{N_2} b_{ij}y(n_1 - i, n_2 - j)$$

Denoting a quantized product $u \cdot v$ by $(u \cdot v)_q$, the output of a finite-wordlength filter $\bar{y}(n_1, n_2)$ for $n_1 \geq 0$, $n_2 \geq 0$ and the corresponding quantization error $f(n_1, n_2)$ can be expressed as

$$\bar{y}(n_1, n_2) = \sum_{i=0}^{N_1} \sum_{j=0}^{N_2} [a_{ij}\bar{x}(n_1 - i, n_2 - j)]_q$$

$$- \sum_{i=0}^{N_1} \sum_{j=0}^{N_2} [b_{ij}y(n_1 - i, n_2 - j)]_q$$

$$f(n_1, n_2) = y(n_1, n_2) - \bar{y}(n_1, n_2)$$

$$= \lambda(n_1, n_2) - \varepsilon(n_1, n_2) - \sum_{i=0}^{N_1} \sum_{j=0}^{N_2} b_{ij}f(n_1 - i, n_2 - j) \quad (11.5)$$

where

$$\lambda(n_1, n_2) = \sum_{i=0}^{N_1} \sum_{j=0}^{N_2} \{a_{ij}\bar{x}(n_1 - i, n_2 - j) - [a_{ij}\bar{x}(n_1 - i, n_2 - j)]_q\}$$

$$\varepsilon(n_1, n_2) = \sum_{i=0}^{N_1} \sum_{j=0}^{N_2} \{b_{ij}y(n_1 - i, n_2 - j)$$

$$- [b_{ij}y(n_1 - i, n_2 - j)]_q\} + y(n_1, n_2)$$

The 2-D z transform of Eq. (11.5) gives

$$F(z_1, z_2) = \frac{\Lambda(z_1, z_2) - E(z_1, z_2)}{D(z_1, z_2)}$$

where $\Lambda(z_1, z_2)$ and $E(z_1, z_2)$ are the z transforms of $\lambda(n_1, n_2)$ and $\varepsilon(n_1, n_2)$, respectively. Consequently, if $h(n_1, n_2)$ for $n_1 \geq 0$, $n_2 \geq 0$ is the impulse response of the filter represented by $1/D(z_1, z_2)$, then

$$f(n_1, n_2) = \sum_{i=0}^{N_1} \sum_{j=0}^{N_2} h(n_1 - i, n_2 - j)[\lambda(i, j) - \varepsilon(i, j)] \quad (11.6)$$

Equation (11.6) can be used in conjunction with Table 11.1 to obtain the mean and variance of the output error for the case of a direct realiza-

Table 11.3 Mean and Variance of the Output Error Due to Product Quantization

Number representation	Quantity	Rounding	Truncation
Sign-magnitude	$\bar{f}$	0	0
	σ_f^2	$\dfrac{E_0^2}{12(k+l)}\sum_{i=0}^{N_1}\sum_{j=0}^{N_2}h^2(i,j)$	$\dfrac{E_0^2}{3(k+l)}\sum_{i=0}^{N_1}\sum_{j=0}^{N_2}h^2(i,j)$
One's complement	$\bar{f}$	0	0
	σ_f^2	$\dfrac{E_0^2}{12(k+l)}\sum_{i=0}^{N_1}\sum_{j=0}^{N_2}h^2(i,j)$	$\dfrac{E_0^2}{3(k+l)}\sum_{i=0}^{N_1}\sum_{j=0}^{N_2}h^2(i,j)$
Two's complement	$\bar{f}$	0	$\dfrac{E_0}{2(k+l)}\sum_{i=0}^{N_1}\sum_{j=0}^{N_2}h(i,j)$
	σ_f^2	$\dfrac{E_0^2}{12(k+l)}\sum_{i=0}^{N_1}\sum_{j=0}^{N_2}h^2(i,j)$	$\dfrac{E_0^2}{12(k+l)}\sum_{i=0}^{N_1}\sum_{j=0}^{N_2}h^2(i,j)$

Source: Ref. 1.

tion of the transfer function (see Sec. 10.4). Formulas for these statistics are given in Table 11.3, where k and l denote the number of coefficients that are different from 0 and 1 in polynomials $N(z_1, z_2)$ and $D(z_1, z_2)$, respectively.

11.2.3 Input Signal Quantization

Let $y(n_1, n_2)$ and $\bar{y}(n_1, n_2)$ be the output of the filter in response to a nonquantized and a quantized input, respectively. It directly follows from Eq. (1.8a) that the output error $f(n_1, n_2)$ of a 2-D digital filter obeys the recursive equation

$$f(n_1, n_2) = \sum_{i=0}^{N_1}\sum_{j=0}^{N_2} a_{ij}e(n_1 - i, n_2 - j) - \sum_{i=0}^{N_1}\sum_{j=0}^{N_2} b_{ij}f(n_1 - i, n_2 - j)$$

where

$$e(i, j) = x(i, j) - \bar{x}(i, j)$$

The z-domain description of the preceding recursive relation is given by

$$F(z_1, z_2) = H(z_1, z_2)E(z_1, z_2)$$

where $E(z_1, z_2)$ denotes the 2-D z transform of $e(i, j)$ and $H(z_1, z_2)$ is given by Eq. (11.1). The mean and variance of the output error are given in Table 11.4 [1].

Table 11.4 Mean and Variance of the Output Error Due to Input Quantization

Number representation	Quantity	Rounding	Truncation
Sign-magnitude	$\bar{f}$	0	0
	σ_f^2	$\dfrac{E_0^2}{12} \displaystyle\sum_{i=0}^{N_1} \sum_{j=0}^{N_2} h^2(i, j)$	$\dfrac{E_0^2}{3} \displaystyle\sum_{i=0}^{N_1} \sum_{j=0}^{N_2} h^2(i, j)$
One's complement	$\bar{f}$	0	0
	σ_f^2	$\dfrac{E_0^2}{12} \displaystyle\sum_{i=0}^{N_1} \sum_{j=0}^{N_2} h^2(i, j)$	$\dfrac{E_0^2}{3} \displaystyle\sum_{i=0}^{N_1} \sum_{j=0}^{N_2} h^2(i, j)$
Two's complement	$\bar{f}$	0	$\dfrac{E_0}{2} \displaystyle\sum_{i=0}^{N_1} \sum_{j=0}^{N_2} h(i, j)$
	σ_f^2	$\dfrac{E_0^2}{12} \displaystyle\sum_{i=0}^{N_1} \sum_{j=0}^{N_2} h^2(i, j)$	$\dfrac{E_0^2}{12} \displaystyle\sum_{i=0}^{N_1} \sum_{j=0}^{N_2} h^2(i, j)$

Source: Ref. 1.

11.2.4 Evaluation of $\Sigma_{i=0}^{\infty} \Sigma_{j=0}^{\infty} h^2(i, j)$

As can be seen in Tables 11.3 and 11.4, the computation of the output errors due to product and input quantization depends on the quantity $\Sigma_{i=0}^{N_1} \Sigma_{j=0}^{N_2} h^2(i, j)$ which is bounded by

$$J_0 = \sum_{i=0}^{\infty} \sum_{j=0}^{\infty} h^2(i, j)$$

$$= \frac{1}{(2\pi j)^2} \oint_{|z_1|=1} \oint_{|z_2|=1} \frac{1}{D(z_1, z_2)D(z_1^{-1}, z_2^{-1})} \frac{dz_2}{z_2} \frac{dz_1}{z_1} \quad (11.7)$$

Here we consider the computation for the more general case

$$J = \frac{1}{(2\pi j)^2} \oint_{|z_1|=1} \oint_{|z_2|=1} H(z_1, z_2)H(z_1^{-1}, z_2^{-1}) \frac{dz_2}{z_2} \frac{dz_1}{z_1} \quad (11.8)$$

where $H(z_1, z_2)$ is given by

$$H(z_1, z_2) = \frac{N(z_1, z_2)}{D(z_1, z_2)}$$

In what follows, it is assumed that

$$D(z_1, z_2) \neq 0 \quad \text{for } |z_1| \leq 1, |z_2| \leq 1 \quad (11.9)$$

which guarantees the BIBO stability of the filter considered.

Writing the integrand of Eq. (11.8) as

$$\frac{H(z_1, z_2)H(z_1^{-1}, z_2^{-1})}{z_1^{-1}z_2^{-1}} = \frac{N(z_1, z_2)N_1(z_1, z_2)}{D(z_1, z_2)D_1(z_1, z_2)} z_1^{k_1} z_2^{k_2}$$

where $N_1(z_1, z_2)$ and $D_1(z_1, z_2)$ are polynomials in z_1 and z_2, and k_1 and k_2 are integers, the integral in Eq. (11.8) becomes

$$J = \frac{1}{2\pi j} \oint_{|z_1|=1} J_1(z_1)z_1^{k_1} \, dz_1$$

where

$$J_1(z_1) = \frac{1}{2\pi j} \oint_{|z_2|=1} \frac{N(z_1, z_2)N_1(z_1, z_2)}{D(z_1, z_2)D_1(z_1, z_2)} z_2^{k_2} \, dz_2$$

The integrand in $J_1(z_1)$ can now be decomposed as

$$\frac{N(z_1, z_2)N_1(z_1, z_2)}{D(z_1, z_2)D_1(z_1, z_2)} z_2^{k_2} = \frac{Q_1(z_2)}{\tilde{D}(z_2)} + \frac{Q_2(z_2)}{\tilde{D}_1(z_2)} z_2^{k_2}$$

where $Q_1(z_2)$, $Q_2(z_2)$, $\tilde{D}(z_2)$ and $\tilde{D}_1(z_2)$ are polynomials in z_2 with coefficients that are polynomials in z_1. Note that the stability assumption given by Eq. (11.9) implies that

$$\tilde{D}(z_2) \neq 0 \qquad \text{for} \quad |z_1| = 1, |z_2| < 1$$

and

$$\tilde{D}_1(z_2) \neq 0 \qquad \text{for} \quad |z_1| = 1, |z_2| > 1$$

Consequently, for a fixed z_1 on $U_1 = \{z_1 : |z_1| = 1\}$,

$$\oint_{|z_2|=1} \frac{Q_1(z_2)}{\tilde{D}(z_2)} \, dz_2 = 0$$

which leads to

$$J_1(z_1) = \frac{1}{2\pi j} \oint_{|z_2|=1} \frac{Q_2(z_2)}{\tilde{D}(z_2)} z_2^{k_2} \, dz_2$$

By the residue theorem, $J_1(z_1)$ is the coefficient of the z_2^{-1} term in the Laurent expansion of $Q_2(z_2)/(\tilde{D}_1(z_2)z_2^{-k_2})$. Once $J_1(z_1)$ is found, the complex integral in Eq. (11.8) can be computed by calculating the residues of $J_1(z_1)z_1^{k_1}$ at all its singularities lying inside the unit disk of the z_1 plane.

Example 11.1 Compute the complex integral J in Eq. (11.8) if

$$H(z_1, z_2) = \frac{(z_1 - 1)(z_2 - 1)}{2 - z_1 - z_2}$$

Solution. $H(z_1, z_2)$ represents a BIBO unstable filter although its impulse response is square summable (see [12]). Therefore, the preceding approach is applicable. We can write

$$\frac{H(z_1, z_2)H(z_1^{-1}, z_2^{-1})}{z_1 z_2} = \frac{(z_1 - 1)(z_2 - 1)}{2 - z_1 - z_2} \frac{(1 - z_1)(1 - z_2)}{2z_1 z_2 - z_1 - z_2} z_1^{-1} z_2^{-1}$$

and so

$$\frac{(z_1-1)(z_2-1)}{2-z_1-z_2} \frac{(1-z_1)(1-z_2)}{2z_1z_2-z_1-z_2} z_2^{-1} = \frac{(1-z_1)z_2-(1-z_1)}{z_2-(2-z_1)} \frac{(z_1-1)z_2+(1-z_1)}{(2z_1-1)z_2^2-z_1z_2}$$

$$= \frac{a_0(z_1)z_2+a_1(z_1)}{z_2-(2-z_1)} + \frac{a_2(z_1)z_2+a_3(z_1)}{(2z_1-1)z_2^2-z_1z_2}$$

where

$$a_2(z_1) = \frac{3(z_1 - 1)^2}{2(z_1 - 1)}$$

Hence

$$J_1(z_1) = \frac{a_2(z_1)}{2z_1 - 1} = \frac{3(z_1 - 1)^2}{2(2z_1 - 1)(z_1 - 2)}$$

Therefore

$$J = \frac{1}{2\pi j} \oint_{|z_1|=1} J_1(z_1)z_1^{-1}\, dz_1 = \frac{1}{2\pi j} \oint_{|z_1|=1} \frac{3(z_1 - 1)^2}{2(2z_1 - 1)(z_1 - 2)z_1}\, dz_1 = \frac{1}{2}$$

11.3 STATE-SPACE STRUCTURES WITH MINIMIZED ROUNDOFF NOISE

Minimum roundoff noise, subject to dynamic-range constraints based on the L_2 norm, can be achieved in 1-D state-space structures by using a method proposed by Mullis and Roberts [4], which is based on a state-space noise model developed by Hwang [5]. This method can be extended to the case of 2-D state-space structures, as will be demonstrated below [6].

11.3.1 Derivation of Noise Model

A 2-D digital filter can be represented by the local state-space model given by

$$\begin{bmatrix} \mathbf{q}^H (i+1, j) \\ \mathbf{q}^V (i, j+1) \end{bmatrix} = \begin{bmatrix} \mathbf{A}_1 & \mathbf{A}_2 \\ \mathbf{A}_3 & \mathbf{A}_4 \end{bmatrix} \mathbf{q}(i, j) + \begin{bmatrix} \mathbf{b}_1 \\ \mathbf{b}_2 \end{bmatrix} x(i, j)$$

$$\equiv \mathbf{A}\mathbf{q}(i, j) + \mathbf{b}x(i, j) \tag{11.10a}$$

$$y(i, j) = [\mathbf{c}_1 \quad \mathbf{c}_2]\mathbf{q}(i, j) + dx(i, j) \equiv \mathbf{c}\mathbf{q}(i, j) + dx(i, j) \tag{11.10b}$$

where $\mathbf{A}_1 \in R^{N_1 \times N_1}$, $\mathbf{A}_4 \in R^{N_2 \times N_2}$ (see Chap. 2). It is assumed that

$$\det \begin{bmatrix} \mathbf{I}_{N_1} - z_1^{-1}\mathbf{A}_1 & -z_1^{-1}\mathbf{A}_2 \\ -z_2^{-1}\mathbf{A}_3 & \mathbf{I}_{N_2} - z_2^{-1}\mathbf{A}_4 \end{bmatrix} \neq 0 \qquad \text{for } (z_1^{-1}, z_2^{-1}) \in \overline{U}^2$$

which implies the BIBO stability of the filter (see Chap. 5).

If we let

$$\mathbf{A}_{00} = \mathbf{I}_{N_1+N_2}, \qquad \mathbf{A}_{10} = \begin{bmatrix} \mathbf{A}_1 & \mathbf{A}_2 \\ 0 & 0 \end{bmatrix}, \qquad \mathbf{A}_{01} = \begin{bmatrix} 0 & 0 \\ \mathbf{A}_3 & \mathbf{A}_4 \end{bmatrix}$$

$$\mathbf{A}_{ij} = \mathbf{A}_{10}\mathbf{A}_{i-1,j} + \mathbf{A}_{01}\mathbf{A}_{i,j-1} \quad \text{for } (i, j) > (0, 0) \tag{11.11}$$

and

$$\mathbf{A}_{-i,j} = \mathbf{A}_{i,-j} = 0 \qquad \text{for } i \geq 1, j \geq 1$$

the transfer function of the filter can be expressed as

$$H(z_1, z_2) = \mathbf{c}(\mathbf{I} - z_1^{-1}\mathbf{A}_{10} - z_2^{-1}\mathbf{A}_{01})^{-1}\begin{bmatrix} z_1^{-1}\mathbf{b}_1 \\ z_2^{-1}\mathbf{b}_2 \end{bmatrix} + d$$

$$= \sum_{i=0}^{\infty} \sum_{j=0}^{\infty} \mathbf{c}\left(\mathbf{A}_{i-1,j}\begin{bmatrix} \mathbf{b}_1 \\ 0 \end{bmatrix} + \mathbf{A}_{i,j-1}\begin{bmatrix} 0 \\ \mathbf{b}_2 \end{bmatrix}\right)z_1^{-i}z_2^{-j} + d$$

$$= \sum_{i=0}^{\infty} \sum_{j=0}^{\infty} \mathbf{c}\mathbf{f}(i, j)z_1^{-i}z_2^{-j} + d \tag{11.12}$$

where

$$\mathbf{f}(i, j) = \mathbf{A}_{i-1,j}\begin{bmatrix} \mathbf{b}_1 \\ 0 \end{bmatrix} + \mathbf{A}_{i,j-1}\begin{bmatrix} 0 \\ \mathbf{b}_2 \end{bmatrix} \tag{11.13}$$

By applying the 2-D similarity transformation

$$\mathbf{T} = \begin{bmatrix} \mathbf{T}_1 & 0 \\ 0 & \mathbf{T}_2 \end{bmatrix} \equiv \mathbf{T}_1 \oplus \mathbf{T}_2 \tag{11.14}$$

where $\mathbf{T}_1 \in R^{N_1 \times N_1}$, $\mathbf{T}_2 \in R^{N_2 \times N_2}$, and $\oplus$ denotes the direct sum, to vector $\mathbf{q}(i, j)$ in realization $(\mathbf{A}, \mathbf{b}, \mathbf{c}, d)$ of Eq. (11.10), an equivalent realization $(\tilde{\mathbf{A}}, \tilde{\mathbf{b}}, \tilde{\mathbf{c}}, \tilde{d})$ where

$$\tilde{\mathbf{A}} = \mathbf{T}^{-1}\mathbf{A}\mathbf{T}, \qquad \tilde{\mathbf{b}} = \mathbf{T}^{-1}\mathbf{b}, \qquad \tilde{\mathbf{c}} = \mathbf{c}\mathbf{T}, \qquad \tilde{d} = d \qquad (11.15)$$

can be obtained. Once matrices $\tilde{\mathbf{A}}_{ij}$ are defined by analogy with Eq. (11.11), it is easy to show that

$$\tilde{\mathbf{A}}_{ij} = \mathbf{T}^{-1}\mathbf{A}_{ij}\mathbf{T} \qquad \text{for all } i, j$$

Now if finite wordlength is used for the representation of internal signals, the state-space model of the filter becomes

$$\begin{bmatrix} \bar{\mathbf{q}}^H(i + 1, j) \\ \bar{\mathbf{q}}^V(i, j + 1) \end{bmatrix} = \mathbf{A}\begin{bmatrix} \bar{\mathbf{q}}^H(i, j) \\ \bar{\mathbf{q}}^V(i, j) \end{bmatrix} + \mathbf{b}x(i, j) + \begin{bmatrix} \tau_1(i + 1, j) \\ \tau_2(i, j + 1) \end{bmatrix} \qquad (11.16a)$$

$$\bar{y}(i, j) = \mathbf{c}\begin{bmatrix} \bar{\mathbf{q}}^H(i, j) \\ \bar{\mathbf{q}}^V(i, j) \end{bmatrix} + dx(i, j) + \gamma(i, j) \qquad (11.16b)$$

where $\tau_1(i + 1, j)$ and $\tau_2(i, j + 1)$ are the random errors generated in the computation of $\mathbf{q}^H(i + 1, j)$ and $\mathbf{q}^V(i, j + 1)$, respectively, and $\gamma(i, j)$ is the random error generated in the computation of $y(i, j)$. If we define the state-error vector as

$$\begin{bmatrix} \Delta\bar{\mathbf{q}}^H(i, j) \\ \Delta\bar{\mathbf{q}}^V(i, j) \end{bmatrix} = \begin{bmatrix} \bar{\mathbf{q}}^H(i, j) \\ \bar{\mathbf{q}}^V(i, j) \end{bmatrix} - \begin{bmatrix} \mathbf{q}^H(i, j) \\ \mathbf{q}^V(i, j) \end{bmatrix}$$

and output error (or noise) as

$$\Delta y(i, j) = \bar{y}(i, j) - y(i, j)$$

then Eqs. (11.10) and (11.16) yield

$$\begin{bmatrix} \Delta\mathbf{q}^H(i, j) \\ \Delta\mathbf{q}^V(i, j) \end{bmatrix} = \mathbf{A}_{10}\begin{bmatrix} \Delta\mathbf{q}^H(i - 1, j) \\ \Delta\mathbf{q}^V(i - 1, j) \end{bmatrix}$$

$$+ \mathbf{A}_{01}\begin{bmatrix} \Delta\mathbf{q}^H(i, j - 1) \\ \Delta\mathbf{q}^V(i, j - 1) \end{bmatrix} + \begin{bmatrix} \tau_1(i, j) \\ \tau_2(i, j) \end{bmatrix} \qquad (11.17a)$$

$$\Delta y(i, j) = \mathbf{c}\begin{bmatrix} \Delta\mathbf{q}^H(i, j) \\ \Delta\mathbf{q}^V(i, j) \end{bmatrix} + \gamma(i, j) \qquad (11.17b)$$

11.3.2 Output-Noise Power

The noise model of Eq. (11.16) can be used to derive an explicit expression for the output-noise power, as will now be demonstrated.

For a fixed point $(i, j) \geq (0, 0)$, Eq. (11.17) gives

$$\Delta y(i, j) = \mathbf{c} \sum_{(0,0) < (l,k) \leq (i,j)} \left(\mathbf{A}_{l-1,k} \left[\begin{matrix} \boldsymbol{\tau}_1(i - l, j - k) \\ 0 \end{matrix} \right] \right.$$

$$\left. + \mathbf{A}_{l,k-1} \left[\begin{matrix} 0 \\ \boldsymbol{\tau}_2(i - l, j - k) \end{matrix} \right] \right) + \gamma(i, j) \qquad (11.18)$$

If the quantization of products is carried out by rounding and the quantization errors are assumed to be independent from error to error and from sample to sample, then the variance of each error is $E_0^2/12$ where E_0 is the quantization step (see Table 11.1). From Eq. (11.18), the expected square error is

$$E[\Delta y^2(i, j)] = \frac{E_0^2}{12} \mathbf{c} \left\{ \sum_{(0,0) < (l,k) \leq (i,j)} \mathbf{A}_{l-1,k} \left[\begin{matrix} \mathbf{Q}_1 & 0 \\ 0 & 0 \end{matrix} \right] \mathbf{A}_{l-1,k}^T \right.$$

$$\left. + \mathbf{A}_{l,k-1} \left[\begin{matrix} 0 & 0 \\ 0 & \mathbf{Q}_2 \end{matrix} \right] \mathbf{A}_{l,k-1}^T \right\} \mathbf{c}^T + \frac{E_0^2}{12} (\mu + \nu)$$

where

$$\frac{E_0^2}{12} \mathbf{Q}_l = E(\boldsymbol{\tau}_l \boldsymbol{\tau}_l^T), \qquad l = 1, 2$$

are positive-definite diagonal matrices of dimensions N_1 and N_2, respectively, and μ and ν are the numbers of constants in $\mathbf{c}$ and d, respectively, which are neither zero nor one.

Therefore, the variance of the output noise $\Delta y(i, j)$ can be calculated as

$$E(\Delta y^2) = \lim_{\substack{i \to \infty \\ j \to \infty}} E \left[\frac{1}{(i + 1)(j + 1) - 1} \sum_{(0,0) \leq (l,k) < (i,j)} \Delta y^2(l, k) \right]$$

$$= \frac{E_0^2}{12} \sum_{i=0}^{\infty} \sum_{j=0}^{\infty} \mathbf{c} \mathbf{A}_{ij} \mathbf{Q} \mathbf{A}_{ij}^T \mathbf{c}^T + \frac{E_0^2}{12} (\mu + \nu)$$

$$= \frac{E_0^2}{12} \mathrm{tr}(\mathbf{Q}\mathbf{W}) + \frac{E_0^2}{12} (\mu + \nu) \qquad (11.19)$$

where

$$\mathbf{Q} = \mathbf{Q}_1 \oplus \mathbf{Q}_2$$

$$\mathbf{W} = \sum_{i=0}^{\infty} \sum_{j=0}^{\infty} \mathbf{A}_{ij}^T \mathbf{c}^T \mathbf{c} \mathbf{A}_{ij} \qquad (11.20)$$

and $\mathrm{tr}(\cdot)$ denotes the trace of a matrix.

Note that the variance of the output noise given by Eq. (11.19) is dependent on the coordinate system. Specifically, if the transformation in Eq. (11.14) is applied, the output-noise power of realization $(\tilde{A}, \tilde{b}, \tilde{c}, \tilde{d})$ is given by

$$E(\Delta \tilde{y}^2) = \frac{E_0^2}{12} \text{tr}(\tilde{Q}\tilde{W}) + \frac{E_0^2}{12}(\mu + \nu) \tag{11.21}$$

where

$$\tilde{W} = T^T W T = \sum_{i=0}^{\infty} \sum_{j=0}^{\infty} \tilde{A}_{ij}^T \tilde{c}^T \tilde{c} \tilde{A}_{ij} \tag{11.22}$$

and $\tilde{Q} = \tilde{Q}_1 \oplus \tilde{Q}_2$.

11.3.3 Dynamic Range Constraints

To prevent overflow in a digital filter, signal scaling based on the L_2 norm can be applied, which leads to a set of dynamic range constraints on the local state variables. The local state at (i, j) due to an input $x(l, k)$ for $(0, 0) \le (l, k) < (i, j)$ can be obtained from Eq. (11.12) as

$$\begin{bmatrix} q^H(i, j) \\ q^V(i, j) \end{bmatrix} = \sum_{(0,0)<(l,k)\le(i,j)} f(l, k)x(i - l, j - k) \tag{11.23}$$

Consequently, if e_p is the pth column of the identity matrix of dimension $(N_1 + N_2)$, then the pth component of the local state in Eq. (11.23) can be estimated as

$$\left| e_p^T \begin{bmatrix} q^H(i, j) \\ q^V(i, j) \end{bmatrix} \right|^2$$

$$= \left[\sum_{(0,0)<(l,k)\le(i,j)} e_p^T f(l, k)x(i - l, j - k) \right]^2$$

$$\le e_p^T \left(\sum_{l=0}^{i} \sum_{k=0}^{j} f(l, k)f^T(l, k) \right) e_p^T \left(\sum_{l=0}^{i} \sum_{k=0}^{j} x^2(i - l, j - k) \right)$$

$$\le e_p^T K e_p \|x\|^2 \tag{11.24}$$

where

$$K = \sum_{l=0}^{\infty} \sum_{k=0}^{\infty} f(l, k)f^T(l, k) \tag{11.25}$$

Notice that $e_p^T K e_p$ is the pth diagonal element of K and, therefore, if all the diagonal elements of K are equal to one, Eq. (11.24) implies that the amplitude of each state component is no more than $\|\mathbf{x}\|$. The dynamic range constraints on the state variables are, therefore, given by

$$
\mathbf{K} = \begin{bmatrix} 1 & & & & & & \\ & \ddots & & & & * & \\ & & 1 & & & & \\ & & & 1 & & & \\ & & * & & & \ddots & \\ & & & & & & 1 \end{bmatrix}
$$

Further, once a similarity transformation $\mathbf{T}$ is used to reduce the output-noise power, the preceding dynamic range constraints should also be satisfied by the transformed realization. Since

$$
\tilde{\mathbf{K}} = \sum_{i=0}^{\infty} \sum_{j=0}^{\infty} \tilde{\mathbf{f}}(i,j)\tilde{\mathbf{f}}^T(i,j) = \mathbf{T}^{-1}\mathbf{K}\mathbf{T}^{-T} \tag{11.26}
$$

the dynamic range constraints on the new variables $\tilde{\mathbf{q}}^H(i,j)$ and $\tilde{\mathbf{q}}^V(i,j)$ are given by

$$
\tilde{\mathbf{K}} = \mathbf{T}^{-1}\mathbf{K}\mathbf{T}^{-T} = \begin{bmatrix} 1 & & & & & & \\ & \ddots & & & & * & \\ & & 1 & & & & \\ & & & 1 & & & \\ & & * & & & \ddots & \\ & & & & & & 1 \end{bmatrix} \tag{11.27}
$$

11.3.4 2-D Second-Order Modes

If we denote

$$
\mathbf{K} = \begin{bmatrix} \mathbf{K}_{11} & \mathbf{K}_{12} \\ \mathbf{K}_{12}^T & \mathbf{K}_{22} \end{bmatrix}, \qquad \mathbf{W} = \begin{bmatrix} \mathbf{W}_{11} & \mathbf{W}_{12} \\ \mathbf{W}_{12}^T & \mathbf{W}_{22} \end{bmatrix}
$$

and

$$
\tilde{\mathbf{K}} = \begin{bmatrix} \tilde{\mathbf{K}}_{11} & \tilde{\mathbf{K}}_{12} \\ \tilde{\mathbf{K}}_{12}^T & \tilde{\mathbf{K}}_{22} \end{bmatrix}, \qquad \tilde{\mathbf{W}} = \begin{bmatrix} \tilde{\mathbf{W}}_{11} & \tilde{\mathbf{W}}_{12} \\ \tilde{\mathbf{W}}_{12}^T & \tilde{\mathbf{W}}_{22} \end{bmatrix}
$$

then Eqs. (11.22) and (11.27) yield

$$
\tilde{\mathbf{K}}_{11}\tilde{\mathbf{W}}_{11} = \mathbf{T}_1^{-1}\mathbf{K}_{11}\mathbf{W}_{11}\mathbf{T}_1 \tag{11.28a}
$$

$$
\tilde{\mathbf{K}}_{22}\tilde{\mathbf{W}}_{22} = \mathbf{T}_2^{-1}\mathbf{K}_{22}\mathbf{W}_{22}\mathbf{T}_2 \tag{11.28b}
$$

In other words, the eigenvalues of $\boldsymbol{\Phi} \equiv \mathbf{K}_{11}\mathbf{W}_{11}$ and $\boldsymbol{\Psi} \equiv \mathbf{K}_{22}\mathbf{W}_{22}$, denoted by $\phi \equiv \{\phi_i: 1 \leq i \leq N_1\}$ and $\psi \equiv \{\psi_j: 1 \leq j \leq N_2\}$, are invariant under a similarity transformation. We call sets $\phi \equiv \{\phi_i: 1 \leq i \leq N_1\}$ and $\psi \equiv \{\psi_j: 1 \leq j \leq N_2\}$ the second-order modes of the filter.

11.3.5 Minimization of Output Noise

For the sake of simplicity, it is assumed that the computation of each component of the new state variables $\tilde{q}^H(i + 1, j)$ and $\tilde{q}^V(i, j + 1)$ always involves $(N_1 + N_2 + 1)$ multiplications; that is, both $\tilde{A}$ and $\tilde{b}$ have neither zero nor one entries. This assumption implies that $\tilde{Q}$ in Eq. (11.21) is equal to $(N_1 + N_2 + 1)I$ and is independent of the similarity transformation used. Thus Eq. (11.21) now becomes

$$E(\Delta \bar{y}^2) = \frac{(N_1 + N_2 + 1)E_0^2}{12}\tilde{G} + \frac{E_0^2}{12}(\mu + \nu) \qquad (11.29)$$

where

$$\tilde{G} = \text{tr}(T^T W T) \qquad (11.30)$$

is referred to as the unit noise of realization $(T^{-1}AT, T^{-1}b, cT, d)$.

By using the singular-value decomposition and noting the invariance of the second-order modes of the filter, it can be shown that a transformation $T = T_1 \oplus T_2$ can be found that minimizes the unit noise $\tilde{G}$ subject to the constraints in Eq. (11.27) [6]. The resulting state-space realization, namely, $\{\tilde{A}, \tilde{b}, \tilde{c}, \tilde{d}\} \equiv \{T^{-1}AT, T^{-1}b, cT, d\}$, is said to be optimal. The required transformation T can be computed through the following steps:

STEP 1: Compute positive-definite matrices W and K using Eqs. (11.20) and (11.25), respectively.

STEP 2: Find matrix $P = P_1 \oplus P_2$ such that

$$P^{-1}KP^{-T} = \begin{bmatrix} I_{N_1} & \Gamma \\ \Gamma^T & I_{N_2} \end{bmatrix}$$

and compute

$$\overline{W} = P^T W P = \begin{bmatrix} \overline{W}_{11} & \overline{W}_{12} \\ \overline{W}_{12}^T & \overline{W}_{22} \end{bmatrix}$$

STEP 3: Find block-orthogonal matrix $R = R_1 \oplus R_2$ such that $R_1^T \overline{W}_{11} R_1 = \text{diag}\{u_1 \cdots u_{N_1}\}$ and $R_2^T \overline{W}_{22} R_2 = \text{diag}\{u_{N_1+1} \cdots u_{N_1+N_2}\}$.

STEP 4: Form matrix Λ^* as

$$\Lambda^* = \Lambda_1^* \oplus \Lambda_2^*$$

where $\Lambda_1^* = \text{diag}\{\lambda_{11}^* \cdots \lambda_{1N_1}^*\}$, $\Lambda_2^* = \text{diag}\{\lambda_{21}^* \cdots \lambda_{2N_2}^*\}$, and

$$\lambda_{1j}^* = \left[\frac{\dfrac{1}{N_1}\sum_{i=1}^{N_1} u_i}{u_j}\right]^{1/2}, \qquad 1 \le j \le N_1$$

$$\lambda_{2j}^* = \left[\frac{\dfrac{1}{N_2}\sum_{i=1}^{N_2} u_{N_1+i}}{u_{N_1+j}}\right]^{1/2}, \qquad 1 \le j \le N_2$$

STEP 5: Find block-orthogonal matrix $S = S_1 \oplus S_2$ such that

$$S_i(\Lambda_i^*)^{-2}S_i^T = \begin{bmatrix} 1 & & * \\ & \ddots & \\ * & & 1 \end{bmatrix}_{N_i \times N_i}, \qquad i = 1, 2$$

using the algorithm given in the appendix of [5].

STEP 6: Form

$$T = PR\Lambda^*S^T$$

This procedure is illustrated by the following example.

Example 11.2 Consider a stable state-space digital filter of order (2,2) represented by Eq. (11.10) where

$$A = \begin{bmatrix} 1.8890 & -0.9122 & \vdots & -1.0 & 0.0 \\ 1.0 & 0.0 & \vdots & 0.0 & 0.0 \\ \hdashline 0.0277 & -0.0258 & \vdots & 1.8890 & 1.0 \\ -0.0258 & 0.0243 & \vdots & -0.9122 & 0.0 \end{bmatrix}$$

$$b^T = [0.2191 \quad 0.0 \quad \vdots \quad -0.0289 \quad 0.0912]$$

$$c = [0.2889 \quad -0.0912 \quad \vdots \quad -0.2191 \quad 0.0]$$

Find the similarity transformation that will lead to an optimal state-space realization.

Solution. As the filter is stable, we can use finite sums

$$\sum_{i=0}^{M_1}\sum_{j=0}^{M_2} A_{ij}^T c^T c A_{ij} \qquad \text{and} \qquad \sum_{i=0}^{M_1}\sum_{j=0}^{M_2} f(i, j)f^T(i, j)$$

with sufficiently large M_1 and M_2 to approximate W and K. Taking $M_1 =$

$M_2 = 240$, numerical computation gives

$$\mathbf{W} = \left[\begin{array}{cc|cc} 1.1336 & -1.0329 & 0.9779 & 1.7744 \\ -1.0329 & 0.9652 & -0.9411 & -1.6723 \\ \hline 0.9779 & -0.9411 & 87.1245 & 85.2572 \\ 1.7744 & -1.6723 & 85.2572 & 87.1724 \end{array} \right]$$

and

$$\mathbf{K} = \left[\begin{array}{cc|cc} 87.1245 & 85.2572 & 1.6398 & -1.5391 \\ 85.2572 & 87.1724 & 1.3212 & -1.2332 \\ \hline 1.6398 & 1.3212 & 1.1336 & -1.0329 \\ -1.5391 & -1.2332 & -1.0329 & 0.9652 \end{array} \right]$$

The unit noise of this filter after scaling is the sum of products of the corresponding diagonal entries in $\mathbf{W}$ and $\mathbf{K}$ and is given by

$$G_0 = \sum_{i=1}^{4} w_{ii} k_{ii} = 365.8049$$

Following these steps, we obtain

$$\mathbf{T}_1 = \begin{bmatrix} -3.3098 & 10.4428 \\ -5.3284 & 10.6095 \end{bmatrix}, \qquad \mathbf{T}_2 = \begin{bmatrix} 0.9153 & 0.2613 \\ -0.9493 & -0.0659 \end{bmatrix}$$

and hence the required similarity transformation $\mathbf{T} = \mathbf{T}_1 \oplus \mathbf{T}_2$ can be formed. The characterization of the optimal state-space realization can now be obtained as

$$\tilde{\mathbf{A}} = \mathbf{T}^{-1}\mathbf{A}\mathbf{T} = \left[\begin{array}{cc|cc} 0.9645 & -0.1190 & -0.4731 & -0.1350 \\ 0.1724 & 0.9245 & -0.2376 & -0.0678 \\ \hline 0.0453 & 0.0107 & 0.8884 & 0.1816 \\ 0.0166 & 0.0220 & -0.1281 & 1.0006 \end{array} \right]$$

$$\tilde{\mathbf{b}} = \mathbf{T}^{-1}\mathbf{b} = [0.1132 \quad 0.0569 \quad | \quad -0.1168 \quad 0.2985]^T$$

$$\tilde{\mathbf{c}} = \mathbf{c}\mathbf{T} = [-0.4703 \quad 2.0493 \quad | \quad -0.2005 \quad -0.0573]$$

The corresponding matrices $\tilde{\mathbf{W}}$ and $\tilde{\mathbf{K}}$ are

$$\tilde{\mathbf{W}} = \begin{bmatrix} 3.3898 & -0.0009 & -1.2564 & 0.2644 \\ -0.0009 & 3.3898 & -0.5394 & 0.0075 \\ -1.2564 & -0.5394 & 3.3889 & 0.0 \\ 0.2644 & 0.0075 & 0.0 & 3.3910 \end{bmatrix}$$

and

$$\tilde{K} = \begin{bmatrix} 1.0 & 0.4757 & 0.1724 & 0.0673 \\ 0.4757 & 1.0 & 0.2047 & 0.0968 \\ 0.1724 & 0.2047 & 1.0 & 0.4754 \\ 0.0673 & 0.0968 & 0.4754 & 1.0 \end{bmatrix}$$

These give the unit noise of realization $\{\tilde{A}, \tilde{b}, \tilde{c}\}$ as

$$\tilde{G} = \sum_{i=1}^{4} \tilde{w}_{ii}\tilde{k}_{ii} = 13.5595$$

11.4 LIMIT CYCLES

Two types of limit cycles may occur in a recursive 2-D filter, namely, quantization and overflow limit cycles. The first type is due to the quantization of products, and the second is due to the finite dynamic range of the filter. In this section, conditions for the absence as well as existence of limit cycles are presented.

11.4.1 Quantization Limit Cycles

A 2-D sequence $\{y_{ij}: i \geq 0, j \geq 0\}$ is said to be periodic if

$$y(i + P_1, j + P_2) = y(i, j) \qquad \text{for all } i, j$$

where the integer pair (P_1, P_2) is called the period of the sequence. A 2-D periodic sequence $\{y(i, j): i \geq 0, j \geq 0\}$ is said to be separable if

$$y(i + P_1, j + P_2) = y(i, j) = y(i + P_1, j) = y(i, j + P_2) \qquad \text{for all } i, j$$

Note that if $\{y(i, j)\}$ is periodic and separable, then it can be generated by repeating the $P_1 \times P_2$ elements of $\{y(i, j)\}$ indefinitely in both i and j directions. If either P_1 or P_2 is equal to zero, then the periodic sequence is called degenerate. A 2-D limit cycle is a 2-D periodic sequence which may or may not be separable.

Now let us consider a 2-D digital filter with zero input as depicted in Fig. 11.1, where Q_l $(1 \leq l \leq m)$ are quantizers under fixed-point arithmetic satisfying the sector conditions

$$0 \leq \frac{Q_l[y(i, j)]}{y(i, j)} \leq k_l \qquad \text{for } y(i, j) \neq 0 \tag{11.31}$$

and

$$Q_l(0) = 0$$

where $Q_l[y(i, j)]$ is the quantized value of $y(i, j)$. Constant k_l in (11.31) depends on the type of quantization used. For example, for truncation $k_l = 1$ and for rounding $k_l = 2$.

Assume that $H(z_1, z_2)$ in Fig. 11.1 has no singularities on the unit bicircle U^2 (see Sec. 5.2, p. 108, for definition), and that the initial conditions for $\{y(i, j)\}$ are given by

$$y(i, 0) = 0 \quad \text{for } 0 \le i \le M_1 \quad (11.32a)$$

and

$$y(0, j) = 0 \quad \text{for } 0 \le j \le M_2 \quad (11.32b)$$

where M_1 and M_2 are some positive integers. The following theorem gives a sufficient condition for the absence of separable limit cycles of period (M_1, M_2) [10].

Theorem 11.1 *If there exists a diagonal matrix* $\mathbf{D} = \text{diag}\{d_1, \ldots, d_m\}$ *with* Re $d_i > 0$ *for* $1 \le i \le m$ *such that for positive integers* k_1 *and* k_2

$$\mathbf{M}(k_1, k_2) \equiv \mathbf{D}^H[\mathbf{H}(e^{j2\pi k_1/M_1}, e^{j2\pi k_2/M_2}) - \mathbf{K}]$$

$$+ [\mathbf{H}^H(e^{j2\pi k_1/M_1}, e^{j2\pi k_2/m_2}) - \mathbf{K}]\mathbf{D} \quad (11.33)$$

is negative definite where $\mathbf{K} = \text{diag}\{k_1^{-1}, \ldots, k_m^{-1}\}$ *and* $\mathbf{D}^H$ *denotes the complex conjugate transpose of* $\mathbf{D}$, *then separable limit cycles of period* (M_1, M_2) *are absent from the filter in Fig. 11.1.*

Corollary 11.1 *The filter depicted in Fig. 11.1 is free from separable limit cycles if there exists a matrix* $\mathbf{D} = \text{diag}\{d_1, \ldots, d_m\}$ *with* Re $d_i > 0$ *for* 1

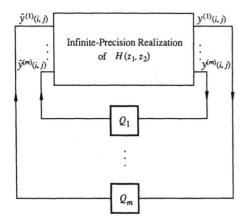

Figure 11.1 A 2-D filter with m quantizers.

$\leq i \leq m$ such that

$$\hat{\mathbf{M}}(\theta_1, \theta_2) \equiv \mathbf{D}^H[\mathbf{H}(e^{j\theta_1}, e^{j\theta_2}) - \mathbf{K}] + [\mathbf{H}^H(e^{j\theta_1}, e^{j\theta_2}) - \mathbf{K}]\mathbf{D} \quad (11.34)$$

is negative definite for $0 \leq \theta_1 \leq 2\pi$, $0 \leq \theta_2 \leq 2\pi$.

More feasible conditions for the absence as well as existence of limit cycles can be derived for some low-order filters. First, we consider 2-D filters of order $(1, 1)$ with one quantizer Q characterized by

$$y(i + 1, j + 1) = Q[ay(i + 1, j) + by(i, j + 1) + cy(i, j)] \quad (11.35)$$

where

$$0 \leq \frac{Q(y)}{y} \leq 1 \qquad \text{for} \quad y \neq 0 \qquad\qquad (11.36)$$

A 2-D filter reaches a zero steady state if there exist integers $K_1 > 0$ and $K_2 > 0$ such that

1. $y(i, j) \neq 0$ for some $i < K_1$ and $j < K_2$
2. $y(i, j) \neq 0$ for $i \geq K_1$ and arbitrary j
   or
   $y(i, j) = 0$ for $j \geq K_2$ and arbitrary i

The following theorem provides sufficient conditions for the absence of limit cycles in the filter represented by Eq. (11.35) [8].

Theorem 11.2 *Under zero initial conditions, the filter described by Eq. (11.35) reaches a zero steady state, that is, the filter is free from limit cycles, if*

$$|b| < \tfrac{1}{2} \qquad and \qquad |a| + |c| < \tfrac{1}{2} \qquad\qquad (11.37)$$

or

$$|a| < \tfrac{1}{2} \qquad and \qquad |b| + |c| < \tfrac{1}{2} \qquad\qquad (11.38)$$

Note that the conditions in Eq. (11.37) or (11.38) imply that

$$|b| < 1, \qquad \left|\frac{1 + b}{a + c}\right| > 1, \qquad and \qquad \left|\frac{1 - b}{a - c}\right| > 1$$

which are the necessary and sufficient conditions for the filter to be BIBO stable (see Sec. 5.5—case 4). It follows that the sufficient conditions in Eq. (11.37) or (11.38) imply that the filter is also BIBO stable.

If in the filter described by Eq. (11.35), $c = 0$ and there are two quantizers involved, then Eq. (11.35) becomes

$$y(i + 1, j + 1) = Q[ay(i + 1, j)] + Q[by(i, j + 1)] \quad (11.39)$$

By Theorem 11.2 we know that with $c = 0$, the filter given by Eq. (11.35) is free from limit cycles if $|a| < 1/2$ and $|b| < 1/2$. The following theorem states that this condition is necessary and sufficient for the filter described by Eq. (11.39) to be free from limit cycles [8].

Theorem 11.3 *The filter represented by Eq. (11.39) under the initial conditions in Eq. (11.32) is free from limit cycles if and only if*

$$|a| < \tfrac{1}{2} \quad \text{and} \quad |b| < \tfrac{1}{2} \tag{11.40}$$

11.4.2 Overflow Limit Cycles

In a fixed-point digital filter, overflow may occur during addition, and its nonlinear effect may result in autonomous oscillations or limit cycles of large amplitude.

Given a stable 1-D recursive digital filter, one can always find a state-space realization that is free of overflow limit cycles for various types of arithmetic [13]. An important feature of such a realization is that the Euclidean norm of its transition matrix is strictly less than one; that is, the transition matrix represents a contraction mapping meaning that for a vector $\mathbf{q}$, the length of its image $\mathbf{Aq}$ is always smaller than that of $\mathbf{q}$. To obtain such a realization, we note that the stability of the filter implies that for a realization $\{\mathbf{A}, \mathbf{b}, \mathbf{c}, d\}$ there exists a positive-definite matrix $\mathbf{W}$ such that

$$\mathbf{W} - \mathbf{A}^T\mathbf{W}\mathbf{A} = \mathbf{I} \tag{11.41}$$

If $\mathbf{W}^{1/2}$ denotes the symmetric square root of matrix $\mathbf{W}$, then the 1-D Lyapunov equation, namely, Eq. (11.41), can be written as

$$\mathbf{I} - (\mathbf{W}^{1/2}\mathbf{A}\mathbf{W}^{-1/2})^T(\mathbf{W}^{1/2}\mathbf{A}\mathbf{W}^{-1/2}) = \mathbf{W}^{-1} \tag{11.42}$$

As $\mathbf{W}^{-1}$ is positive definite, Eq. (11.42) implies that

$$\|\mathbf{W}^{1/2}\mathbf{A}\mathbf{W}^{-1/2}\| < 1$$

Consequently, by taking $\mathbf{T} = \mathbf{W}^{1/2}$ as a similarity transformation, an equivalent realization $\{\hat{\mathbf{A}}, \hat{\mathbf{b}}, \hat{\mathbf{c}}, \hat{d}\} = \{\mathbf{TAT}^{-1}, \mathbf{Tb}, \mathbf{CT}^{-1}, d\}$ is obtained where $\|\hat{\mathbf{A}}\| < 1$. In other words, the new realization will be free of overflow limit cycles. This Lyapunov approach can be extended to the 2-D case, as is demonstrated below. Let us consider a 2-D state-space digital filter with zero input described by

$$\begin{bmatrix} \mathbf{q}^H(i+1, j) \\ \mathbf{q}^V(i, j+1) \end{bmatrix} = Q\left\{ \begin{bmatrix} \mathbf{A}_1 & \mathbf{A}_2 \\ \mathbf{A}_3 & \mathbf{A}_4 \end{bmatrix} \begin{bmatrix} \mathbf{q}^H(i, j) \\ \mathbf{q}^V(i, j) \end{bmatrix} \right\} \equiv Q\{\mathbf{Aq}(i, j)\} \tag{11.43a}$$

where

$$Q(\mathbf{q}) = \begin{bmatrix} Q_1(q_1) \\ \vdots \\ Q_N(q_N) \end{bmatrix}$$

is a nonlinear operator satisfying the conditions

$$|Q_l(q_l)| \le |q_l| \qquad 1 \le l \le N \tag{11.43b}$$

and

$$Q_l(0) = 0 \tag{11.43c}$$

The following theorem provides a sufficient condition for the filter described by Eq. (11.43) to be free of overflow limit cycles [9].

Theorem 11.4   *The filter described by Eq. (11.43) is asymptotically stable, that is,*

$$\lim_{i \text{ and/or } j \to \infty} x(i, j) = 0$$

if there exists a diagonal positive-definite matrix **D** *such that*

$$\mathbf{D} - \mathbf{A}^T \mathbf{D} \mathbf{A} = \mathbf{Q}$$

where **Q** *is a positive-definite matrix.*

Note that if $\{\hat{\mathbf{A}}, \hat{\mathbf{b}}, \hat{\mathbf{c}}, \hat{d}\}$ represents a realization of a 2-D filter such that $\|\hat{\mathbf{A}}\| < 1$, then

$$\mathbf{I} - \hat{\mathbf{A}}^T \hat{\mathbf{A}} > 0$$

Therefore, by Theorem 11.4, the realization $\{\hat{\mathbf{A}}, \hat{\mathbf{b}}, \hat{\mathbf{c}}, \hat{d}\}$ is free of overflow limit cycles. Furthermore, if $\{\mathbf{A}, \mathbf{b}, \mathbf{c}, d\}$ is a state-space realization of a stable 2-D filter such that **A** satisfies the 2-D Lyapunov equation [14]

$$\mathbf{W} - \mathbf{A}^T \mathbf{W} \mathbf{A} = \mathbf{Q} \tag{11.44}$$

with $\mathbf{W} = \mathbf{W}_1 \oplus \mathbf{W}_2 > 0$ and $\mathbf{Q} > 0$, then by taking

$$\mathbf{T} = \mathbf{W}^{1/2} = \mathbf{W}_1^{1/2} \oplus \mathbf{W}_2^{1/2}$$

as a 2-D similarity transformation, Eq. (11.44) yields

$$\mathbf{I} - (\mathbf{TAT}^{-1})^T (\mathbf{TAT}^{-1}) = \mathbf{W}^{-1/2} \mathbf{Q} \mathbf{W}^{-1/2} > 0$$

which implies that $\|\mathbf{TAT}^{-1}\| < 1$. We have, in effect, demonstrated the following corollary.

Corollary 11.2   *If matrix* **A** *in a state-space realization of a stable 2-D filter satisfies the Lyapunov equation, namely, Eq. (11.44), for some* $\mathbf{W} = \mathbf{W}_1 \oplus$

$W_2 > 0$ and $Q > 0$, *then the realization* $\{TAT^{-1}, Tb, cT^{-1}, d\}$, *where* T *is the symmetric square root of matrix* W, *is free of overflow limit cycles.*

The BIBO stability of a 2-D state-space realization does not, in general, imply Eq. (11.44), as was demonstrated in Sec. 5.4. Certain special cases where the BIBO stability of a 2-D filter does imply Eq. (11.44) can be found in [10] and [14]. One of these cases is described by the following theorem.

Theorem 11.5 *In 2-D state-space digital filters of order (1, 1), BIBO stability implies the absence of overflow limit cycles.*

REFERENCES

1. S. H. Mneney and A. N. Venetsanopoulos, Finite register length effects in two-dimensional digital filters, *Proc. 22nd Midwest Symp. on Circuits and Systems*, pp. 669–676, June 1979.
2. L. M. Roytman and M. N. S. Swamy, Determination of quantization error in two-dimensional digital filters, *Proc. IEEE*, vol. 69, pp. 832–834, July 1981.
3. P. Agathoklis, E. I. Jury, and M. Mansour, Evaluation of quantization error in two-dimensional digital filters, *IEEE Trans. Acoust., Speech, Signal Process.*, vol. ASSP-28, pp. 273–279, June 1980.
4. C. T. Mullis and R. Roberts, Synthesis of minimum roundoff noise fixed point digital filters, *IEEE Trans. Circuits Syst.*, vol. CAS-23, pp. 551–562, Sept. 1976.
5. S. Y. Hwang, Minimum uncorrelated unit noise in state-space digital filtering, *IEEE Trans. Acoust., Speech, Signal Process.*, vol. ASSP-25, pp. 256–262, June 1976.
6. W.-S. Lu and A. Antoniou, Synthesis of 2-D state-space fixed-point digital-filter structures with minimum roundoff noise, *IEEE Trans. Circuits Syst.*, vol. CAS-33, pp. 965–973, Oct. 1986.
7. T.-L. Chang, Limit cycles in a two-dimensional first-order digital filter, *IEEE Trans. Circuits Syst.*, vol. CAS-24, pp. 15–19, Jan. 1977.
8. N. G. El-Agizi and M. M. Fahmy, Sufficient conditions for the nonexistence of limit cycles in two-dimensional digital filters, *IEEE Trans. Circuits Syst.*, vol. CAS-26, pp. 402–406, June 1979.
9. N. G. El-Agizi and M. M. Fahmy, Two-dimensional digital filters with no overflow oscillations, *IEEE Trans. Acoust., Speech, Signal Process.*, vol. ASSP-27, pp. 465–469, Oct. 1979.
10. P. Agathoklis, E. I. Jury, and M. Mansour, Criteria for the absence of limit cycles in two-dimensional discrete systems, *IEEE Trans. Acoust., Speech, Signal Process.*, vol. ASSP-32, pp. 432–434, April 1984.
11. A. Antoniou, *Digital Filters: Analysis and Design*, New York: McGraw-Hill, 1979 (2nd ed. in press).

12. D. M. Goodman, Some stability properties of two-dimensional linear shift-invariant digital filters, *IEEE Trans. Circuits Syst.*, vol. CAS-24, pp. 201–208, April 1977.

13. C. W. Barnes and A. T. Fam, Minimum norm recursive digital filters that are free of overflow limit cycles, *IEEE Trans. Circuits Syst.*, vol. CAS-24, pp. 569–574, Oct. 1977.

14. B. D. O. Anderson, P. Agathoklis, E. I. Jury, and M. Mansour, Stability and the matrix Lyapunov equation for discrete two-dimensional systems, *IEEE Trans. Circuits Syst.*, vol. CAS-33, pp. 261–267, March 1986.

PROBLEMS

11.1 Verify the entries in Table 11.2.

11.2 Verify Eq. (11.5).

11.3 Verify the entries in Table 11.3.

11.4 Find upper bounds of the means and variances of the output error due to product quantization for the transfer function $H(z_1, z_2) = N(z_1, z_2)/D(z_1, z_2)$ where $D(z_1, z_2)$ is given by
(a) $D(z_1, z_2) = 2 - z_1 - z_2$.
(b) $D(z_1, z_2) = 2 - z_1z_2 - z_1 - z_2$.

11.5 A 2-D digital filter is characterized by the allpass transfer function

$$H(z_1, z_2) = \frac{a_3 + a_2z_1^{-1} + a_1z_2^{-1} + z_1^{-1}z_2^{-1}}{1 + a_1z_1^{-1} + a_2z_2^{-1} + a_3z_1^{-1}z_2^{-1}}$$

(a) Show that if $H(z_1, z_2)$ is BIBO stable, the filter can be represented in the local state space by Eq. (11.10) with all A_i $(1 \le i \le 4)$ scalar.
(b) Verify that if $a_1 = a_2 = 0$, then $H(z_1, z_2)$ can be realized by the state-space model in Eq. (11.10) where

$$\mathbf{A} = \begin{bmatrix} 0 & -a_3 \\ 1 & 0 \end{bmatrix}, \qquad \mathbf{b} = \begin{bmatrix} 1 \\ 0 \end{bmatrix}, \qquad \mathbf{c} = [0 \quad 1 - a_3^2], \qquad d = a_3$$

11.6 (a) Show that for the state-space model given in Problem 11.5b,

$$\mathbf{K} = \frac{1}{1 - a_3^2}\mathbf{I} \quad \text{and} \quad \mathbf{W} = (1 - a_3^2)\mathbf{I}$$

(b) Find a realization of $H(z_1, z_2)$ with $a_1 = a_2 = 0$ that minimizes the output roundoff-noise power subject to L_2 dynamic-range constraints.

11.7 Consider a BIBO stable state-space 2-D filter of order (1, 1). The filter is described by Eq. (11.10) where

$$\mathbf{A} = \begin{bmatrix} a_1 & a_2 \\ a_3 & a_4 \end{bmatrix}, \qquad \mathbf{b} = \begin{bmatrix} b_1 \\ b_2 \end{bmatrix}, \qquad \mathbf{c} = [c_1 \quad c_2], \qquad d = 0$$

Assume that matrices $\mathbf{K}$ and $\mathbf{W}$ have been computed using Eqs. (11.25) and (11.20), respectively, and are given by

$$\mathbf{K} = \begin{bmatrix} k_1 & k_2 \\ k_2 & k_4 \end{bmatrix} \qquad \text{and} \qquad \mathbf{W} = \begin{bmatrix} w_1 & w_2 \\ w_2 & w_4 \end{bmatrix}$$

Find a realization with minimum output-noise power.

11.8 Prove Theorem 11.4.

11.9 Prove Theorem. 11.5.

12

Implementation

12.1 INTRODUCTION

Once a transfer function that satisfies the requirements imposed by the application at hand is obtained, the implementation of the filter must be undertaken.

As was stated in the introduction to Chap. 6, a 2-D digital filter can be implemented in terms of software or hardware. In applications where a record of the data to be processed is available, a software implementation may be entirely acceptable, whereas in applications where real-time processing is required or where a massive amount of data needs to be processed, a hardware implementation may be the only available choice.

In a software implementation, the difference equation, a signal-flow graph, a digital-filter network, or a state-space representation is converted into a computer program that can be run on a general-purpose computer or workstation or on a general-purpose digital signal processing chip. A software implementation is deemed to be good if it entails low computational complexity, needs a small amount of memory, and is insensitive to quantization effects and aliasing errors. Low computational complexity can be achieved through the application of fast-Fourier transforms [1–4], whereas low sensitivity to coefficient quantization can be achieved by emulating a wave digital-filter structure of the type described in Sec. 10.6.1. On the other hand, if low output roundoff noise is required, a state-space structure of the type described in Sec. 11.3.5 can be used.

In a hardware implementation, the digital filter is implemented in terms of a collection of specialized very-large-scale integrated (VLSI) circuit chips. Such an implementation is deemed to be good if it is fast, reliable, and economical. Although the advantages associated with wave digital-filter and state-space structures (e.g., low sensitivity and low roundoff noise) are important, the network complexity of these structures often renders them unsuitable for VLSI implementation. To achieve high speed of operation, the filter structure must allow a high degree of concurrency, that is, a large number of processing elements must operate simultaneously. In such a case, communication among processing elements becomes critical. Since the cost, performance, and speed of a chip depend heavily on the delay and area of the interconnection network, processing elements should be interconnected by simple, short, and regular communication paths. A class of digital-filter structures that are suitable for VLSI implementation is the class of systolic structures [5–6]. These are highly regular structures of simply interconnected processing elements that pass data from one processing element to the next in a fashion resembling the rhythmical systolic operation of the heart and arteries.

This chapter consists of two parts. In the first part, comprising Secs. 12.2 and 12.3, the 2-D discrete Fourier transform (DFT) is described and methods for its efficient computation in terms of 1-D and 2-D fast-Fourier transforms are considered. These techniques are important tools for the software implementation of nonrecursive filters. In the second part, namely, Sec. 12.4, systolic structures for the implementation of 2-D nonrecursive and recursive filters are examined.

12.2 2-D DISCRETE FOURIER TRANSFORM

In many applications, the 2-D discrete signals under consideration are of finite extent or periodic. In such cases, the 2-D z transform and 2-D Fourier transform described in Chap. 3 are well defined and, consequently, the signal can be represented in terms of the DFT. In what follows, the DFT is defined and its salient properties are described. Its application for the implementation of 2-D nonrecursive digital filters is then considered in some detail.

12.2.1 Definition and Properties

The DFT of a 2-D discrete and periodic signal $f_p(n_1, n_2)$ with periods N_1 and N_2 is defined by

$$F_p(k_1, k_2) = \sum_{n_1=0}^{N_1-1} \sum_{n_2=0}^{N_2-1} f_p(n_1, n_2) e^{-j(2\pi/N_1)k_1 n_1} e^{-j(2\pi/N_2)k_2 n_2} \quad (12.1)$$

We can write

$$\sum_{k_1=0}^{N_1-1} \sum_{k_2=0}^{N_2-1} F_p(k_1, k_2)e^{j(2\pi/N_1)k_1n_1} e^{j(2\pi/N_2)k_2n_2}$$

$$= \sum_{k_1=0}^{N_1-1} \sum_{k_2=0}^{N_2-1} \left[\sum_{\bar{n}_1=0}^{N_1-1} \sum_{\bar{n}_2=0}^{N_2-1} f_p(\bar{n}_1, \bar{n}_2)e^{-j(2\pi/N_1)k_1\bar{n}_1}e^{-j(2\pi/N_2)k_2\bar{n}_2} \right]$$
$$\times\ e^{j(2\pi/N_1)k_1n_1}e^{j(2\pi/N_2)k_2n_2}$$

$$= \sum_{\bar{n}_1=0}^{N_1-1} \sum_{\bar{n}_2=0}^{N_2-1} f_p(\bar{n}_1, \bar{n}_2)\left[\sum_{k_1=0}^{N_1-1} \sum_{k_2=0}^{N_2-1} e^{j(2\pi/N_1)(n_1-\bar{n}_1)k_1}e^{j(2\pi/N_2)(n_2-\bar{n}_2)k_2} \right]$$

$$= \sum_{\bar{n}_1=0}^{N_1-1} \sum_{\bar{n}_2=0}^{N_2-1} f_p(\bar{n}_1, \bar{n}_2)\left[\sum_{k_1=0}^{N_1-1} e^{j(2\pi/N_1)(n_1-\bar{n}_1)k_1} \right]\left[\sum_{k_2=0}^{N_2-1} e^{j(2\pi/N_2)(n_2-\bar{n}_2)k_2} \right]$$

$$(12.2)$$

and since

$$\sum_{k_i=0}^{N_i-1} e^{j(2\pi/N_i)(n_i-\bar{n}_i)k_i} = \begin{cases} N_i & \text{if } \bar{n}_i = n_i \\ 0 & \text{otherwise} \end{cases}$$

for $i = 1, 2$, Eq. (12.2) implies that

$$f_p(n_1, n_2) = \frac{1}{N_1N_2} \sum_{k_1=0}^{N_1-1} \sum_{k_2=0}^{N_2-1} F_p(k_1, k_2)e^{j(2\pi/N_1)k_1n_1}e^{j(2\pi/N_2)k_2n_2} \quad (12.3)$$

The function at the right-hand side in Eq. (12.3) is called the inverse DFT of $F_p(k_1, k_2)$ and it can be used to recover the signal $f_p(n_1, n_2)$ from its DFT.

If $f(n_1, n_2)$ is a finite-extent, first-quadrant signal with the region of support $R_{N_1N_2} = \{0 \leq n_1 \leq N_1 - 1, 0 \leq n_2 \leq N_2 - 1\}$, a periodic extension $f_p(n_1, n_2)$ with periods N_1 and N_2 can be formed as

$$f_p(n_1, n_2) = \sum_{r_2=-\infty}^{\infty} \sum_{r_1=-\infty}^{\infty} f(n_1 + r_1N_1, n_2 + r_2N_2) \quad (12.4)$$

which has a DFT $F_p(n_1, n_2)$ given by Eq. (12.1).

In what follows, we assume that the signals of interest are periodic. If they are not periodic, they are forced to become periodic through the periodic extension in Eq. (12.4). For the sake of simplicity, subscript p in $f_p(n_1, n_2)$ and $F_p(k_1, k_2)$ is omitted.

The properties of the 2-D DFT are essentially the same as those of its 1-D counterpart and can be summarized in terms of a number of theorems. If $f(n_1, n_2)$ and $g(n_1, n_2)$ are periodic signals with periods N_1 and N_2, respectively, such that

$$f(n_1, n_2) \leftrightarrow F(k_1, k_2) \quad \text{and} \quad g(n_1, n_2) \leftrightarrow G(k_1, k_2)$$

and a, b are arbitrary constants, then the following theorems hold.

Theorem 12.1 Linearity

$$af(n_1, n_2) + bg(n_1, n_2) \leftrightarrow aF(k_1, k_2) + bG(k_1, k_2)$$

Theorem 12.2 Periodicity $F(k_1, k_2)$ *is a periodic function of* (k_1, k_2) *with periods* N_1 *and* N_2; *that is, for any integers* r_1 *and* r_2

$$F(k_1 + r_1N_1, k_2 + r_2N_2) = F(k_1, k_2)$$

Theorem 12.3 Separability *If* $F_1(k_1)$ *is the* N_1-*point DFT of* $f_1(n_1)$ *and* $F_2(k_2)$ *is the* N_2-*point DFT of* $f_2(n_2)$, *then*

$$f(n_1, n_2) = f_1(n_1)f_2(n_2) \leftrightarrow F(k_1, k_2) = F_1(k_1)F_2(k_2)$$

Theorem 12.4 Parseval's Theorem *If* $g^*(n_1, n_2)$ *is the complex conjugate of* $g(n_1, n_2)$, *then*

$$\sum_{n1=0}^{N_1-1} \sum_{n2=0}^{N_2-1} f(n_1, n_2)g^*(n_1, n_2)$$

$$= \frac{1}{N_1N_2} \sum_{k1=0}^{N_1-1} \sum_{k2=0}^{N_2-1} F(k_1, k_2)G^*(k_1, k_2) \quad (12.5)$$

Theorem 12.5 Duality

$$F^*(n_1, n_2) \leftrightarrow N_1N_2f^*(k_1, k_2)$$

The circular convolution of $f(n_1, n_2)$ and $g(n_1, n_2)$ is defined by

$$f(n_1, n_2) \circledast g(n_1, n_2)$$

$$= \sum_{m1=0}^{N_1-1} \sum_{m2=0}^{N_2-1} f(m_1, m_2)g[(n_1 - m_1)_{N_1}, (n_2 - m_2)_{N_2}] \quad (12.6)$$

where $(n_i - m_i)_{N_i}$ represents an integer p_i that satisfies the constraints $0 \leq p_i \leq N_i - 1$ and $p_i = (n_i - m_i) + l_iN_i$ for some integer l_i. An important property of the circular convolution is given by the following theorem.

Theorem 12.6 Circular Convolution

$$f(n_1, n_2) \circledast g(n_1, n_2) \leftrightarrow F(k_1, k_2)G(k_1, k_2) \quad (12.7)$$

12.2.2 Implementation of Nonrecursive Filters Using the 2-D DFT

Consider a 2-D nonrecursive digital filter characterized by the transfer function

$$H(z_1, z_2) = \sum_{m_1=0}^{N_1-1} \sum_{m_2=0}^{N_2-1} h(m_1, m_2) z_1^{-m_1} z_2^{-m_2} \qquad (12.8)$$

The impulse response of the filter $h(m_1, m_2)$ has a finite region of support in the first quadrant given by $R_h = \{0 \le m_1 \le N_1 - 1, 0 \le m_2 \le N_2 - 1\}$. If $x(n_1, n_2)$ and $y(n_1, n_2)$ are the input and output signals of the filter, respectively, then $y(n_1, n_2)$ can be computed through the convolution

$$y(n_1, n_2) = \sum_{m_1=0}^{N_1-1} \sum_{m_2=0}^{N_2-1} h(m_1, m_2) x(n_1 - m_1, n_2 - m_2) \qquad (12.9)$$

If the input signal $x(n_1, n_2)$ has a finite region of support $R_x = \{0 \le n_1 \le M_1 - 1, 0 \le n_2 \le M_2 - 1\}$, then Eq. (12.9) implies that the output signal $y(n_1, n_2)$ has a region of support $R_y = \{0 \le n_1 \le N_1 + M_1 - 1, 0 \le n_2 \le N_2 + M_2 - 1\}$.

In order to facilitate the application of the DFT for the implementation of the filter, the impulse response, input, and output must have a common region of support. Since the regions of support of $h(n_1, n_2)$ and $x(n_1, n_2)$ are subsets of that of $y(n_1, n_2)$, we can use $R_a = \{0 \le n_1 \le P_1, 0 \le n_2 \le P_2\}$, where $P_1 \ge N_1 + M_1 - 1$ and $P_2 \ge N_2 + M_2 - 1$, as the common region of support. Augmented versions of $x(n_1, n_2)$ and $h(n_1, n_2)$ over region R_a can be defined as

$$x_a(n_1, n_2) = \begin{cases} x(n_1, n_2) & \text{for } (n_1, n_2) \in R_x \\ 0 & \text{for } (n_1, n_2) \notin R_x \text{ but } (n_1, n_2) \in R_a \end{cases}$$

$$(12.10)$$

and

$$h_a(n_1, n_2) = \begin{cases} h(n_1, n_2) & \text{for } (n_1, n_2) \in R_h \\ 0 & \text{for } (n_1, n_2) \notin R_h \text{ but } (n_1, n_2) \in R_a \end{cases}$$

$$(12.11)$$

From Eqs. (12.9) and (12.10), we can now write

$$y(n_1, n_2) = \sum_{m_1=0}^{P_1-1} \sum_{m_2=0}^{P_2-1} h_a(m_1, m_2) x_a(n_1 - m_1, n_2 - m_2)$$

$$= h_a(n_1, n_2) \circledast x_a(n_1, n_2) \qquad \text{for } (n_1, n_2) \in R_a$$

and from Theorem 12.6

$$Y(k_1, k_2) = H_a(k_1, k_2)X_a(k_1, k_2)$$

where $Y(k_1, k_2)$, $H_a(k_1, k_2)$, and $X_a(k_1, k_2)$ are the $(P_1 \times P_2)$-point DFTs of $y(n_1, n_2)$, $h_a(n_1, n_2)$, and $x_a(n_1, n_2)$, respectively. Once $Y(k_1, k_2)$ is found, $y(n_1, n_2)$ for $(n_1, n_2) \in R_y$ can be computed by using the $(P_1 \times P_2)$-point inverse DFT of $Y(k_1, k_2)$. The steps involved are as follows:

STEP 1: Identify the regions of support of $h(n_1, n_2)$ and $x(n_1, n_2)$.

STEP 2: Identify the region of support R_a.

STEP 3: Augment signals $h(n_1, n_2)$ and $x(n_1, n_2)$ to $h_a(n_1, n_2)$ and $x_a(n_1, n_2)$ using Eqs. (12.10) and (12.11), respectively.

STEP 4: Compute the $(P_1 \times P_2)$-point DFTs for $h_a(n_1, n_2)$ and $x_a(n_1, n_2)$.

STEP 5: Compute the $(P_1 \times P_2)$-point inverse DFT of $H_a(k_1, k_2)X_a(k_1, k_2)$ to obtain $y(n_1, n_2)$.

12.3 COMPUTATION OF 2-D DFT

It follows from Eq. (12.1) that computing the DFT of a signal $f(n_1, n_2)$ over the region $R_f = \{0 \le k_1 \le N_1 - 1, 0 \le k_2 \le N_2 - 1\}$ requires totally $N_1^2 N_2^2$ complex multiplications and $N_1 N_2(N_1 N_2 - 1)$ complex additions. More efficient computation can be achieved (1) by decomposing the DFT into a number of 1-D DFTs that can be computed using 1-D fast-Fourier transforms (FFTs) or (2) by computing the 2-D DFT using 2-D FFTs.

12.3.1 Row–Column Decomposition

Let

$$W_N = e^{-j2\pi/N}$$

and write Eq. (12.1) as

$$F(k_1, k_2) = \sum_{n_1=0}^{N_1-1} \sum_{n_2=0}^{N_2-1} f(n_1, n_2)W_{N_1}^{k_1 n_1} W_{N_2}^{k_2 n_2}$$

$$= \sum_{n_2=0}^{N_2-1} \left[\sum_{n_1=0}^{N_1-1} f(n_1, n_2)W_{N_1}^{k_1 n_1} \right] W_{N_2}^{k_2 n_2}$$

$$= \sum_{n_2=0}^{N_2-1} \hat{F}(k_1, n_2)W_{N_2}^{k_2 n_2} \tag{12.12}$$

where

$$\hat{F}(k_1, n_2) = \sum_{n_1=0}^{N_1-1} f(n_1, n_2) W_{N_1}^{k_1 n_1} \tag{12.13}$$

Each row of the 2-D signal $f(n_1, n_2)$, namely, $f(n_1, n_2)$ for $0 \le n_1 \le N_1 - 1$ with n_2 fixed, can be regarded as a 1-D signal with DFT $\hat{F}(k_1, n_2)$. Once $\hat{F}(k_1, n_2)$ is found for all n_2 in the range $0 \le n_2 \le N_2 - 1$, a column comprising N_2 elements is formed whose DFT is $F(k_1, k_2)$ as per Eq. (12.12). Therefore, the 2-D DFT of $f(n_1, n_2)$ can be obtained by computing N_2 1-D N_1-point DFTs and N_1 1-D N_2-point DFTs. If all the 1-D DFTs involved are computed directly, then it can easily be shown that the preceding row–column decomposition method requires a total of $N_1 N_2 (N_1 + N_2)$ multiplications and about the same number of additions. For example, the 2-D DFT of a 512×512 signal would require approximately 2.685×10^8 multiplications. By comparison, the direct evaluation of the 2-D DFT would require approximately 6.872×10^{10} multiplications.

A more substantial reduction in the amount of computation can be achieved by employing 1-D FFT algorithms [3, Ch. 13] for the computation of each 1-D DFT involved. Since the total number of multiplications required in the FFT of an N-point signal is $(N/2) \log_2 N$, the use of the row–column decomposition approach in conjunction with a 1-D FFT would require a total of $(N_1 N_2/2) \log_2(N_1 N_2)$ multiplications and $N_1 N_2 \log_2(N_1 N_2)$ additions. Consequently, the DFT of a 512×512 signal would require only 2.36×10^6 multiplications, which represents a 99.65 percent reduction in the number of multiplications relative to direct evaluation.

12.3.2 Vector–Radix FFT

An alternative approach to the efficient evaluation of a 2-D DFT is to use a 2-D FFT algorithm. The basic idea in the various 1-D FFT algorithms [3, Ch. 13], [4, Ch. 8] is that an N-point DFT can be evaluated by computing two $(N/2)$-point DFTs or by computing four $(N/4)$-point DFTs, and so on. The same idea applies to the 2-D case. In what follows, we describe the so-called decimation-in-place FFT algorithm, which is the 2-D counterpart of the well-known decimation-in-time 1-D FFT algorithm.

Consider the 2-D DFT of $f(n_1, n_2)$ given in Eq. (12.1) and assume that

$$N_1 = N_2 = N = 2^r$$

A radix-(2×2) FFT algorithm can be obtained by writing Eq. (12.1) as

$$F(k_1, k_2) = F_{ee}(k_1, k_2) + W_N^{k_1} F_{oe}(k_1, k_2)$$
$$+ W_N^{k_2} F_{eo}(k_1, k_2) + W_N^{k_1+k_2} F_{oo}(k_1, k_2) \tag{12.14}$$

where

$$F_{ee}(k_1, k_2) = \sum_{n_1=0}^{(N/2)-1} \sum_{n_2=0}^{(N/2)-1} f(2n_1, 2n_2) W_N^{2k_1 n_1} W_N^{2k_2 n_2}$$

$$F_{oe}(k_1, k_2) = \sum_{n_1=0}^{(N/2)-1} \sum_{n_2=0}^{(N/2)-1} f(2n_1 + 1, 2n_2) W_N^{2k_1 n_1} W_N^{2k_2 n_2}$$

$$F_{eo}(k_1, k_2) = \sum_{n_1=0}^{(N/2)-1} \sum_{n_2=0}^{(N/2)-1} f(2n_1, 2n_2 + 1) W_N^{2k_1 n_1} W_N^{2k_2 n_2}$$

$$F_{oo}(k_1, k_2) = \sum_{n_1=0}^{(N/2)-1} \sum_{n_2=0}^{(N/2)-1} f(2n_1, + 1, 2n_2 + 1) W_N^{2k_1 n_1} W_N^{2k_2 n_2}$$

Since F_{ee}, F_{oe}, F_{eo}, and F_{oo} are periodic with periods $N/2$, it can readily be shown that

$$F\left(k_1 + \frac{N}{2}, k_2\right) = F_{ee}(k_1, k_2) - W_N^{k_1} F_{oe}(k_1, k_2)$$
$$+ W_N^{k_2} F_{eo}(k_1, k_2) - W_N^{k_1+k_2} F_{oo}(k_1, k_2) \quad (12.15)$$

$$F\left(k_1, k_2 + \frac{N}{2}\right) = F_{ee}(k_1, k_2) + W_N^{k_1} F_{oe}(k_1, k_2)$$
$$- W_N^{k_2} F_{eo}(k_1, k_2) - W_N^{k_1+k_2} F_{oo}(k_1, k_2) \quad (12.16)$$

$$F\left(k_1 + \frac{N}{2}, k_2 + \frac{N}{2}\right) = F_{ee}(k_1, k_2) - W_N^{k_1} F_{oe}(k_1, k_2)$$
$$- W_N^{k_2} F_{eo}(k_1, k_2) - W_N^{k_1+k_2} F_{oo}(k_1, k_2) \quad (12.17)$$

Equations (12.14)–(12.17) can be represented by the "radix-(2 × 2) butterfly" flow graph of Fig. 12.1.

Since $N/2$ is a power of 2, F_{ee}, F_{oe}, F_{eo} and F_{oo} can be expressed in terms of $(N/4)$-point DFTs using the same approach. By repeating this decimation process $\log_2 N$ times, only a number of 2 × 2 DFTs need to be performed, which do not require complex multiplications. As an example, Fig. 12.2 shows a complete vector–radix FFT for a 4 × 4 array. Note that each cycle of the FFT entails $N^2/4$ butterflies and each butterfly needs three multiplications and eight additions. Consequently, the total number of multiplications required in a radix-(2 × 2) FFT is $(3N^2/4) \log_2 N$. The use of the preceding 2-D FFT for the computation of the DFT of a 512 × 512 signal would require only 1.77×10^6 multiplications, which represents a 25 percent reduction in the number of multiplications relative to that required by the row–column decomposition approach.

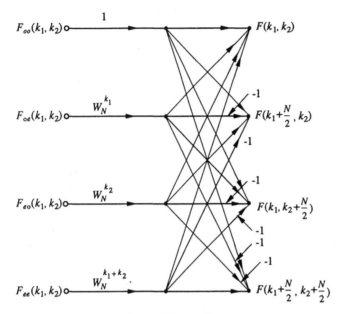

Figure 12.1 A radix-(2×2) butterfly.

12.4 SYSTOLIC IMPLEMENTATION

Although a software implementation is entirely satisfactory for many applications, the highest processing rate that can be achieved is limited by the use of general-purpose hardware. Consequently, in applications where a software implementation cannot provide the required processing rate, an implementation in terms of specialized hardware should be considered. Of the numerous structures that have been proposed for the realization of 2-D digital filters (see Chap. 10), systolic structures appear to offer some significant advantages when it comes to VLSI implementation. In this section, some basic systolic structures for the implementation of 1-D nonrecursive filters are first described and are then applied for the implementation of 2-D digital filters.

12.4.1 Systolic Implementation of 1-D Nonrecursive Filters

Let

$$H(z) = \sum_{m=0}^{N} h_m z^{-m} \qquad (12.18)$$

be the transfer function of a 1-D nonrecursive digital filter. The output of the filter in response to input signal $x(n)$, $n = 0, 1, \ldots$ is given by the

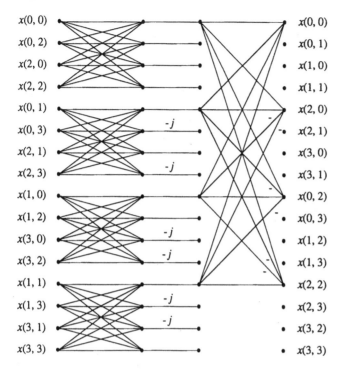

Figure 12.2 A vector–radix FFT for a 4 × 4 2-D array.

convolution

$$y(n) = \sum_{m=0}^{N} h_m x(n - m) \tag{12.19}$$

There are a number of realizations for transfer function $H(z)$ that can be transformed directly into corresponding systolic structures. A realization of this type, which can readily be obtained by using the direct realization method [3], is illustrated in Fig. 12.3a. This realization can easily be rearranged into a pipelined structure comprising a number of cascade processing elements (PEs) as shown in Fig. 12.3b. The unit delay in the rightmost PE and the adder in the leftmost PE are included in order to achieve a regular structure with identical PEs. An obvious disadvantage of this structure is that, unlike the N multiplications, which can be performed concurrently, the N additions have to be carried out sequentially. Therefore, the processing time required is $\tau_m + N\tau_a$, where τ_m and τ_a denote the time needed to perform one multiplication and one addition, respectively.

The processing rate in the preceding structure can be improved by adding a pair of unit delays in each PE, as shown in Fig. 12.4. The addition of

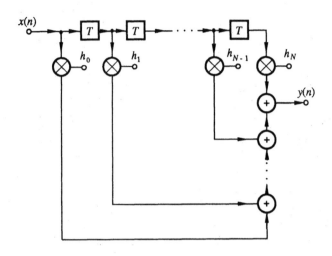

(a)

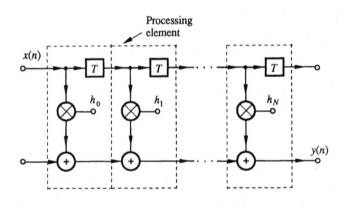

(b)

Figure 12.3   (a) A direct realization of $H(z)$; (b) corresponding pipelined structure.

the N delay pairs will delay the signals in the top and bottom paths in Fig. 12.3b by N sampling periods without destroying their relative timing. Consequently, the output will be delayed by N sampling periods. In other words, if $y_d(n)$ is the output of the modified structure and $y(n)$ is the output of the original structure, then $y_d(n) = y(n - N)$. It is noted that in the modified structure only one multiplication and one addition are required

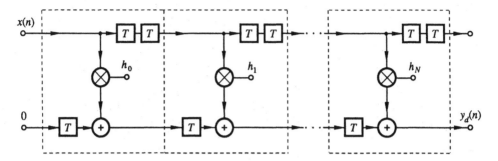

Figure 12.4 A systolic structure based on the structure of Fig. 12.3(b).

during each digital-filter cycle. Hence the processing time is now independent of the order of the filter and is given by $\tau_m + \tau_a$.

An alternative direct realization of $H(z)$ is depicted in Fig. 12.5a [3]. This can be rearranged as a pipelined structure, as shown in Fig. 12.5b. A drawback of this structure is that the input signal must be communicated to all the PEs simultaneously. When the order of the filter is high, the propagation delays due to long signal paths may decrease the processing rate significantly. Global communication can be eliminated by using padding delays, as depicted in Fig. 12.5c.

12.4.2 Systolic Implementation of 1-D Recursive Filters

Consider now the systolic implementation of 1-D recursive digital filters characterized by the transfer function

$$H(z) = \frac{\sum_{i=0}^{N} a_i z^{-i}}{\sum_{i=0}^{N} b_i z^{-i}}, \qquad b_0 = 1 \qquad (12.20)$$

The output of the filter in response to input signal $x(n)$, $n = 0, 1, \ldots$ is given by

$$y(n) = \sum_{i=0}^{N} a_i x(n - i) - \sum_{i=1}^{N} b_i y(n - i) \qquad (12.21)$$

Equation (12.21) implies that output $y(n)$ can be obtained by computing two convolutions in which the N most recent input and output values are used. Therefore, systolic structures for the implementation of $H(z)$ can be obtained by properly combining the structures discussed in Sec. 12.4.1. Figure 12.6a shows one such combination [6]. In this structure, two types of PEs are used, as depicted in Fig. 12.6b. The implementation requires totally $2N$ adders, $2N$ multipliers, and $2N$ unit delays, and the minimum

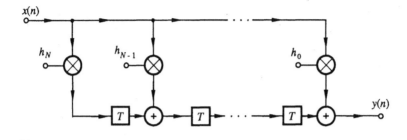

(a)

Figure 12.5 (a) A direct realization of $H(z)$; (b) corresponding pipelined structure; (c) systolic structure.

cycle time is $\tau_m + 2\tau_a$ as the multiplications in each PE can be performed in parallel.

12.4.3 Systolic Implementation of 2-D Filters

Consider a nonrecursive 2-D filter with a transfer function given by

$$H(z_1, z_2) = \sum_{n_1=0}^{N_1-1} \sum_{n_2=0}^{N_2-1} h(n_1, n_2) z_1^{-n_1} z_2^{-n_2} \tag{12.22}$$

From Sec. 10.3.1, it follows that the filter can be realized by first realizing r_c pairs of 1-D nonrecursive transfer functions $L_i(z_1)$ and $U_i(z_2)$ and then connecting pairs of cascaded 1-D subfilters in parallel, as depicted in Fig. 10.4. Parameter r_c is the rank of the coefficient matrix of the transfer function, and

$$L_i(z_1) = \sum_{k=i}^{N_1} l_{k,i} z_1^{-(k-1)} \tag{12.23}$$

$$U_i(z_2) = \sum_{k=i}^{N_2} u_{k,i} z_2^{-(k-1)} \tag{12.24}$$

A systolic implementation of the 2-D filter can be obtained by realizing the preceding transfer functions in terms of the systolic structures discussed in Sec. 12.4.1. It is noted that the ith pair of the 1-D transfer functions given by Eqs. (12.23) and (12.24) can be written as

$$L_i(z_1) = z_1^{-(i-1)} \sum_{k=0}^{N_1-i} l_{k+i,i} z_1^{-k} \tag{12.25}$$

$$U_i(z_2) = z_2^{-(i-1)} \sum_{k=0}^{N_2-i} u_{i,k+i} z_2^{-k} \tag{12.26}$$

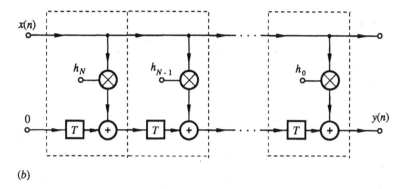

(b)

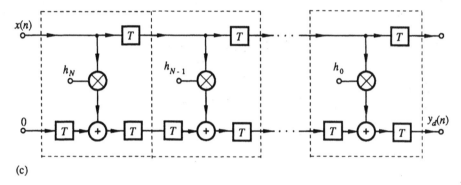

(c)

respectively and, therefore, the implementation of $L_i(z_1)$ and $U_i(z_2)$ requires only $N_j - i$ cells plus $i - 1$ unit-delay elements for $j = 1$ and 2, respectively.

The systolic implementation obtained above can be used as a building block for the implementation of 2-D recursive filters. The transfer function

$$H(z_1, z_2) = \frac{\sum_{i=0}^{N_1} \sum_{j=0}^{N_2} a_{ij} z_1^{-i} z_2^{-j}}{\sum_{i=0}^{N_1} \sum_{j=0}^{N_2} b_{ij} z_1^{-i} z_2^{-j}}, \qquad b_{00} = 1$$

can be written as

$$H(z_1, z_2) = \frac{\overline{N}(z_1^{-1}, z_2^{-1})}{1 + \tilde{D}(z_1^{-1}, z_2^{-1})}$$

where $\overline{N}(z_1^{-1}, z_2^{-1})$ and $\tilde{D}(z_1^{-1}, z_2^{-1})$ are given by Eqs. (10.25) and (10.26), respectively, and hence the recursive filter can be realized using one of the two-block feedback schemes depicted in Fig. 10.9a and b.

An alternative approach to the implementation of recursive 2-D filters is to use a systolic implementation for 1-D recursive filters as a building

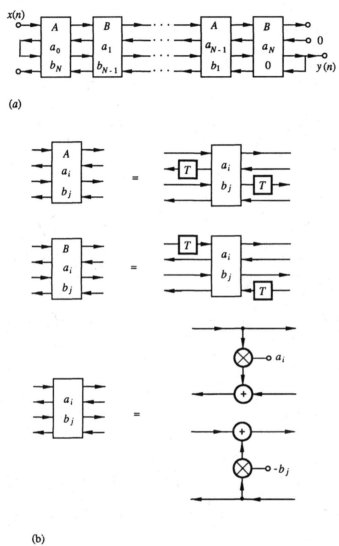

(a)

(b)

Figure 12.6 (a) Systolic implementation for 1-D recursive filters; (b) structures of processing elements A and B.

block. This approach is especially useful if the 2-D filter under consideration is quadrantally symmetric [7], as is demonstrated below.

It can be shown that the transfer function of a quadrantally symmetric filter has a separable denominator (see Ref. 6, Chap. 8). If we apply an LU decomposition to its numerator, then it follows from Sec. 10.3.1 that

$$\sum_{i=0}^{N_1} \sum_{j=0}^{N_2} a_{ij} z_1^{-i} z_2^{-j} = \sum_{k=0}^{r_a} L_k(z_1) U_k(z_2)$$

where $r_a = \text{rank}(A)$ and

$$A = \begin{bmatrix} a_{00} & a_{01} & \cdots & a_{0N_2} \\ a_{10} & a_{11} & \cdots & a_{1N_2} \\ \vdots & \vdots & & \vdots \\ a_{N_10} & a_{N_11} & \cdots & a_{N_1N_2} \end{bmatrix}$$

Hence the transfer function can be expressed as

$$H(z_1, z_2) = \frac{\sum_{k=0}^{r_a} L_k(z_1) U_k(z_2)}{d_1(z_1) d_2(z_2)} = \sum_{k=0}^{r_a} f_k(z_1) g_k(z_2)$$

where

$$f_k(z_1) = \frac{L_k(z_1)}{d_1(z_1)} \quad \text{and} \quad g_k(z_2) = \frac{U_k(z_2)}{d_2(z_2)}$$

represent the kth pair of recursive 1-D subfilters. Consequently, a systolic implementation of $H(z_1, z_2)$ can be constructed by connecting r_a pairs of cascaded 1-D systolic structures in parallel.

REFERENCES

1. D. E. Dudgeon and R. M. Mersereau, *Multidimensional Digital Signal Processing,* Englewood Cliffs, N.J.: Prentice-Hall, 1984.
2. J. S. Lim and A. V. Oppenheim (eds.), *Advanced Topics in Signal Processing,* Englewood Cliffs, N.J.: Prentice-Hall, 1988.
3. A. Antoniou, *Digital Filters: Analysis and Design,* New York: McGraw-Hill, 1979 (2nd ed. in press).
4. A. V. Oppenheim and R. W. Schafer, *Discrete-Time Signal Processing,* Englewood Cliffs, N.J.: Prentice-Hall, 1989.
5. H. T. Kung, Why systolic architectures?, *IEEE Computer Mag.,* vol. 15, pp. 37–46, Jan. 1982.
6. W. Luk and G. Jones, Systolic recursive filters, *IEEE Trans. Circuits Syst.,* vol. 35, pp. 1067–1068, Aug. 1988.
7. W.-S. Lu and E. B. Lee, An efficient implementation scheme for quadrantally symmetric 2-D digital filters, *IEEE Tran. Circuits Syst.,* vol. 35, pp. 239–241, Feb. 1983.

PROBLEMS

12.1 Prove Theorems 12.1–12.3.

12.2 Prove Theorems 12.4–12.6.

12.3 Demonstrate the computational efficiency of the row–column decomposition for the 2-D DFT by producing a table listing the total numbers of multiplications and additions required for the 2-D DFT of an $N \times N$ signal versus N where N varies from 2^6 to 2^{10}.

12.4 Repeat Problem 12.3 for the vector-radix FFT approach.

12.5 A nonrecursive 2-D filter can be implementated by using the SVD–LUD realization method (see Sec. 10.3.2) in conjunction with 1-D DFT. Compare the computational efficiency of this approach to those of the row–column decomposition approach and the vector-radix FFT approach.

12.6 Find a systolic implementation for the transfer function $H(z_1, z_2) = \mathbf{z}_1^T \mathbf{C} \mathbf{z}_2$, where $\mathbf{z}_i = [1 \quad z_i^{-1} \cdots z_i^{-4}]$, $i = 1, 2$, for the following cases:

$$(a)\ \mathbf{C} = \begin{bmatrix} 0.0224 & 0.2186 & 0.6043 & 0.2186 & 0.0224 \\ 0.3212 & 0.4143 & 0.5258 & 0.4143 & 0.3212 \\ 0.5051 & 0.6137 & 0.7045 & 0.6137 & 0.5051 \\ 0.3212 & 0.4143 & 0.5258 & 0.4143 & 0.3212 \\ 0.0224 & 0.2186 & 0.6043 & 0.2186 & 0.0224 \end{bmatrix}.$$

$$(b)\ \mathbf{C} = \begin{bmatrix} -0.0414 & 0.0117 & 0.2217 & 0.0117 & -0.0414 \\ 0.2072 & 0.3114 & 0.4792 & 0.3114 & 0.2072 \\ 0.4558 & 0.6111 & 0.7367 & 0.6111 & 0.4558 \\ 0.2072 & 0.3114 & 0.4792 & 0.3114 & 0.2072 \\ -0.0414 & 0.0117 & 0.2217 & 0.0117 & -0.0414 \end{bmatrix}.$$

12.7 Obtain a systolic implementation for each of the following transfer functions:

(a) $H(z_1, z_2) = \dfrac{z_1^2 z_2^2}{2z_1^2 z_2^2 + z_1^2 z_2 + z_1 z_2^2 + 0.4z_1 + 0.5z_2 + 0.5}.$

(b) $H(z_1, z_2)$

$$= 0.00895\ \dfrac{[1 \quad z_1^{-1} \quad z_1^{-2}] \begin{bmatrix} 1.0 & -1.6215 & 1.0 \\ -1.6215 & 2.6370 & 1.6213 \\ 1.0 & 1.6213 & 1.0 \end{bmatrix} \begin{bmatrix} 1 \\ z_2^{-1} \\ z_2^{-2} \end{bmatrix}}{[1 \quad z_1^{-1} \quad z_1^{-2}] \begin{bmatrix} 1.0 & -1.7881 & 0.8293 \\ -1.7881 & 3.2064 & -1.4927 \\ 0.8293 & -1.4927 & 0.6982 \end{bmatrix} \begin{bmatrix} 1 \\ z_2^{-1} \\ z_2^{-2} \end{bmatrix}}.$$

12.8 Find a systolic implementation of the state-space filter described by Eq. (11.10) where

$$\mathbf{A} = \begin{bmatrix} 1.8890 & -0.9122 & \vdots & -1.0 & 0.0 \\ 1.0 & 0.0 & \vdots & 0.0 & 0.0 \\ \cdots & \cdots & + & \cdots & \cdots \\ 0.0277 & -0.0258 & \vdots & 1.8890 & 1.0 \\ -0.0258 & 0.0243 & \vdots & -0.9122 & 0.0 \end{bmatrix}$$

$$\mathbf{b} = [0.2191 \quad 0.0 \quad \vdots \quad -0.0289 \quad 0.0912]^T$$

$$\mathbf{c} = [0.2889 \quad -0.0912 \quad \vdots \quad -0.2191 \quad 0.0]$$

$$d = 0$$

13

Applications

13.1 INTRODUCTION

Two-dimensional digital filters have been used in many different areas in the past, ranging from the digital processing of satellite photographs, radar maps, and medical x-ray images to the planning of power systems [1–9]. In this chapter, some of the applications of linear 2-D digital filters in the areas of image and seismic data processing are briefly reviewed. In Sec. 13.2, two techniques for the enhancement of images are examined. In Sec. 13.3, a well-known constrained least-squares estimation approach [3] to image restoration is described. The approach is formulated in terms of the 2-D FFT and is, therefore, efficient. In Sec. 13.4, the application of 2-D fan filters for the processing of seismic data is discussed.

13.2 APPLICATIONS OF LINEAR 2-D FILTERS TO IMAGE ENHANCEMENT

Image enhancement is the process by which an image is manipulated or transformed in order to extract useful information from it, to improve the visual effect of the image, or to change it into some form that is more suitable for further analysis. For example, in robotics a vision system is often required simply because the determination of the geometric shape of the object to be handled by the robot is critical to successful robot motion. In this case, a highpass filter can be used to strengthen the contour of the object. In what follows, two image enhancement techniques based on the use of 2-D linear filters are described. Section 13.2.1 deals with a technique for the removal of noise from a digitized image by using lowpass filters. Section 13.2.2, on the other hand, deals with a technique for the enhancement of edges by using highpass filters [1].

13.2.1 Noise Removal

Transducer and transmission noise in an image tends to be spatially independent, and, therefore, its energy content is usually concentrated at frequencies higher than that of the image. Consequently, processing an image by using a lowpass filter tends to remove a large amount of the noise content without changing the image significantly.

The basic requirements imposed on the filter are as follows:

1. A record of the image is usually available. Hence the filter need not be causal.
2. Phase distortion is usually objectionable in image processing (see Ref. [6] of Chap. 6), and hence the filter should have zero or linear phase response.
3. The sum of the impulse-response values should, if possible, be equal to unity to prevent bias in the light intensity during the processing. This requirement ensures that the image background will not be affected by the noise removal process.

A 2-D nonrecursive filter of order (N_1, N_2) where N_1 and N_2 are odd can be represented by

$$y(n_1, n_2) = \sum_{k_1=-C_1}^{C_1} \sum_{k_2=-C_2}^{C_2} h(n_1 - k_1, n_2 - k_2)x(k_1, k_2) \quad (13.1)$$

where $C_1 = (N_1 - 1)/2$, $C_2 = (N_2 - 1)/2$ and $\{x(n_1, n_2): 0 \le n_1 \le M_1 - 1, 0 \le n_2 \le M_2 - 1\}$ represents a discretized version of the image to be processed. The quantity $x(n_1, n_2)$ is a measure of the light intensity at coordinate $(n_1 T_1, n_2 T_2)$ of the image. The transfer function of the filter is given by

$$H(z_1, z_2) = \sum_{n_1=-C_1}^{C_1} \sum_{n_1=-C_2}^{C_2} h(n_1, n_2)z_1^{-n_1}z_2^{-n_2} \quad (13.2)$$

and if

$$h(n_1, n_2) = h(-n_1, -n_2)$$
$$\text{for } -C_1 \le n_1 \le C_1, \ -C_2 \le n_2 \le C_2 \tag{13.3}$$

a zero phase response is achieved (see Sec. 6.3.1).

The impulse responses of three 3×3 nonrecursive filters that are commonly used for noise removal are as follows [1].

$$\mathbf{h}_1 = \{h(n_1, n_2): n_1 = -1, 0, 1, n_2 = 1, 0, -1\} = \frac{1}{9}\begin{bmatrix} 1 & 1 & 1 \\ 1 & 1 & 1 \\ 1 & 1 & 1 \end{bmatrix} \tag{13.4}$$

$$\mathbf{h}_2 = \{h(n_1, n_2): n_1 = -1, 0, 1, n_2 = 1, 0, -1\} = \frac{1}{10}\begin{bmatrix} 1 & 1 & 1 \\ 1 & 2 & 1 \\ 1 & 1 & 1 \end{bmatrix} \tag{13.5}$$

$$\mathbf{h}_3 = \{h(n_1, n_2): n_1 = -1, 0, 1, n_2 = 1, 0, -1\} = \frac{1}{16}\begin{bmatrix} 1 & 2 & 1 \\ 2 & 4 & 2 \\ 1 & 2 & 1 \end{bmatrix} \tag{13.6}$$

These filters satisfy Eq. (13.3) and, therefore, have zero phase response, as required. In addition

$$\sum_{n_1=-C_1}^{C_1} \sum_{n_2=-C_2}^{C_2} h(n_1, n_2) = 1 \tag{13.7}$$

that is, requirement 3 above is satisfied. The amplitude response of the filter represented by Eq. (13.6) is shown in Fig. 13.1. The application of this filter for the removal of noise is illustrated in Fig. 13.2.

An alternative approach to image smoothing is to use the DFT in conjunction with analog-filter approximations [2]. A normalized Butterworth lowpass filter of order N has the transfer function

$$H(s) = \frac{1}{D(s)}$$

where $D(s)$ is the Nth-order stable polynomial given by

$$D(s)D(-s) = 1 + s^{2N}$$

(see Chap. 5 of [10]). We can write

$$|H(j\omega)|^2 = \frac{1}{1 + \omega^{2N}}$$

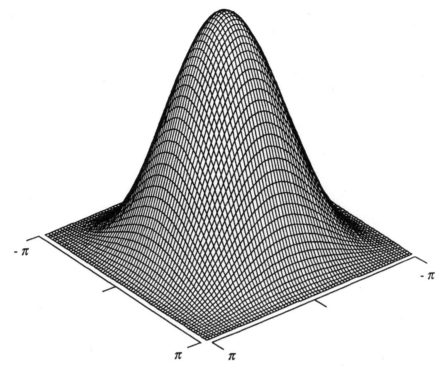

Figure 13.1 Amplitude response of filter represented by Eq. (13.6).

and hence a 2-D zero-phase frequency response can be obtained as

$$H_L(\omega_1, \omega_2) = \frac{1}{1 + c[D(\omega_1, \omega_2)/\omega_c]^{2N}}$$

where

$$D(\omega_1, \omega_2) = \sqrt{(\omega_1 T_1)^2 + (\omega_2 T_2)^2} \qquad (13.8)$$

Constants c and ω_c depend on the desired cutoff frequency. If $T_1 = T_2 = 1$ and $c = \sqrt{2} - 1$, then ω_c is the 3-dB cutoff frequency of the filter and $H_L(\omega_1, \omega_2)$ can be expressed as

$$H_L(\omega_1, \omega_2) = \frac{\omega_c^{2N}}{\omega_c^{2N} + 0.4142(\omega_1^2 + \omega_2^2)^N} \qquad (13.9)$$

In this way, a 2-D lowpass circularly symmetric frequency response is achieved. Figure 13.3 shows a 3-D plot of $H_L(\omega_1, \omega_2)$ for the case where $N = 4$ and $\omega_c = 2.0$ rad/s. If $X(k_1, k_2)$ is the discrete Fourier transform

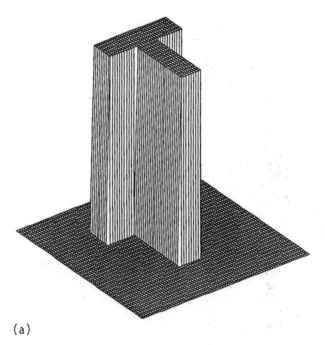

(a)

Figure 13.2 (a) Ideal image; (b) image of (a) contaminated by high-frequency noise; (c) the processed image.

(DFT) of the original noise-contaminated image and $H_L(k_1, k_2) = H_L(k\Omega_1, k\Omega_2)$ where $\Omega_1 = 2\pi/T_1$ and $\Omega_2 = 2\pi/T_2$, then the filtered image can be obtained as the inverse DFT of $H_L(k_1, k_2)X(k_1, k_2)$ as

$$y(n_1, n_2) = \mathcal{F}^{-1}[H_L(k_1, k_2)X(k_1, k_2)]$$

Figure 13.4 shows the output obtained by processing the image of Fig. 13.2b with $N = 4$ and $\omega_c = 2$ rad/s.

Note that if the processing is to be carried out by using the DFT, the frequency response of the filter need not correspond to a *rational* transfer function since it is unnecessary to realize the filter in terms of unit delays, adders, and multipliers. As a matter of fact, it is not possible to find a rational transfer function whose frequency response is given by Eq. (13.9).

13.2.2 Edge Enhancement

Many image-related applications such as pattern recognition and machine vision require images of objects with enhanced edges. In an optical image, the pixel intensity changes rapidly across edges, which means that the frequency content due to edges tends to be concentrated at the high end

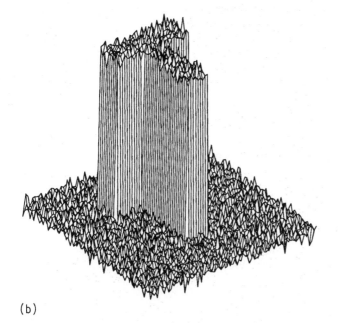

(b)

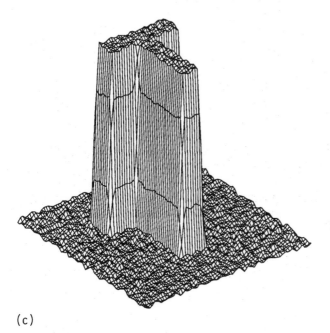

(c)

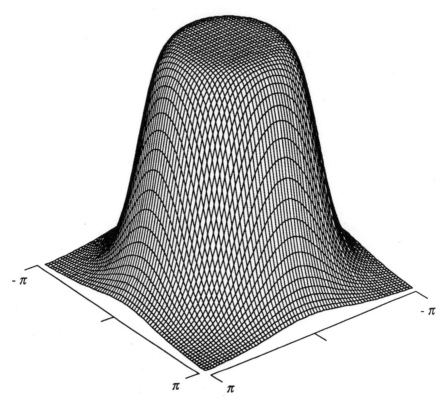

Figure 13.3 Amplitude response of filter represented by Eq. (13.9) ($n = 4$ and $\omega_c = 2$ rad/s).

of the spectrum. As a consequence, edge enhancement can be achieved by using highpass filters.

Three 3×3 nonrecursive filters that have been used for edge enhancement in the past [1] are characterized by the following impulse responses:

$$\mathbf{h}_1 = \{h(n_1, n_2): n_1 = -1, 0, 1, n_2 = 1, 0, -1\}$$

$$= \begin{bmatrix} 0 & -1 & 0 \\ -1 & 5 & -1 \\ 0 & -1 & 0 \end{bmatrix} \tag{13.10}$$

$$\mathbf{h}_2 = \{h(n_1, n_2): n_1 = -1, 0, 1, n_2 = 1, 0, -1\}$$

$$= \begin{bmatrix} -1 & -1 & -1 \\ -1 & 9 & -1 \\ -1 & -1 & -1 \end{bmatrix} \tag{13.11}$$

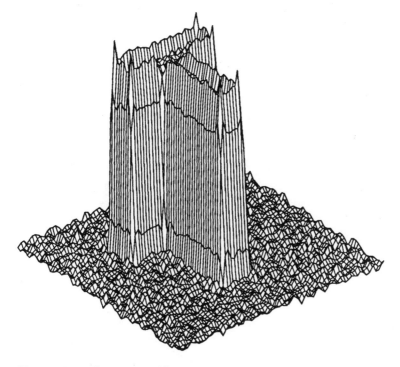

Figure 13.4 The processed image.

$$\mathbf{h}_3 = \{h(n_1, n_2)\colon n_1 = -1, 0, 1, n_2 = 1, 0, -1\}$$

$$= \begin{bmatrix} 1 & -2 & 1 \\ -2 & 5 & -2 \\ 1 & -2 & 1 \end{bmatrix} \tag{13.12}$$

Like the lowpass filters of Sec. 13.2.1, the preceding highpass filters have zero phase and satisfy the constraint in Eq. (13.7). Figure 13.5 shows the amplitude response of the filter represented by Eq. (13.10). To demonstrate the effect of a highpass filter on an image, the filter represented by Eq. (13.10) was used to process the image of Fig. 13.6a. The image obtained, depicted in Fig. 13.6b, demonstrates that edges are enhanced.

As in the case of noise elimination, edge enhancement can be accompished by using the DFT in conjunction with analog-filter approximations. A highpass zero-phase frequency response can be obtained from the Butterworth approximation as [2]:

$$H_H(\omega_1, \omega_2) = \begin{cases} \dfrac{1}{1 + c[\omega_c/D(\omega_1, \omega_2)]^{2N}} & \text{if } D(\omega_1, \omega_2) \neq 0 \\ 0 & \text{otherwise} \end{cases}$$

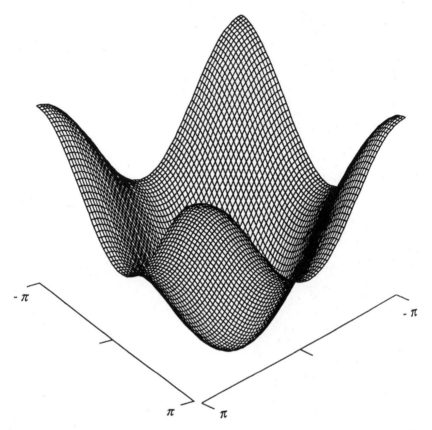

Figure 13.5 Amplitude response of highpass filter represented by Eq. (13.10).

where $D(\omega_1, \omega_2)$ is given by Eq. (13.8) and c and ω_c depend on the desired cutoff frequency. If $T_1 = T_2 = 1$ and $c = \sqrt{2} - 1$, then ω_c is the 3dB cutoff frequency and $H_H(\omega_1, \omega_2)$ can be expressed as

$$H_H(\omega_1, \omega_2) = \frac{(\omega_1^2 + \omega_2^2)^N}{0.4142\omega_c^{2N} + (\omega_1^2 + \omega_2^2)^N} \tag{13.13}$$

The frequency response achieved is depicted in Fig. 13.7.

As in Sec. 13.2.1, we first compute the DFT of the image $X(k_1, k_2)$, then multiply it by the sampled frequency response $H_H(k_1, k_2)$, and then compute the inverse DFT of the product to obtain the processed image

$$y(n_1, n_2) = \mathscr{F}^{-1}[H_H(k_1, k_2)X(k_1, k_2)]$$

Figure 13.8 shows how the edges in the image of Fig. 13.6a are enhanced by using a Butterworth highpass filter with $n = 4$ and $\omega_c = 2$ rad/s. The

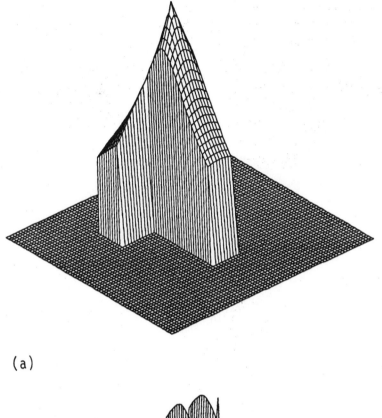

(a)

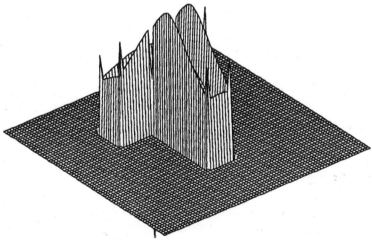

(b)

Figure 13.6   (a) The original image; (b) the processed image.

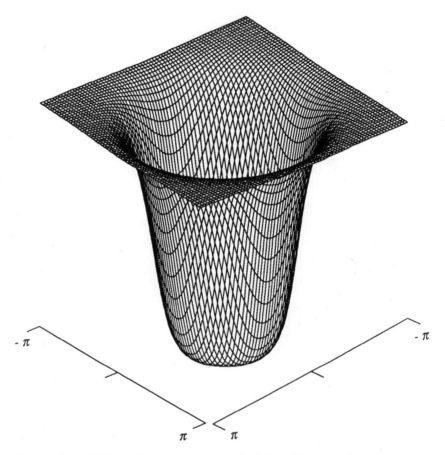

$-\pi$ $-\pi$

π π

Figure 13.7 Highpass frequency response based on Butterworth approximation ($n = 4$ and $\omega_c = 2$ rad/s).

spikes are due to the increased high frequency content which is brought about by the steep transitions at the edges.

13.3 IMAGE RESTORATION

Images are often corrupted by bandwidth reduction during the formation of the image, nonlinearity of the recording medium (e.g., photographic film), and noise introduced during the transmission, recording, measurement, digitization, etc. [4]. The purpose of image restoration is to improve a recorded image according to some norm or criterion. In this section, we

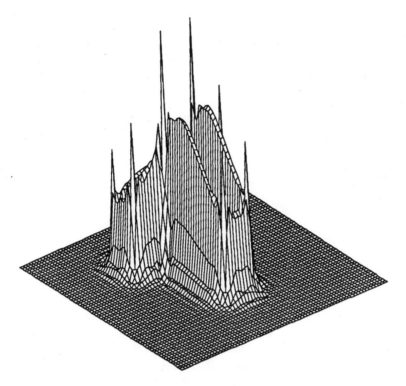

Figure 13.8 The processed version of the image in Fig. 13.6a.

present a constrained least-squares estimation approach to the image restoration problem proposed by Hunt [3].

13.3.1 The Image Restoration Problem

The basic model that describes the process of image formation can be presented as [4]

$$g(x, y) = \mathscr{S}\left[\int_{-\infty}^{\infty} \int_{-\infty}^{\infty} h(x, y; s, t)f(s, t) \, ds \, dt\right] + n(x, y)$$

where $f(x, y)$ represents the ideal image, $h(x, y; s, t)$ is the point-spread function of the image formation system, $\mathscr{S}[\cdot]$ represents the sensor nonlinearity, $n(x, y)$ is a noise term, and $g(x, y)$ is the recorded image. If the image formation system is stationary across the image and the sensor nonlinearity is negligible, then the image model becomes

$$g(x, y) = \int_{-\infty}^{\infty} \int_{-\infty}^{\infty} h(x - s, y - t)f(s, t) \, ds \, dt + n(x, y) \quad (13.14)$$

If the image $f(x, y)$ is defined over the rectangle $R_f = \{0 \leq x \leq A, 0 \leq y \leq B\}$ and the point-spread function can be well represented by its values over the rectangle $R_h = \{0 \leq x \leq C, 0 \leq y \leq D\}$, then the recorded image $g(x, y)$ is a finite-extent, first-quadrant 2-D signal with the region of support $R_g = \{0 \leq x \leq A + C, 0 \leq y \leq B + D\}$. A discrete version of the preceding process can be obtained by approximating the integral in Eq. (13.14) by a double summation. In this way, the light intensity of the recorded image at point $(n_1\Delta x, n_2\Delta y)$ can be evaluated as

$$g(n_1\Delta x, n_2\Delta y) = \Delta x \Delta y \sum_{m_1=0}^{n_1} \sum_{m_2=0}^{n_2} h[(n_1 - m_1)\Delta x, (n_2 - m_2)\Delta y]$$

$$\times f(m_1\Delta x, m_2\Delta y) + n(n_1\Delta x, n_2\Delta y)$$

that is,

$$g(n_1, n_2) = \sum_{m_1=0}^{n_1} \sum_{m_2=0}^{n_2} h(n_1 - m_1, n_2 - m_2)f(m_1, m_2)$$

$$+ n(n_1, n_2) \qquad (13.15)$$

where $g(n_1, n_2) \equiv g(n_1\Delta x, n_2\Delta y)$, $h(n_1 - m_1, n_2 - m_2) \equiv h[(n_1 - m_1)\Delta x, (n_2 - m_2)\Delta y] \Delta x \Delta y$, $n(n_1, n_2) \equiv n(n_1\Delta x, n_2\Delta y)$. If $A = (M_1 - 1)\Delta x$, $B = (M_2 - 1)\Delta y$, $C = (N_1 - 1)\Delta x$, and $D = (N_2 - 1)\Delta y$, then Eq. (13.14) implies that $\{g(n_1, n_2)\}$ and $\{n(n_1, n_2)\}$ are matrices of dimension $(M_1 + N_1 - 1) \times (M_2 + N_2 - 1)$. Now define the augmented matrices $\{f_e(m_1, m_2)\}$ and $\{h_e(n_1, n_2)\}$ as

$f_e(m_1, m_2)$

$$= \begin{cases} f(m_1, m_2) & \text{for } 0 \leq m_1 \leq M_1 - 1 \text{ and } 0 \leq m_2 \leq M_2 - 1 \\ 0 & \text{for } M_1 \leq m_1 \leq P_1 - 1 \text{ or } M_2 \leq m_2 \leq P_2 - 1 \end{cases}$$

$$h_e(n_1, n_2) = \begin{cases} h(n_1, n_2) & \text{for } 0 \leq n_1 \leq N_1 - 1 \text{ and } 0 \leq n_2 \leq N_2 - 1 \\ 0 & \text{for } N_1 \leq n_1 \leq P_1 - 1 \text{ or } N_2 \leq n_2 \leq P_2 - 1 \end{cases}$$

where P_1 and P_2 are integers satisfying the conditions $P_1 \geq M_1 + N_1 - 1$ and $P_2 \geq M_2 + N_2 - 1$. In addition, define three column vectors $\mathbf{f}$, $\mathbf{g}$, and $\mathbf{n}$ by lexicographically ordering the matrices $\{f_e(m_1, m_2)\}$, $\{g(n_1, n_2)\}$, and $\{n(n_1, n_2)\}$ as

$$\mathbf{f} = [\mathbf{f}_{e0} \ \mathbf{f}_{e1} \ldots \mathbf{f}_{e,P_1-1}]^T$$

$$\mathbf{g} = [\mathbf{g}_0 \ \mathbf{g}_1 \ldots \mathbf{g}_{P_1-1}]^T$$

$$\mathbf{n} = [\mathbf{n}_0 \ \mathbf{n}_1 \ldots \mathbf{n}_{P_1-1}]^T$$

where $\mathbf{f}_{ei}$, $\mathbf{g}_i$, $\mathbf{n}_i$ are the ith row vector of matrices $\{f_e(m_1, m_2)\}$, $\{g(n_1, n_2)\}$, and $\{n(n_1, n_2)\}$, respectively. Under these circumstances, Eq. (13.14) can be expressed in a vector-matrix form as

$$\mathbf{g} = \mathbf{Hf} + \mathbf{n} \tag{13.16}$$

where $\mathbf{H}$ is a matrix of dimension $P_1 P_2 \times P_1 P_2$ given by

$$\mathbf{H} = \begin{bmatrix} \mathbf{H}_0 & \mathbf{H}_{P_1-1} & \cdots & \mathbf{H}_1 \\ \mathbf{H}_1 & \mathbf{H}_0 & \cdots & \mathbf{H}_2 \\ \vdots & \vdots & & \vdots \\ \mathbf{H}_{P_1-1} & \mathbf{H}_{P_1-2} & \cdots & \mathbf{H}_0 \end{bmatrix} \tag{13.17}$$

with each submatrix $\mathbf{H}_j$ formed by the jth row of matrix $\{h_e(n_1, n_2)\}$ as a *circulant* matrix

$$\mathbf{H}_j = \begin{bmatrix} h_e(j, 0) & h_e(j, P_2 - 1) & \cdots & h_e(j, 1) \\ h_e(j, 1) & h_e(j, 0) & \cdots & h_e(j, 2) \\ \vdots & \vdots & & \vdots \\ h_e(j, P_2 - 1) & h_e(j, P_2 - 2) & \cdots & h_e(j, 0) \end{bmatrix} \tag{13.18}$$

In terms of the preceding formulation, the image restoration problem can be stated as: Given an image formation system charactlerized by matrix $\mathbf{H}$ and a recorded image represented by vector $\mathbf{g}$, find the column vector $\hat{\mathbf{f}}$ that optimally approximates the representation of the original image $\mathbf{f}$ under some norm or criterion.

There are two problems in image restoration that need to be carefully addressed. The first is associated with the ill-posedness of the image restoration process. As was stated earlier, Eq. (13.16) is a discrete approximation of the integral in Eq. (13.14), which is often called the integral equation of the first kind [11]. Solving Eq. (13.14) for a given $g(x, y)$ often leads to numerical difficulties because the unboundedness of the inverse of the integral operator associated with Eq. (13.14) tends to magnify the noise component $n(x, y)$ and, therefore, leads to an unacceptable numerical solution. As a result, $\mathbf{H}$ is always an ill-conditioned matrix; that is, the ratio of the largest to the smallest singular value is very large and, therefore, a small noise component $\mathbf{n}$ can lead to large restoration errors. Consequently, physically meaningful constraints need to be imposed on the approximate solution of Eq. (13.16) to reduce numerical instability. The second problem is directly related to the computational complexity associated with the solution of Eq. (13.16). The dimension of an image in most cases is at least 512×512. If the point-spread function of the degradation system can be represented by a matrix $\{h(n_1, n_2)\}$ of dimension 32×32, then the dimension of matrix $\mathbf{H}$ is $295{,}936 \times 295{,}936$. Obviously, solving such a large linear system of equations is unrealistic for most existing computer systems.

Solutions to the preceding problems have been obtained by Hunt in his classic work on this subject [3]. The first problem is solved by using a constrained least-squares estimation approach. The second problem is solved by noting that matrix **H** is block-circulant (i.e., **H** as a block matrix is circulant and each block matrix $\mathbf{H}_j$ is also circulant) and then using a frequency-domain approach. In this way, the required computation is carried out by computing the DFTs of several 2-D signals of dimension $P_1 \times P_2$. These solutions are described in some detail below.

13.3.2   The Constrained Least-Squares Estimation Approach

If **f** is the solution to Eq. (13.16), then **f** satisfies the equation

$$\|\mathbf{g} - \mathbf{Hf}\|^2 = \|\mathbf{n}\|^2 \tag{13.19}$$

The constrained least-squares estimation approach [3] yields an approximate solution $\hat{\mathbf{f}}$ for Eq. (13.16) that minimizes

$$\|\mathbf{C}\hat{\mathbf{f}}\|^2 \tag{13.20}$$

subject to the constraint

$$\|\mathbf{g} - \mathbf{H}\hat{\mathbf{f}}\|^2 = \|\mathbf{n}\|^2 \tag{13.21}$$

where **C** is a matrix of dimension $P_1 \times P_2$ representing a measure of smoothness. On comparing Eq. (13.21) with Eq. (13.19), we observe that Eq. (13.21) provides a reasonable constraint on the solution $\hat{\mathbf{f}}$. To identify matrix **C** in Eq. (13.20), the Laplacian of function $f(x, y)$ defined by

$$\nabla^2 f = \frac{\partial^2 f(x, y)}{\partial x^2} + \frac{\partial^2 f(x, y)}{\partial y^2}$$
$$\approx f(n_1 + 1, n_2) + f(n_1 - 1, n_2) + f(n_1, n_2 + 1)$$
$$+ f(n_1, n_2 - 1) - 4f(n_1, n_2) \tag{13.22}$$

with $f(n_1, n_2) \equiv f(n_1 \Delta x, n_2 \Delta y) = f(x, y)$ is employed as a measure of smoothness since $\Delta^2 f$ responds to transitions in image light intensity.

The discrete analogy of the Laplacian described by Eq. (13.22) can be realized by the 2-D convolution of $f(x, y)$ with signal $p(x, y)$ given by

$$p(n_1, n_2) = \begin{bmatrix} 0 & 1 & 0 \\ 1 & -4 & 1 \\ 0 & 1 & 0 \end{bmatrix}$$

To avoid wraparound error in the discrete convolution, the augmented matrix $p_e(n_1, n_2)$ is defined as

$$p_e(n_1, n_2) = \begin{cases} p(n_1, n_2) & 0 \le n_1 \le 2 \text{ and } 0 \le n_2 \le 2 \\ 0 & 3 \le n_1 \le P_1 - 1 \text{ or } 0 \le n_2 \le P_2 - 1 \end{cases}$$

where $P_1 \ge M_1 + 2$, $P_2 \ge M_2 + 2$, and $M_1 \times M_2$ is the dimension of $\{f(m_1, m_2)\}$. The discrete Laplacian can now be expressed as

$$\sum_{k_1=0}^{P_1-1} \sum_{k_2=0}^{P_2-1} f_e(k_1, k_2) p_e(n_1 - k_1, n_2 - k_2)$$

Proceeding as in Sec. 13.3.1, a block-circulant matrix $\mathbf{C}$ can be formed as

$$\mathbf{C} = \begin{bmatrix} \mathbf{C}_0 & \mathbf{C}_{P_1-1} & \cdots & \mathbf{C}_1 \\ \mathbf{C}_1 & \mathbf{C}_0 & \cdots & \mathbf{C}_2 \\ \vdots & \vdots & & \vdots \\ \mathbf{C}_{P_1-1} & \mathbf{C}_{P_1-2} & \cdots & \mathbf{C}_0 \end{bmatrix} \tag{13.23}$$

where each submatrix $\mathbf{C}_j$ is a $P_1 \times P_2$ circulant matrix generated by the jth row of matrix $\{p_e(n_1, n_2), 0 \le n_1 \le P_1 - 1, 0 \le n_2 \le P_2 - 1\}$, that is,

$$\mathbf{C}_j = \begin{bmatrix} p_e(j, 0) & p_e(j, P_2 - 1) & \cdots & p_e(j, 1) \\ p_e(j, 1) & p_e(j, 0) & \cdots & p_e(j, 2) \\ \vdots & \vdots & & \vdots \\ p_e(j, P_2 - 1) & p_e(j, P_2 - 2) & \cdots & p_e(j, 0) \end{bmatrix} \tag{13.24}$$

Having defined matrix $\mathbf{C}$ of Eq. (13.20), we can now form the objective function

$$J(\hat{\mathbf{f}}) = \|\mathbf{C}\hat{\mathbf{f}}\|^2 + \alpha(\|\mathbf{g} - \mathbf{H}\hat{\mathbf{f}}\|^2 - \|\mathbf{n}\|^2)$$

where α is the Lagrange multiplier. It is known from calculus that the vector $\hat{\mathbf{f}}$ that minimizes $\|\mathbf{C}\hat{\mathbf{f}}\|^2$ can be obtained by solving

$$\frac{\partial J(\hat{\mathbf{f}})}{\partial \hat{\mathbf{f}}} = 2\mathbf{C}^H\mathbf{C}\hat{\mathbf{f}} - 2\alpha\mathbf{H}^H(\mathbf{g} - \mathbf{H}\hat{\mathbf{f}}) = 0$$

for $\hat{\mathbf{f}}$, that is,

$$\hat{\mathbf{f}} = (\mathbf{H}^H\mathbf{H} + \gamma\mathbf{C}^H\mathbf{C})^{-1}\mathbf{H}^H\mathbf{g} \tag{13.25}$$

where $\gamma = 1/\alpha$ has to be chosen so that $\hat{\mathbf{f}}$ given by Eq. (13.25) satisfies constraint (13.21). To find such a γ, define scalar function $\phi(\gamma)$ as

$$\phi(\gamma) = \|\mathbf{g} - \mathbf{H}\hat{\mathbf{f}}\|^2$$

where $\hat{\mathbf{f}}$ is given by (13.25) so that it depends on γ. It can be shown [3] that $d\phi(\gamma)/d\gamma$ is nonnegative and, therefore, $\phi(\gamma)$ is monotonically increasing with respect to γ. Since the constraint in Eq. (13.21) can be expressed as

$$\phi(\gamma) = \|\mathbf{n}\|^2$$

the monotonicity property of $\phi(\gamma)$ leads immediately to the following algorithm for the computation of the optimizing $\hat{\mathbf{f}}$.

STEP 1:   Choose an initial value of γ that is greater than zero.

STEP 2:   Compute $\hat{\mathbf{f}}$ using Eq. (13.25) and compute $\phi(\gamma)$.

STEP 3:   If $\phi(\gamma) \in [\|n\|^2 - \varepsilon, \|n\|^2 + \varepsilon]$ where ε is a prescribed tolerance, then take output $\hat{\mathbf{f}}$ as the solution and stop.

STEP 4:   If $\phi(\gamma) < \|n\|^2 - \varepsilon$, increase γ using an appropriate algorithm such as the Newton–Raphson algorithm and return to Step 2.

STEP 5:   If $\phi(\gamma) > \|n\|^2 + \varepsilon$, decrease γ using an appropriate algorithm and return to Step 2.

13.3.3   Implementation of the Constrained Least-Squares Estimation

As can be seen from Eq. (13.25), the constrained least-square estimation of the image being restored requires the inversion of matrix $\mathbf{H}^T\mathbf{H} + \gamma\mathbf{C}^T\mathbf{C}$. The dimension of this matrix is very large, as was demonstrated in Sec. 13.3.1 (see p. 373), and its inversion is, therefore, not practical.

The key fact that enabled Hunt [3] to obtain a very efficient implementation of the algorithm is that matrices $\mathbf{H}$ and $\mathbf{C}$, defined by Eqs. (13.17)–(13.18) and (13.23)–(13.24), respectively, are block-circulant. Consider first a circulant matrix $\mathbf{S}$ of dimension $N \times N$ given by

$$\mathbf{S} = \begin{bmatrix} s(0) & s(N-1) & \cdots & s(1) \\ s(1) & s(0) & \cdots & s(2) \\ \vdots & \vdots & & \vdots \\ s(N-1) & s(N-2) & \cdots & s(0) \end{bmatrix}$$

It can readily be shown that

$$\mathbf{S}\mathbf{w}_{NK} = \lambda_{NK}\mathbf{w}_{NK} \qquad \text{for } k = 0, 1, \ldots, N-1 \qquad (13.26)$$

where

$$\mathbf{w}_{NK} = [1 \; w_N^k \; w_N^{2k} \cdots w_N^{(N-1)k}]^T \qquad (13.27)$$

$$w_N = e^{j2\pi/N}$$

$$\lambda_{NK} = \mathbf{s}_1\mathbf{w}_{Nk} \qquad (13.28)$$

and $\mathbf{s}_1$ denotes the first row of matrix $\mathbf{S}$. Equation (13.26) implies that $\mathbf{w}_{Nk}$

for $k = 0, 1, \ldots, N - 1$ are N eigenvectors of $\mathbf{S}$ with the corresponding eigenvalues λ_{Nk} for $k = 0, 1, \ldots, N - 1$, that is,

$$\mathbf{S} = \mathbf{W}_N \mathbf{D}_N \mathbf{W}_N^{-1}$$

where

$$\mathbf{W}_N = [\mathbf{w}_{N0} \ \mathbf{w}_{N1} \cdots \mathbf{w}_{N,N-1}] \tag{13.29}$$

$$\mathbf{D}_N = \begin{bmatrix} \lambda_{N0} & & & \mathbf{0} \\ & \lambda_{N1} & & \\ & & \ddots & \\ \mathbf{0} & & & \lambda_{NN-1} \end{bmatrix}$$

Note that the eigenvectors $\mathbf{w}_{Nk}$ specified by Eq. (13.27) are independent of matrix $\mathbf{S}$. In other words, $\mathbf{w}_{Nk}$ for $k = 0, 1, \ldots, N - 1$ can be regarded as the complete set of eigenvectors for *any* circulant matrix of dimension $N \times N$. In addition, it can easily be verified that

$$\mathbf{W}_N^{-1} = \frac{1}{N} \overline{\mathbf{W}}_N \tag{13.30}$$

where $\overline{\mathbf{W}}_N$ is the complex conjugate of $\mathbf{W}_N$. Furthermore, the eigenvalues λ_{Nk} given by Eq. (13.28) can be written as

$$\begin{aligned} \lambda_{Nk} &= s(0) + s(N - 1)e^{j2\pi k/N} + \cdots + s(1)e^{j2\pi(N-1)k/N} \\ &= s(0) + s(N - 1)e^{-j2\pi(N-1)k/N} + \cdots + s(1)e^{-j2\pi k/N} \\ &= \sum_{l=0}^{N-1} s(l)e^{-j2\pi lk/N} \\ &= S(k) \end{aligned} \tag{13.31}$$

where $S(k)$ is the kth component of the DFT of the signal $\{s(l): 0 \le l \le N - 1\}$.

Similarly, for the block-circular matrix $\mathbf{H}$ given by Eq. (13.17), one can find a transformation matrix $\mathbf{W}$ and a diagonal matrix $\mathbf{D}$ such that

$$\mathbf{H} = \mathbf{W}\mathbf{D}_h\mathbf{W}^{-1} \tag{13.32}$$

where

$$\mathbf{W} = \begin{bmatrix} \mathbf{W}_{00} & \mathbf{W}_{01} & \cdots & \mathbf{W}_{0,P_1-1} \\ \mathbf{W}_{10} & \mathbf{W}_{11} & \cdots & \mathbf{W}_{1,P_1-1} \\ \vdots & \vdots & & \vdots \\ \mathbf{W}_{P_1-1,0} & \mathbf{W}_{P_1-1,1} & \cdots & \mathbf{W}_{P_1-1,P_1-1} \end{bmatrix} \tag{13.33}$$

$$\mathbf{W}_{i,m} = w_{P_1}(i, m)\mathbf{W}_{P_2} \qquad (13.34)$$

$$\mathbf{W}_{P_2} = \{w_{P_2}(k, l)\} \quad k = 0, 1, \dots, P_2 - 1, \quad l = 0, 1, \dots, P_2 - 1 \qquad (13.35)$$

$$w_{P_1}(i, m) = e^{j2\pi i m/P_1} \qquad (13.36)$$

$$w_{P_2}(k, l) = e^{j2\pi k l/P_2} \qquad (13.37)$$

and the diagonal elements of matrix $\mathbf{D}_h$, namely, the eigenvalues of matrix $\mathbf{H}$, are given by

$$d_h(k, k) = H_e\left(\left[\frac{k}{P_2}\right], k \bmod P_2\right) \qquad \text{for } k = 0, 1, \dots, P_1 - 1 \qquad (13.38)$$

where H_e is the 2-D DFT of $h_e(n_1, n_2)$ defined in Sec. 13.3.1, $[k/P_2]$ denotes the biggest integer not exceeding k/P_2, and $k \bmod P_2$ is the remainder obtained by dividing k by P_2. It is observed that the transformation matrix $\mathbf{W}$ given by Eqs. (13.33) to (13.37) does not depend on the entries of matrix $\mathbf{H}$, that is, it can be used as the transformation matrix for diagonalizing *any* block-circulant matrices with the same block structure as matrix $\mathbf{H}$ of Eqs. (13.17) and (13.18). Furthermore, by using Eqs. (13.33) to (13.37), it can be shown that

$$\mathbf{W}^{-1} = \begin{bmatrix} \mathbf{U}_{00} & \mathbf{U}_{01} & \cdots & \mathbf{U}_{0,P_1-1} \\ \mathbf{U}_{10} & \mathbf{U}_{11} & \cdots & \mathbf{U}_{1,P_1-1} \\ \vdots & \vdots & & \vdots \\ \mathbf{U}_{P_1-1,0} & \mathbf{U}_{P_1-1,1} & \cdots & \mathbf{U}_{P_1-1,P_1-1} \end{bmatrix} \qquad (13.39)$$

where

$$\mathbf{U}_{i,m} = \frac{1}{P_1} w_{P_1}^{-1}(i, m)\mathbf{W}_{P_2}^{-1}$$

$$= \frac{1}{P_1} e^{-j2\pi i m/P_1}\mathbf{W}_{P_2}^{-1}$$

and

$$\mathbf{W}_{P_2}^{-1} = \left\{\frac{1}{P_2} e^{-j2\pi k l/P_2}\right\} \quad k = 0, 1, \dots, P_2 - 1, \quad l = 0, 1, \dots, P_2 - 1$$

Now let us consider the evaluation of the restored signal $\hat{\mathbf{f}}$. Equation (13.25) can be written as

$$(\mathbf{H}^H\mathbf{H} + \gamma\mathbf{C}^H\mathbf{C})\hat{\mathbf{f}} = \mathbf{H}^H\mathbf{g} \qquad (13.40)$$

where matrix **H** can be replaced by $\mathbf{WD}_h\mathbf{W}^{-1}$ and matrix **C** can be replaced by $\mathbf{WD}_c\mathbf{W}^{-1}$. Since

$$\mathbf{D}_h^H = \overline{\mathbf{D}}_h$$

$$\mathbf{D}_c^H = \overline{\mathbf{D}}_c$$

$$\mathbf{W}^H = \overline{\mathbf{W}} = P_1P_2\mathbf{W}^{-1}$$

Eq. (13.40) implies that

$$(\overline{\mathbf{D}}_h\mathbf{D}_h + \gamma\overline{\mathbf{D}}_c\mathbf{D}_c)\mathbf{W}^{-1}\hat{\mathbf{f}} = \overline{\mathbf{D}}_h\mathbf{W}^{-1}\mathbf{g}$$

that is,

$$\mathbf{W}^{-1}\hat{\mathbf{f}} = (\overline{\mathbf{D}}_h\mathbf{D}_h + \gamma\overline{\mathbf{D}}_c\mathbf{D}_c)^{-1}\overline{\mathbf{D}}_h\mathbf{W}^{-1}\mathbf{g}$$

which leads to

$$\hat{F}(k_1, k_2) = \left[\frac{\overline{H_e(k_1, k_2)}}{|H_e(k_1, k_2)|^2 + \gamma|P_e(k_1, k_2)|^2}\right]G(k_1, k_2) \quad (13.41)$$

where $H_e(k_1, k_2)$, $P_e(k_1, k_2)$, $G(k_1, k_2)$, and $\hat{F}(k_1, k_2)$ are the 2-D DFTs of $\{h_e(n_1, n_2)\}$, $\{p_e(n_1, n_2)\}$, $\{g(n_1, n_2)\}$, and $\{\hat{f}_e(n_1, n_2)\}$. Once $\{\hat{F}(k_1, k_2)\}$ is found from Eq. (13.41), the restored signal $\hat{\mathbf{f}}$ can then be found by computing the inverse DFT of $\hat{F}(k_1, k_2)$.

To illustrate the implementation approach, consider the 64 × 64 ideal image shown in Fig. 13.9a. Figure 13.9b shows an image that was blurred by linear motion and was then contaminated by additive zero-mean white noise with variance $\sigma = 0.1$. The application of Hunt's algorithm leads to $\gamma = 0.36$. The restored image obtained by employing the DFT implementation approach is depicted in Fig. 13.9c.

13.4 APPLICATIONS OF LINEAR 2-D FAN FILTERS TO SEISMIC SIGNAL PROCESSING

The structure of subsurface ground formations can be explored by using the so-called seismic reflection method [12]. In this method, a seismic wave generated by an explosion of dynamite near the surface travels through different wave paths to different horizontal interfaces and returns to the surface after being reflected by the interfaces. If a detector is placed on the earth surface near the point of the explosion, the time of arrival of each reflection can be recorded and hence the depth of each reflecting subsurface formation can be determined. If a trace of the ground motion

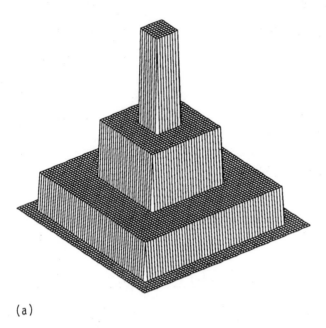

(a)

Figure 13.9 (a) Ideal image; (b) an image blurred by linear motion and contaminated by additive white noise with $\sigma = 0.1$; (c) the restored image.

versus time is recorded, distinct ground interfaces can be identified from the intensity of the trace. If a linear array comprising N_1 equally spaced detectors is used, as depicted in Fig. 13.10, the waveforms collected can be used to construct a 3-D plot in which the x, y, and z axes represent distance along the array, time, and ground motion, respectively. In plots of this type, which are often referred to as seismic images, subsurface formations can be identified as well-defined ridges.

Seismic waveforms consist of two distinct components: a component due to reflections from the ground and subsurface formations and a surface component often referred to as ground roll. The required information is embedded in the reflected component. The ground roll is believed to consist largely of Rayleigh waves [12], which are known to travel along free surfaces of elastic solids. Its presence in a seismic image is highly undesirable since it distorts the patterns produced by the reflected wave and thus renders the image less discernible. An interesting property of the surface wave is that its velocity is much lower than that of the reflected wave, as can be seen in Fig. 13.11, and as a result its spectrum is concentrated at low frequencies.

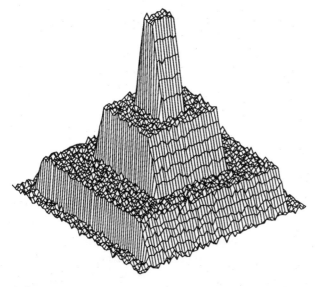

(b)

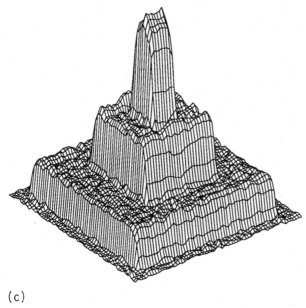

(c)

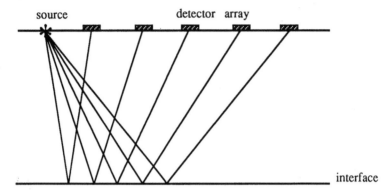

Figure 13.10 Measurement of seismic data.

The ground motion can be represented by the 2-D discrete signal $v(n_1, n_2) = v(n_1 L, n_2 T)$ where L is the distance between adjacent detectors and $1/T$ is the sampling rate used to sample waveforms. An actual seismic image of this type comprising 48 horizontal spatial points and 1000 vertical temporal points is illustrated in Fig. 13.12a. The well-defined ridges inclined at approximately 45° to the n_1 axis are due to the ground roll.

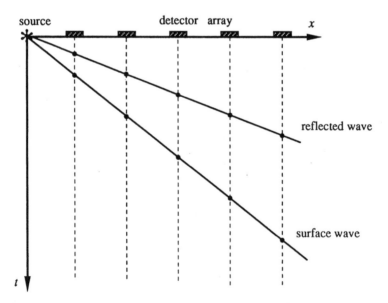

Figure 13.11 Travel time for reflected and surface waves.

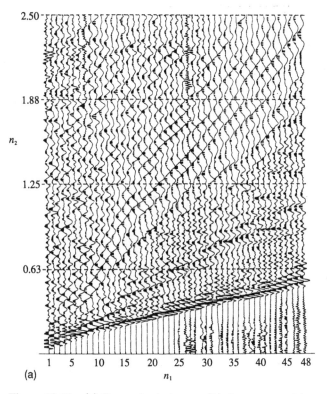

n_2

1 5 10 15 20 25 30 35 40 45 48

(a)

n_1

Figure 13.12 (a) Raw seismic image; (b) the processed image (from [6]).

In order to study the effect of the ground roll on a seismic image, assume that

$$f_0(y) = \sum_{k=1}^{60} \frac{1}{1 + \log k} \{\sin (k\omega_0 y) + \cos (k\omega_0 y)\}$$

is the signal generated by the explosion and neglect any reflections from subsurface formations. The signal will propagate along the surface at some velocity μ and will arrive at distance x along the axis of the array x/μ seconds later. Consequently, the signal received at point x will be

$$f(x, y) = \sum_{k=1}^{60} \frac{1}{1 + \log k} \left\{ \sin \left[\left(y - \frac{x}{\mu} \right) k\omega_0 \right] + \cos \left[\left(y - \frac{x}{\mu} \right) k\omega_0 \right] \right\}$$

Obviously, $f(x, y)$ contains a wide range of frequencies with respect to line

$$y = \frac{x}{\mu} + c$$

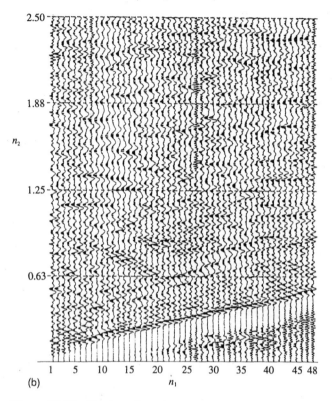

(b)

Figure 13.12 Continued

of the (x, y) plane. Consequently, the frequency spectrum of $f(x, y)$ is concentrated on a line

$$\omega_2 = \mu\omega_1 \tag{13.42}$$

as illustrated in Fig. 13.13a and b. In practice, the line in Eq. (13.42) forms an angle of 3° to 5° [6] and, therefore, the ground roll can largely be eliminated by using a fan bandstop filter with an amplitude response of the type shown in Fig. 13.14. Methods for the design of nonrecursive and recursive fan filters have been described in Chaps. 6, 8, and 9 and by Bruton et al. [6]. The effect of a fan filter with $\theta_s = 10°$ [6] on the seismic image in Fig. 13.12a is illustrated in Fig. 13.12b. As can be seen, the ground-roll ridges have been removed.

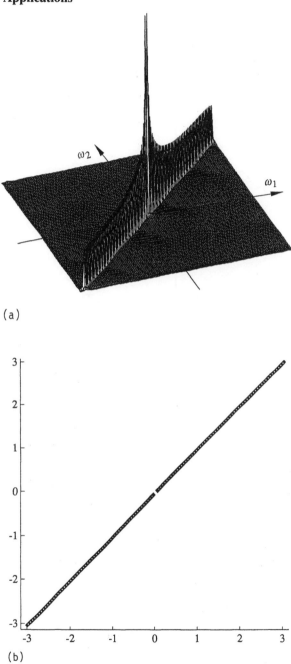

(a)

(b)

Figure 13.13 (a) Amplitude spectrum of signal $f(x, y)$; (b) contour plot of the amplitude spectrum.

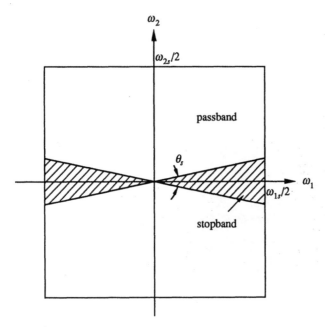

Figure 13.14   Amplitude response of an ideal fan filter.

REFERENCES

1. W. K. Pratt, *Digital Image Processing*, New York: Wiley, 1978.
2. R. C. Gonzalez and P. Wintz, *Digital Image Processing*, 2nd ed., Reading, Mass.: Addison-Wesley, 1987.
3. B. R. Hunt, The application of constrained least squares estimation to image restoration by digital computer, *IEEE Trans. Computers*, vol. C-22, pp. 805–812, Sept. 1973.
4. J. Biemond, R. L. Lagendijk, and R. M. Mersereau, Iterative methods for image deblurring, *Proc. IEEE*, vol. 78, pp. 856–883, May 1990.
5. G. Garibotto, 2-D recursive phase filters for the solution of two-dimensional wave equations, *IEEE Trans. Acoust., Speech, Signal Process.*, vol. ASSP-27, pp. 367–373, Aug. 1979.
6. L. T. Bruton, N. R. Bartley, and R. A. Stein, Stable 2-D recursive filter design using equi-terminated lossless *N*-port structures, *Proc. European Conf. Circuit Theory and Design*, pp. 300–307, Aug. 1981.
7. K. L. Peacock, On the practical design of discrete velocity filters for seismic data processing, *IEEE Trans. Acoust., Speech, Signal Process.*, vol. ASSP-30, pp. 52–60, Feb. 1982.
8. A. N. Venetsanopoulos, A survey of multidimensional digital signal processing and applications, *Alta Frequenza*, vol. LVI, pp. 315–326, Oct. 1987.

9. H. L. Willis and J. V. Aanstoos, Some unique signal processing applications in power system planning, *IEEE Tran. Acoust., Speech, Signal Process.*, vol. ASSP-27, pp. 685–697, Dec. 1979.

10. A. Antoniou, *Digital Filters: Analysis and Design*, New York: McGraw-Hill, 1979 (2nd ed. in press).

11. R. Kress, *Linear Integral Equations*, New York: Springer-Verlag, 1989.

12. M. B. Dobrin, *Introduction to Geophysical Prospecting*, New York: McGraw-Hill, 1976.

PROBLEMS

13.1 (a) Using MATLAB generate a 128 × 128 binary image as the original image. The 3-D plot of the 2-D array generated should display an image with clear images, which can be used to evaluate edge detection or enhancement algorithms. For example, images of letters are suitable for this purpose.

(b) Add a 128 × 128 random noise array to the array obtained in part (a) to form a noisy image. This can be done in MATLAB by using the commands *rand('normal')* and *r = rand*(128, 128). The *r* is a zero-mean normally distributed 2-D array with variance = 1.0. Modify *r* by multiplying it by a small constant to reduce the variance of the noise to an acceptable level.

13.2 (a) Use a lowpass nonrecursive filter of order higher than 3 × 3 to remove the noise of the image generated in Problem 13.1b.

(b) Use one of the 3 × 3 lowpass filters described by Eqs. (13.4)–(13.6) to filter the same noisy image.

13.3 (a) Use the 3 × 3 highpass nonrecursive filters described in Eqs. (13.10)–(13.12) to enhance the edges of the image generated in Problem 13.1a.

(b) Does the same filter work if only the noisy image obtained in Problem 13.1b is available?

(c) Use an appropriate combination of different kinds of filters to enhance the edges of the noisy image.

13.4 Apply the image restoration technique proposed by Hunt to remove the noise from the image obtained in Problem 13.1b.

Index